It Must be Beautiful

It Must be Beautiful

Great Equations of Modern Science

Edited by
Graham Farmelo

Granta Books
London · New York

Granta Publications, 2/3 Hanover Yard, London N1 8BE

First published in Great Britain by Granta Books 2002

A CIP catalogue record for this book is available from the British Library.

    5 7 9 10 8 6 4

ISBN 1 86207 479 8

Typeset by M Rules
Printed and bound in Great Britain by Mackays of Chatham plc

Dietary note

I have arranged the essays according to the chronological order of their respective equations, in the absence of an obviously better scheme. If you choose to read them in that order, one consequence is that the diet of physics is especially rich in the first half of the book. If this is not to your taste, I suggest you depart from chronology and assuage your personal appetite in whatever way takes your fancy. There should be no loss in enjoyment, as the essays have been written to be read independently of one another.

Now that I have the floor, I should like to thank all the staff at Granta Books for all the work they have done to bring this collection to fruition. I am especially grateful to Sajidah Ahmad, Neil Belton, Louise Campbell, Angela Rose and Sarah Wasley, all of whom have contributed way beyond duty's call.

Bon appétit.

GF

Contents

Foreword

It Must be Beautiful

Science is for those who learn; poetry, for those who know.
                    Joseph Roux in *Meditations of a Parish Priest*,
                                        part 1, number 71 (1886)

During a radio interview Philip Larkin gave in May 1974 to promote his *High Windows* collection, he pointed out that a good poem is like an onion. On the outside, both are pleasingly smooth and intriguing, and they become more and more so, as their successive layers of meaning are revealed. His aim was to write the perfect onion.

The poetry of science is in some sense embodied in its great equations and, as the essays in this book demonstrate, these equations can also be peeled. But their layers represent their attributes and consequences, not their meanings.

Despite the best efforts of poets and literary critics, no one has ever come up with an uncontroversial definition of a poem. No such problems beset mathematicians asked to define the term 'equation'. An equation is fundamentally an expression of perfect balance. For the pure mathematician – usually unconcerned with science – an equation is an abstract statement, having nothing to do with the concrete realities of the real world. So when mathematicians see an equation like $y = x + 1$, they think of y and x as completely abstract symbols, not as representing things that actually exist.

It is perfectly possible to imagine a universe in which mathematical equations have nothing to do with the workings of nature. Yet the marvellous thing is that they do. Scientists routinely cast their laws in the form of equations featuring symbols that each represent a quantity experimenters can measure. It is through this symbolic representation that the mathematical equation has become one of the most powerful weapons in the scientists' armory.

Best known of all scientific equations is $E = mc^2$, first suggested by Einstein in 1905. Like most great equations, it asserts an equality between things that superficially appear to be quite different[1] – energy, mass and the speed of light in a vacuum. Through this equation Einstein predicted that, for any mass you like (m), if you multiply it twice by the speed of light in a vacuum (denoted by the letter c) the result is *exactly* equal to its corresponding energy (E). Like every other equation, $E = mc^2$ balances two quantities, in the same way as a pair of weighing scales, with the = sign serving as the pivot. But whereas the scales balance weights, most equations balance other quantities; $E = mc^2$, for example, balances energies. This celebrated equation began its life as a confident Einsteinian speculation, and only decades later became part of the corpus of scientific knowledge, after experimenters had shown that it does indeed concur with nature. Now a twentieth-century icon, $E = mc^2$ is one of the few things about science that every TV quiz participant is expected to know.[2]

In common with all great scientific equations, $E = mc^2$ is in many ways similar to a great poem. Just as a perfect sonnet is spoiled if so much as a word or an item of punctuation is changed, not a single detail of a great equation such as $E = mc^2$ can be altered without rendering it useless. $E = 3mc^2$, for example, has nothing whatever to do with nature.

Great equations also share with the finest poetry an extraordinary power – poetry is the most concise and highly charged form of language, just as the great equations of science are the most succinct form of understanding of the aspect of physical reality they describe. $E = mc^2$ is itself enormously powerful: its few symbols encapsulate knowledge that can be applied to every energy conversion, from ones in every cell of every living thing on Earth, to the most distant cosmic explosion. Better yet, it seems to have held good since the beginning of time.

In the same way as close study of a great equation gradually enables scientists to see things that they initially missed, so repeated readings of a great poem invariably stir new emotions and associations. The great equations are just as rich a stimulus as poetry to the prepared imagination. Shakespeare could no more have foreseen the multiple meanings readers have perceived in 'Shall I compare thee to a summer's day?' (sonnet 18) than Einstein could have predicted the myriad consequences of his equations of relativity.

None of this is to imply that poetry and scientific equations are the same. Every poem is written in a particular language and loses its magic in translation, whereas an equation is expressed in the universal language of mathematics: $E = mc^2$ is the same in English as it is in Urdu. Also, poets seek multiple meanings and interactions between words and thoughts whereas scientists intend their equations to convey a single logical meaning.[3]

The meaning a great scientific equation usually furnishes us with what is called a law of nature. An analogy popularized by the physicist Richard Feynman helps to clarify this relationship between equations and laws.[4] Imagine people watching a game being played on a chessboard. If they had never been taught the rules of chess, they could soon work them out simply by observing how the players move various pieces. Now imagine that the players are not playing chess on an ordinary chequered board, but are moving the pieces according to a very much more complicated set of rules on a hugely extended board. For the observers to be able to work out the rules of the game, they would have to watch parts of it extremely carefully, looking for patterns and any other clues they could muster. That, in essence, is the predicament of scientists. They closely observe nature – the movements of the pieces on the board – and try to glean the underlying laws.

Armies of thinkers have been defeated by the enigma of why most fundamental laws of nature can be written down so conveniently as equations. Why is it that so many laws can be expressed as an absolute imperative, that two apparently unrelated quantities (the equation's left and right sides) are exactly equal? Nor is it clear why fundamental laws exist at all.[5] A popular, tongue-in-cheek explanation is that God is a mathematician, an idea that unhelpfully replaces profound questions with a doubly unverifiable proposition. Yet divine design has long been a popular explanation of the efficacy of equations in science. Witness the quotation on the memorial bust of America's first professional woman astronomer Maria Mitchell (1818–89) in the Bronx Hall of Fame: 'Every formula which expresses a law of nature is a hymn of praise to God', words written by Mitchell in 1866.

Even more contentious than the provenance of scientific equations is the question of whether they are invented or discovered.[6] The Indian-American astrophysicist Subrahmanyan Chandrasekhar probably spoke for most great theoreticians when he remarked that when he found some new fact or insight, it appeared to him to be something 'that had always been there and that I had chanced to pick up'. According to this view, the equations that underlie the workings of the universe are in some sense 'out there', independent of human existence, so that scientists are cosmic archaeologists, trying to unearth laws that have lain hidden since time began. The origin of the laws remains a complete mystery.

Of the hundreds of thousands of research scientists who have ever lived, very few have an important scientific equation to their name. Two scientists who were adept at discovering fundamental equations and especially perceptive about the role of mathematics in science were Albert Einstein and the almost comparably brilliant English theoretical physicist Paul Dirac. Neither was a mathematician per se, but both were remarkable in their ability to write down new equations that were as fecund as the greatest poetry. And both men were captivated by the belief that the fundamental equations of physics must be beautiful.[7]

This may sound strange. The subjective concept of beauty is unwelcome in polite intellectual circles, and certainly has no place in academic critiques of high art.[8] Yet it's a word that readily comes to the lips of all of us – even to the most pedantic critics – when we are moved by the sight of a smiling baby, a mountain vista, an exquisitely formed orchid. What does it mean to say that an equation is beautiful?[9] Fundamentally, it means that the equation can evoke the same rapture as other things that many of us describe as beautiful. Much like a great work of art, a beautiful equation has among its attributes much more than mere attractiveness – it will have universality, simplicity, inevitability and an elemental power. Think of masterpieces like Cézanne's *Apples and Pears*, Buckminster Fuller's geodesic dome, Judi Dench's interpretation of Lady Macbeth, Ella Fitzgerald's recording of 'Manhattan'. During my first experience of each of them, I soon realized that I was in the presence of something monumental in conception, fundamentally pure, free of excrescence and crafted so carefully that its power would be diminished if anything in it were changed.

An additional quality of a good scientific equation is that it has utilitarian beauty. It must tally with the results of every relevant experiment and, even better, make predictions that no one has made before. This aspect of an equation's effectiveness is akin to the beauty of a finely engineered machine of the kind we hear about in Stanley Kubrick's film *Full Metal Jacket*, when marine recruit Gomer Pyle starts talking to his rifle. The besotted Pyle praises its meticulous construction, delighting in the qualities that make it supremely fit for its lethal purpose. It would not be nearly so beautiful if it didn't work.

The concept of beauty was especially important to Einstein, the twentieth century's quintessential scientific aesthete. According to his elder son Hans, 'He had a character more like that of an artist than of a scientist as we usually think of them. For instance, the highest praise for a good theory or a good piece of work was not that it was correct nor that it was exact but that it was beautiful.' He once went so far as to say that 'the *only* physical theories that we are willing to accept are the beautiful ones', taking it for granted that a good theory must concur with experiment. Dirac was even more emphatic than Einstein in his belief in mathematical beauty as a criterion for the quality of fundamental theories[10] and even averred that this was for him 'a kind of religion'. In the latter part of his career, he spent a good deal of time touring the world, giving packed-out lectures on the origins of the great equation that bears his name, continually stressing that the pursuit of beauty had always been a lodestar and an inspiration. During a seminar in Moscow University in 1955, when asked to summarize his philosophy of physics, he wrote on the blackboard in capital letters, 'Physical laws should have mathematical beauty.' This piece of blackboard is still on display.

For lesser mortals, such aestheticism is a tough and unproductive credo. The prosaic fact is that for most scientists, beauty is neither a concept that much concerns them nor is it a helpful guide in their day-to-day work. True, the equations they use have an underlying beauty and the correct solutions to these equations are much more likely to be beautiful than ugly. But beauty can be misleading. Science is littered with the remains of theories that were once perceived as beautiful but turned out to be wrong – not what nature had in mind. For most scientists exploring a new theory, the number one criterion of its validity is whether it agrees with experiment.

The idea that science advances through a combination of experiment and mathematically based theory is relatively new. It originated in Florence only 350 years ago, yesterday in comparison with the span of human history. The originator was Galileo, the first modern scientist, who saw that science proceeds best by considering a narrow range of phenomena, and that the result will be laws that can be described in precise mathematical terms.[11] This was among the greatest and most productive discoveries in the entire history of ideas.

Since Galileo's time, science has become steadily more mathematical. Equations are now a hugely important scientific tool and it is virtually an article of faith for most theoreticians – certainly for most physicists – that there exists a fundamental equation to describe the phenomenon they are studying or that someone some day will find a suitable equation. Yet, as Feynman was fond of speculating, it may eventually turn out that fundamental laws of nature do not need to be stated mathematically and that they are better expressed in other ways, like the rules governing a game of chess.

For now, it seems that equations offer the most effective way of expressing most fundamental scientific laws. But equations are not the preoccupation of all scientists, many of whom do well with only a rudimentary acquaintance with mathematics. This point is made in the joke about a mathematician, a physicist, an engineer and a biologist when someone asks each of them the numerical value of π. The mathematician responds crisply that it's 'equal to the circumference of a circle divided by its diameter'. The physicist counters that it's '3.141593, give or take 0.000001'. The engineer says that it's 'about 3'. The biologist asks, 'What's π?'

This is of course a caricature. Some physicists have little mathematics, some engineers are brilliant in their ability to apply mathematics to their work, and some theoretical biologists are hotshot mathematicians. However, like all caricatures, it has a kernel of truth. Engineers tend to have a utilitarian attitude to mathematics, and place a high premium on making good approximations. And of all the sciences, physics is the most mathematical, biology the least. Since Galileo's time, physicists have flourished mainly by keeping things simple, by breaking down the complexities of the everyday world into their simplest component parts. Such reductionism is not always an option for biologists, whose bailiwick is the hugely complicated living world, with its interrelated communities of organisms, every one of which has a hugely complex structure in molecular terms. And let us not forget that the unifying theory of biology is superficially at least non-mathematical – *The Origin of Species*, Darwin's account of his theory of evolution by natural selection, doesn't contain a single equation. The same is true for the geologists' theory of continental drift, whose early papers (published soon after World War I) were virtually an equation-free zone.

The essays in this collection reflect the importance of mathematics in the various overlapping areas of science from 1900 onwards. Physics is appropriately well represented. Three of Einstein's great contributions (including $E = mc^2$ and his equation of general relativity) are discussed and there are contributions on other important equations that have led to our current understanding of the subatomic world. The Dirac equation has a special place: not only did the equation do its intended job of describing the behaviour of the electron, it also unexpectedly predicted the very existence of antimatter, which at one time constituted half of the entire universe. No wonder Dirac commented, 'My equation is smarter than I am.'

The equations of subatomic physics form the basis of the Standard Model, the deeply prosaic name for the current theory of fundamental particles and their interactions (leaving aside the most familiar force of all, gravity, which is ironically beyond the model's scope). This book's afterword draws together the strands that have contributed to the Model, one of the intellectual triumphs of the twentieth century.

Two essays look at some of the equations of modern biology. The first explains how evolutionary ideas can be expressed mathematically to give richly diverse insights into the living world, from the mating behavior of red deer to the ratios of males to females in communities of wasps. The second essay concerns the so-called quadratic map, a deceptively simple equation in theoretical ecology that can be used to understand the varying population of fish in a garden pond and the fluctuating numbers of grouse on a moor, as well as a host of similar problems. This equation played a crucial role in the history of chaos, for it turned out that the equation strikingly embodies chaotic behaviour – behaviour that is extremely sensitive to initial conditions. It was in large part thanks to this equation, one so simple that children can study it at school, that scientists in the 1970s came to see that some equations that appear to predict the future in terms of the past are completely unable to makes such predictions, contrary to what most scientists had thought.

Two other equations covered here concern information science and the search for extraterrestrial intelligence. The essay on information science addresses the equations of the late doyen of information theorists, Claude Shannon, who pioneered the mathematical apparatus that underlies what we now call the communications revolution. Shannon's equations apply to every type of information transfer, including the Internet, radio and television.

The search for extraterrestrial intelligence (SETI) may not be a topic for which you would expect there to be an equation. How can there be an equation for something that may not exist? The answer is that SETI's key equation – first written down by the American astronomer Frank Drake – does not make predictions; rather, it disciplines our thinking about the probability that there exist civilizations that can communicate with us. Not an equation that is beautiful in the sense understood by Einstein and Dirac, Drake's formula has nonetheless brought some coherence to a field potentially rife with woolliness.

Mathematical equations are not the only type of equation used by scientists. Chemists, for example, use equations that are not written solely in terms of mathematical symbols but in terms of letters representing atoms, molecules and their submicroscopic relatives. A huge amount of industrial activity is based on chemical equations like this, each one describing an interaction whose details can be inferred but almost never observed with the naked eye. We have chosen one special set of chemical reactions for inclusion here, to represent the power of chemical thinking. These wonderfully simple equations formed the basis of scientific understanding of the thinning of the ozone layer and its cause, the presence of chemicals known as CFCs in the Earth's atmosphere. In the early 1980s, these simple equations helped to awaken humanity's sense of impending environmental apocalypse.

The authors of this book are leading scientists, historians and writers. They have looked at whichever aspects of their equation – Larkin's layers – strike them as being most intriguing, for the most part eschewing the eye-watering mathematical details. The result is a set of uniquely personal meditations on some of the seminal equations of modern science, equations that through their concision, power and fundamental simplicity can be regarded as some of the most beautiful poetry of the twentieth century.

Among my own collection of poetry, on the shelf above my desk, sits a dustless copy of *High Windows*. I first read it when I was a greenhorn student of subatomic physics, struggling to understand its fundamental equations and to appreciate their beauty. The collection was given to me by a Larkin-loving friend, a student of English literature, just a few days after the collection was published. Her message to me was the same as mine is now to you: 'Enjoy the onions.'

Graham Farmelo
August 2001

A Revolution with no Revolutionaries

The Planck–Einstein Equation for the Energy of a Quantum

Graham Farmelo

I

'Revolutions are celebrated when they are no longer dangerous.'
Pierre Boulez, 13 January 1989, on the bicentennial
celebrations of the French Revolution

The twentieth century chose some undeserving characters to be its celebrities, but it selected its favourite scientist with excellent taste. Albert Einstein, as brilliant at identifying fruitful scientific problems as he was at solving them, did more than anyone else in the most scientifically productive of centuries to advance human knowledge. What a pity that his most truly revolutionary work is now so widely forgotten.

If you were to ask people on the street to name Einstein's most famous contribution to science, they would probably cite his theory of relativity. This was a brilliant piece of work, to be sure, but it was not revolutionary, as Einstein often stressed. He had positioned himself squarely on the shoulders of Newton and Galileo to produce a new theory of space, time and matter that meshed smoothly with their theories. Only once did Einstein depart radically from his predecessors' thinking, when he put forward a remarkable new idea about the energy of light.

Common sense tells us that light enters our eyes in a continuous stream. Scientists at the end of the nineteenth century seemed to confirm this intuitive belief using their universally agreed wave picture of light, which says that the energy of light is delivered smoothly, like the energy of water waves sloshing against a harbour wall. But, as Einstein observed, 'common sense is that body of prejudice acquired before the age of eighteen'. He proposed in 1905, when he was working as a patent examiner in Bern, that this picture of light is wrong and that the energy of light is delivered not continuously but in discrete amounts, which he called quanta. Shortly afterwards, he speculated that the energies of atoms in a solid are also quantized – only certain values of energy are possible. Again, this quantization of energy was contrary to common sense. The energy of motion of the apple that fell in Newton's garden appeared to increase gradually, not in a series of jumps.

Einstein saw more clearly than anyone else that the submicroscopic world is replete with quanta: nature is fundamentally granular, not smooth. Although he was working alone when he came to these conclusions, he did not pluck them from out of the blue. He was inspired by papers written by a physicist twenty-one years his senior, Max Planck, then the dean of German physicists, working in Berlin. Planck had been the first to introduce the idea of energy quanta in the closing weeks of 1900, although it is not clear whether he fully understood the implications of what he had done.

One deceptively simple equation was especially perplexing to the quantum pioneers. First written down by Planck but properly interpreted only later by Einstein, the equation relates the energy E of each quantum to its frequency f: $E = hf$, where h is a fixed quantity that others later named after Planck. This was the first important new scientific equation of the century (Kaiser Wilhelm II had decreed that 1900 was the first year of the twentieth century, not the last of the nineteenth). High-school students now learn it by rote and few of them puzzle over it, but it took the first quantum physicists almost twenty-five years to tease out its meaning. During this time, Einstein's work on the ideas behind the $E = hf$ equation led him to become the first person successfully to predict the existence of a fundamental particle. In addition, he and others laid the foundations of a fully fledged quantum theory, arguably the century's most revolutionary scientific idea.

Albert Einstein and Max Planck dominate the story of this most intellectually productive of equations. The two men were superficially very different. Planck was tall, gaunt and bald, whereas Einstein was muscular, just above average height and blessed with a resplendent mane of hair; Planck was convivial with his peers, Einstein kept an intellectual distance from them; Planck was nationalistic, Einstein avowedly cosmopolitan and liberal; Planck's politics were to the right, Einstein's to the left; Planck was a punctilious administrator, Einstein took every opportunity to avoid paperwork; Planck was a family man, Einstein's home life was dysfunctional.

But the two also had much in common. They were both theoretical physicists, a relatively new breed of scientist with an overwhelming interest in understanding nature in terms of universal, overarching principles. The two men, both workaholics, heeded new experimental results, but they were happiest when they were working in the laboratories inside their heads. Both believed that scientific principles existed independent of human beings and that new principles were out there waiting to be discovered. Like all good scientists, Planck and Einstein approached their work conservatively. They were cautious about new experimental results, chary of innovations that contradicted well-established theories, and mindful that if a new theory is to be taken seriously, it must reproduce every success of its predecessors and ideally make new predictions of its own.

For both men, physics was their first love, music their second. Einstein was fond of Bach, Mozart and Haydn, and loved to play the violin, which he carried everywhere on his travels. Opinions differ on the quality of his technique: his tone was 'beautifully delicate', according to the great violin teacher Shinichi Suzuki, but another authority said that he 'bowed like a lumberjack'. Whatever the truth about his musical gifts, Einstein did not take kindly to criticisms of his playing: 'Einstein got a lot more excited over musical disputes than scientific disputes,' one of his acquaintances observed. Planck was a much finer and more equable musician, a pianist good enough in later life to play duets with the great violinist Joseph Joachim. Planck loved the music of Joachim's friend and collaborator Brahms, and was also especially fond of Schubert and Bach.

When Planck, Einstein and their colleagues were laying the foundations of quantum theory, they were also in the vanguard of the wider movement of modernism, consciously reinventing their subject, exploring the means and limits of classical techniques.[1] In this sense, they were similar to Igor Stravinsky in St Petersburg, Virginia Woolf in London, Pablo Picasso in Paris, Antonio Gaudí in Barcelona. But, unlike the artists, Planck and Einstein were modernists despite themselves: neither set out to challenge the foundations of their subject for its own sake. Whereas the artists were free to create new forms to replace ones that seemed to them outdated, the scientists had no choice but to create new theories to replace those that had been found irredeemably wanting. It was a tiny but troubling disparity between experiment and theory that led to the quantum revolution. The trouble started in a few ovens in Berlin.

II

Berlin has never been a favourite destination for *fins gourmets*. They will concede, however, that it is at least now possible to buy a decent espresso there, for example in one of the Einstein® coffee bars springing up all over the city. Although these smart establishments were not named after the great physicist, the name above their front doors reminds us of the time a century ago when Berlin was not only Europe's fastest-growing and most opulent city,[2] but the capital of physics.

Shortly before the close of the Franco-Prussian War in 1871, Bismarck made Berlin the capital of the victorious new Reich. The city – a bubbling cultural stew since the glory days of the polymathic Prussian despot Frederick the Great a century before – was home to some of the world's leading experimenters. Berlin was also the headquarters of an elite group of theoretical physicists, members of a new discipline that the more prestigious experimenters had begun to employ in the late 1860s to teach the forbiddingly mathematical theories that were becoming increasingly fashionable. This community was at first exclusively male – women were admitted to universities in Berlin only from the summer of 1908.[3] A century later, little has changed: the overwhelming majority of theoretical physicists continue to be men.

It was as a leading member of Berlin's new scientific community that Planck conceived the concept of the energy quantum and wrote down the $E = hf$ equation. To understand Planck's work, we need to look at the two great theories that seized the imagination of physicists in the latter half of the nineteenth century. The first was a unified mathematical treatment of electricity, magnetism and optics, set out in 1864 by James Clerk Maxwell, a Scottish physicist celebrated for his brilliance and versatility. Using a set of equations that now bears his name, he demonstrated that visible light is an electromagnetic wave that travels through an all-pervasive ether, in much the same way as a sound wave zips through air. Like any other wave, an electromagnetic wave has a wavelength and a corresponding frequency. The wavelength is simply the distance between any two consecutive peaks of the wave, and the frequency is the number of times it jiggles up and down every second. At the red end of the rainbow spectrum, light waves have a wavelength of seven ten-thousandths of a millimetre and move up and down 430 trillion times a second, while at the violet end its wavelength is rather shorter and its frequency is rather higher. Maxwell's theory correctly explained why there exist electromagnetic waves outside the visible range, with higher and lower frequencies. Light is simply part of the spectrum of electromagnetic radiation.

Another of Maxwell's many interests was thermodynamics, the second great theory of physics to come of age towards the end of the nineteenth century. The theory dealt with different forms of energy and the extent to which they can be converted into one another, for example, from the motion of a fly-wheel into heat; it was concerned only with bulk matter and said nothing about the behaviour of the individual constituent atoms. The steam engines that powered the industrialization of western Europe were initially responsible for stimulating theoretical work on thermodynamics. By the middle of the nineteenth century, developments in the theory led to improvements in the technology, which itself led to refinements in the apparatus designed to test the theory.

Thermodynamics and electromagnetism, together with Newton's work on forces, is part of what we now call 'classical physics'. Not that its inventors thought they were working in a classical tradition – they believed they were simply doing physics. It was the emergence of the quantum theory of Planck, Einstein and their colleagues that gave rise to this retrolabelling.

Among the leading classical physicists in Germany was Rudolf Clausius, arguably the first theoretical physicist.[4] This quarrelsome man was the pioneering master of a mathematical approach to thermodynamics that concentrated on seeking a few grand, overarching principles or axioms. It was crucial, he argued, that they should be logically consistent and that they lead to results that agree with experiment. This top-down approach contrasted sharply with the traditional piecemeal style of doing mathematical physics, which involved writing down equations to describe a phenomenon before seeing how well they accounted for experimental results.

Others had already established that energy can neither be created nor destroyed – the first law of thermodynamics. But in 1850 Clausius was among the first to fashion what became known as the second law, which says, roughly speaking, that heat does not flow spontaneously from something cold to something hotter. This is plausible enough: a cold cappuccino never warms up if it's left alone. Both thermodynamic laws appeared to be absolute – to be universally valid, wherever and whenever they are checked.

Although the two laws are superficially simple, Clausius had to deploy an enormous amount of intellectual artillery to state them rigorously. His mathematical and linguistic precision, together with the diamond-hard clarity of his reasoning, fascinated Max Planck when he was an impressionable graduate student.[5] Born in 1858 into a patriotic and affluent family of scholars, lawyers and public servants, he was deeply imbued with conservative values. Some of his early memories of political unrest were to stay with him for his entire life, not least his witnessing as an eight-year-old the sight of victorious Prussian and Austrian troops marching into his native town of Kiel after their defeat of Denmark. A diligent undergraduate, if not an especially brilliant one, Planck had enjoyed a broad education in mathematics, physics, philosophy and history. He also studied music, his principal hobby, and distinguished himself by composing a tuneful operetta performed at musical evenings in the homes of his professors.

Unsure of which subject to pursue, he chose physics, no thanks to his professor at the University of Munich, Philip von Jolly. In what ranks among the most egregious howlers in the history of career counselling, von Jolly advised the twenty-year-old Planck against entering physics because after the discovery of the two laws of thermodynamics, all that was left for theoretical physicists to do was to tidy up the loose ends. With his trademark conservatism, Planck touchingly replied that he wished only to deepen the foundations laid by his predecessors and that he had no wish to make any new discoveries. As we shall see, the first wish was to be granted at the expense of the second.

Planck fell in love with thermodynamics when he was working on his doctoral thesis and he became intrigued with the power and generality of its laws. He was, however, deeply uncomfortable about two aspects of the thermodynamic being championed by the leading Austrian theoretical physicist, the passionate and depressive Ludwig Boltzmann. First, Planck was not convinced that matter was ultimately made of atoms: no one had actually observed one, so perhaps they were nothing more than a convenient fiction? He was also sceptical of Boltzmann's argument that the second law of thermodynamics was true only statistically: that heat is overwhelmingly likely – but not certain – to flow spontaneously from something hot to something colder. This lack of certainty was inimical to Planck's passion for absolutes, for incontestability, for certainty.

One absolute did catch his eye. It concerned a problem so subtle that few scientists outside Berlin were concerned with it – and anyone outside science could reasonably have dismissed it as laughably obscure. Imagine a completely sealed cavity, like an electric oven but with neither vents nor windows. Now suppose that the cavity is at a steady, uniform temperature. The walls of the cavity give out electromagnetic radiation, which bounces around inside the cavity, continually being reflected from the walls or being absorbed and then re-emitted.

Experimenters observe this cavity radiation by making a small hole in the side of the cavity, enabling a small amount of the radiation to escape. Some radiation from the surroundings enters the cavity, but it is soon absorbed, re-emitted and reflected around the cavity so that it takes on the same characteristics as the other radiation in the cavity. Because all of the radiation from outside that passes through the hole is 'absorbed', the hole looks black when it is viewed at room temperature, and the emerging radiation – a sample of the cavity radiation – is often called black-body radiation.[6] The question that fascinated physicists was: at any given temperature of the cavity, what is the intensity of its radiation at each colour or, more rigorously, at each wavelength? It was by answering this question that Planck was led to the equation $E = hf$.

One of Planck's research advisers had already proved that whatever the law for the radiation's intensity turned out to be, it would depend on neither the size of the cavity nor its shape, nor on the material from which its walls were made. Such a law would be a classic example of what Planck called an 'absolute', something that 'will necessarily retain its importance for all times and cultures, even for non-terrestrial and non-human ones'. Cavity radiation was not just of academic interest: it was important for Germany's lighting industry, one of the many flourishing branches of the country's economy at a time when electrical and chemical technologies were revolutionizing capitalism. Always looking for sources of illumination that gave out as much visible light and as little heat as possible, engineers who were trying to design increasingly efficient electric lamps needed to know how much radiation their filaments emitted. The more they knew about cavity radiation, the better equipped they would be to produce better lamps of the kind invented in 1897 by the American Thomas Edison.

This was one of the problems under investigation at the lavishly equipped Physikalisch-Technische Reichsanstalt (the Imperial Institute of Physics and Technology) in Charlottenburg, just outside Berlin, three miles from the university where Planck had been working since 1889. Funded jointly by the German government and the industrialist Werner von Siemens, the Reichsanstalt was founded in the aftermath of the Franco-Prussian War (1870–71) to refine the art of making precise measurements and to set standards with which scientists and engineers could work. Mindful of the potential economic benefits for the new Reich, the Reichsanstalt's founders set out to provide nonpareil research facilities that would be of practical benefit to German industry.[7] Even its classically designed yellow-brick buildings, located in almost nine acres of immaculate parkland, spoke of the institution's imperial ambitions.

German physicists had been working on cavity radiation for thirty years. Their understanding could be handily summarized by simple mathematical laws that predicted the intensity of the radiation for every wavelength. While two teams of Reichsanstalt experimenters were working on the problem, Planck was trying to understand the most successful of the radiation laws, written down in March 1896 by his close friend Wilhelm Wien, one of the finest physicists at the Reichsanstalt. Wien was a character: the country-loving son of an East Prussian land owner, he had hoped to divide his time between physics and farming until a disastrous harvest forced his father to sell the farm and made it necessary for the young Wien to take up science as a full-time career. He was also a chauvinist and an anti-Semitic reactionary. A few days after the end of World War I he led a group of volunteers, mainly war veterans, to shoot at communists and other leftists in the streets of Würzburg and Munich, to prevent what he described as 'the Bolshevization of Germany'.

Wien's cavity-radiation law successfully accounted for the intensity of the radiation at each colour, for a wide range of temperatures. Planck wanted to understand his colleague's law using thermodynamics and electromagnetism, and he began with high hopes that he could understand cavity radiation without having to assume the existence of atoms or by having to use a version of the second law of thermodynamics that involved probabilities rather than certainties. By the early summer of 1899, however, he had given up both of these preconceptions. He reluctantly concluded that he could understand cavity radiation only if he accepted that atoms exist and if he embraced Boltzmann's statistical way of thinking. As he was correcting the proof of a paper setting out the theory, the experimenters – 'the shock troops of science', as he called them – brought him some disturbing news: Wien's law appeared suddenly to be in trouble. It consistently underestimated the intensity of cavity radiation, especially at long wavelengths, which new equipment had only just enabled them to investigate (Figure 1.1).

On Sunday 7 October 1900 the Reichsanstalt experimenter Heinrich Rubens and his wife visited Planck and his family at their handsome oak-panelled villa in the Grünewald, the smart Berlin suburb favoured by the professoriate. The two physicists talked shop, and soon after the Rubenses left, Planck set to work to find a better law. That evening in his study – no doubt standing up at his tall writing desk, as was his wont – he produced a modified version of Wien's law that could account for all the experimenters'

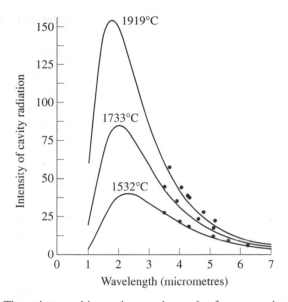

Figure 1.1 The points on this graph were the seeds of quantum theory. Planck observed that the points did not agree with the predictions of the Wien law, shown here as continuous lines, each corresponding to a different temperature. Planck soon came up with a law that could fit *all* the data and, as a result of trying to derive his formula theoretically, was led to propose the idea of energy quanta.

data. Planck dashed off a postcard to Rubens telling him of his new cavity-radiation law and, twelve days later, presented the law for the first time in public to a formal meeting of his Berlin colleagues, including Rubens. Afterwards, Rubens returned to his laboratory and was able the next morning to give Planck a pleasant start to his weekend by confirming that his new law had successfully accounted for his new data. To this day, no one has come up with a law that better predicts the intensity of cavity radiation.

On the very day that Planck first wrote down his cavity-radiation law, he began to try to understand it in terms of what was actually going on inside the ovens, initially using the laws of classical physics.[8] He soon found that he had no choice but to use Boltzmann's statistical reasoning, which he had previously abominated, to understand how radiation interacts with the atoms of the ovens' walls. He accepted the standard picture of the walls, namely that they consisted – like any solid – of atoms vibrating about fixed positions, with an average energy that increased as the oven warmed up. But in this case Boltzmann's way of treating the energies of the atoms didn't work, so Planck had no choice but to abandon some of the assumptions of classical physics that he had thought were the bedrock of his subject and do something that was desperately uncongenial to him – extemporize. It was as if Artur Rubinstein suddenly had to riff like Earl Hines.

During some eight weeks of the most strenuous work of his life, he found that he could derive his law only if he drastically modified Boltzmann's statistical techniques and if he took one particularly strange step. He had to divide the total energy of all the atoms vibrating in the oven's walls at each frequency into discrete amounts, each with an energy given by the equation $E = hf$. Here was the first appearance of the energy quantum, the first suggestion that energy at the molecular level is fundamentally different from energy on the everyday scale.

The notion of energy quanta flew in the face of what was at the time every scientist's understanding of energy. Energy was, like water, supposed to be available in any quantity – you can take water from the sea, or put it back, in any amount you like. The idea that water could come only in definite quantal amounts, say cupfuls, contradicts everyday experience, yet this is how energy apparently behaves at the molecular level. Could it be that, just as water ultimately comes in units of water molecules, energy fundamentally comes in discrete quanta, in lumps?

Planck first publicly presented his $E = hf$ equation in a lecture. On Friday 14 December, shortly after five o'clock in the afternoon, he stood up to read a short paper about his derivation of his cavity law to Berlin physicists at one of the fortnightly meetings of the German Physical Society. It was with no fanfare or excitement that Planck first mentioned the $E = hf$ equation during this talk. His colleagues were, it appears, respectfully interested but underwhelmed.

According to the conventional view, this presentation was Planck's unveiling of the quantum idea to the world.[9] Many quantum historians have been persuaded, however, that this account is simplistic, following the writings of the late philosopher and historian of science Thomas Kuhn. Planck wrote that he considered energy quantization 'a purely formal assumption, and I did not give it much thought except for this: that I had to obtain a positive result, under any circumstances and at whatever cost'. Statements like this convinced Kuhn that in 1900 Planck did not appreciate the significance of energy quanta and that he did not believe that energies are quantized. Rather, Kuhn argued, Planck believed along with everyone else that the atoms could have any energy they liked, and that he was dividing these energies into quanta simply as a mathematical device to make his calculations work out satisfactorily.[10]

But all scholars agree that Planck did correctly seize on the importance of his new constant h.[11] He worked for several years to try to understand the constant in terms of classical physics and he later wrote that many of his colleagues mistakenly regarded his failure as a tragedy. He eventually came to accept that he had discovered the latest of only a handful of truly fundamental constants – including the speed of light and the constant in Newton's law of gravitation – which figure in the equations of physics, with values that cannot be derived. Such a discovery is extremely rare in the history of science: since Planck saw the need for h, not a single new fundamental constant has been identified.

Planck's theory also featured a second constant k that related to Boltzmann's statistical theory, so Planck named it after him, with a generosity that he was to regret, as Boltzmann neither introduced it nor thought of investigating its value. By comparing the predictions of his mathematical law with the Reichsanstalt cavity-radiation data, Planck found the value of each of the two constants. The measurement of the Boltzmann constant was especially pleasing for Planck, as it enabled him to make what was then the most accurate measurement of the mass of an atom.[12] The value of the constant also subsequently enabled scientists to calculate the average energy of an atom in any substance anywhere in the universe, whatever its temperature.[13]

As strange as it might now seem, Planck was less excited by his quantum theory and his $E = hf$ equation than by the possibility the theory raised of measuring length, time and mass in terms of a new set of units that would be natural to use anywhere in the universe. Tradition and convenience leads us here on earth to measure length in metres or feet, to measure time in seconds and mass in kilograms or pounds, but there is no fundamental reason why these units are better than any others. If history had turned out differently, we might now be measuring length in units of the extent of Julius Caesar's little finger, and measuring mass and time in terms of the weight of his crown and the time it took his heart to beat.

Planck quickly realized that the new constant h enabled him to set up units that are not at all arbitrary but emerge from the laws of nature. He saw that he could calculate unique values of length, mass and time using special combinations of the new universal constant with two others, the speed of light and Newton's gravitational constant.[14] Planck reasoned that if these three constants have always been the same everywhere, then his calculated values of mass, length and time gave units that are also valid everywhere in the universe and are therefore more natural than any set up by any earthly authority, no matter how august. Planck found that the unique value for mass that emerged is about the mass of a giant amoeba (10^{-8} kilogrammes), for length it's about a trillionth of a trillionth of the width of an atom (10^{-35} metres) and for time it's about 10^{-43} second, about a millionth of a trillionth of a trillionth of a trillionth of the time it takes you to blink. None of these is convenient to use in everyday life, of course, but Planck saw that the essential new point was that it was not only some laws that had absolute, universal validity – a unique set of units did, too.

Most of Planck's colleagues regarded his cavity-radiation law as little more than a mathematical formula that happened to fit the data. Not one of the Berlin physics Brahmins saw clearly the implications of Planck's work and, in particular, of his new equation $E = hf$. That was left to a young graduate student, working mainly on his own, in Switzerland.

III

The poet Paul Valéry made a point of keeping a notebook in his pocket to jot down his ideas. When he asked Einstein if he did the same, Einstein replied, 'Oh, that's not necessary,' before adding wistfully, 'It's so seldom I have one.' He made that comment in the 1920s, when the most creative part of his career was drawing to a close. When it was beginning, twenty years before, one wonders how much stationery he consumed.

In the autumn of 1900, when Planck first wrote down his $E = hf$ equation, Einstein was in Zurich, eking out a living by giving private tutorials.[15] Several years before, he had read about the problem of understanding cavity radiation and by the spring of 1901 he was familiar with Planck's work. Einstein's reaction to it shows that he was already a special talent. No one else saw beyond the virtuosity of Planck's mathematical analysis and the superficial success of his formula in explaining the Reichsanstalt data, but Einstein swiftly realized that revolution was afoot. He later wrote that, soon after the appearance of Planck's work, 'It was as if the ground had been pulled out from under one, with no firm foundation anywhere, upon which one could have built.'

Einstein thought about the implications of Planck's work for almost four years before he published his revolutionary ideas about light – or, more generally, radiation – and what happens when it interacts with matter. The confident and cheerful young physicist had secured a well-paid job as a patent examiner 'Expert Class III' at the Swiss patent office in Bern and had married a former student colleague. By the early autumn of 1904, their son

Hans had been born, they had moved to a two-room apartment and Einstein's job had been made permanent. Somehow, in the interstices of his life at home and at the office, he was developing his ideas about light, relativity and also the molecular structure of matter, which he chose as the subject of his PhD thesis. In what we now realize was one of the most spectacular flowerings of talent in the history of science, all three of Einstein's lines of research bore fruit in 1905.[16] The first paper he published that year, today recognized as his first great contribution to science, was about light quanta, an idea he described in a letter to a friend as 'very revolutionary'.

The paper was a gem. Although its language is temperate to the point of understatement, its reasoning has the audacity of which only the most brilliant and unfettered young minds are capable. Einstein came straight to the point by gently asserting that, contrary to Maxwell's wave theory of light, 'the energy of [a light ray emitted from a point source] is not distributed continuously over ever-increasing volumes of space but consists of a finite number of energy quanta localized at points of space that move without dividing, and can be absorbed or generated only as complete units.' To followers of Maxwell – and that included every leading physicist at the time – this was abject heresy.

Einstein then moved on to give Planck what amounted to a physics lesson, by demonstrating that the reasoning Planck had used to study his cavity radiation was fatally flawed. Using simple mathematics, Einstein showed that the total energy of radiation in a cavity was infinite according to classical physics. Had Planck realized this, he probably would have discarded his theory and not made his great discovery of energy quanta, as Einstein later pointed out. Wary of Planck's formula for the intensity of cavity radiation, Einstein instead used Wien's earlier formula, which gave an excellent account of cavity radiation with short wavelengths or, equivalently, high frequencies. Einstein noticed that the equation for the density of this radiation is exactly the same as the corresponding equation for a gas of quanta, when they are all bouncing around independent of one another. So – and here is his especially audacious step – Einstein suggested that the radiation described by Wien's law behaves as if it *is* a gas, *whether or not* it is inside a cavity. The comparison also gave him a simple equation for the energy of each 'quantum' in the gas of radiation: it was $E = hf$, where h is Planck's constant and f is the frequency of the radiation.

Although Einstein's formula looks identical to Planck's, they say quite different things: Einstein's applied to the energy of every light quantum, whereas Planck's was about the special case of the energy of the atoms in a cavity when they interact with light. But Einstein had something else to say about this interaction: he proposed that matter absorbs or gives out radiation not in a continuous stream – as classical theory implied – but 'as if' (Einstein's words) the radiation consisted of quanta. So matter could absorb or give out one, thirty-seven or any other whole number of radiation quanta, but not two and a half or any other fractional number.

If radiation is delivered in quanta, why aren't we conscious of individual packets of energy entering our eyes? Although Einstein didn't address this question explicitly, he well knew the answer – each quantum has such a minuscule energy that the number of quanta entering our eyes is normally so vast that our brains cannot differentiate between their separate arrivals, so they appear to arrive in a continuous stream. Each quantum of visible light has only about a trillionth of the energy of the beat of a fly's wing and, as a result, an ordinary candle gives out about a billion trillion quanta every second – far too many for our eyes to be able to distinguish between them.

Einstein concluded his paper by pointing out how his light-quantum idea could be tested experimentally. His most striking suggestion concerned another superficially obscure problem: of understanding what happens when radiation is shone on a metal, the so-called photoelectric effect. The metal reflects and absorbs the radiation, but Einstein knew that experimenters had found that radiation could eject some of the electrons in the metal. If the quantum picture of radiation is correct, Einstein argued, then it is reasonable to suppose that each ejected electron is displaced by a single radiation quantum with energy $E = hf$. What is more, if the quantum surrenders all this energy to the electron, then the energy of the emergent electron is simply equal to the energy of the radiation, quantum, less the energy needed to fish out the electron from the metal. Einstein expressed this idea mathematically in what became known as his photoelectric law.

Einstein called his view of light 'heuristic' – something put forward as an aid to learning – so he appeared to be hedging his bets about whether light quanta are real. It is easy to see why he did this – he was arguing that, in this case, the hallowed Maxwell theory of radiation was wrong and that radiation behaved as quanta, not as waves. Yet if there was one thing that physicists thought they knew about radiation, it was that it *did* behave as waves. And they had incontrovertible proof: if an experimenter shines a beam of light on a sufficiently narrow slit, the light is diffracted, or – to use a less formal term – spread out. Diffracted light often has a characteristic pattern of peaks and troughs in brightness, and it is easy to explain this pattern if radiation is a wave. Otherwise, it is impossible.

Einstein was well aware of this and he understood the force of the argument, but he was not going to be intimidated by inconvenient experimental facts. He completed his paper in mid-March and sent it to the world's leading journal of physics research, *Annalen der Physik*, edited by Planck and Wien. Planck was a particularly effective editor: he rejected second-rate submissions but was happy to publish papers that were either orthodox or heretical, provided they were well argued and logically consistent. Einstein's revolutionary paper was duly published a few months later.

The majority will always reject a truly revolutionary idea, because it is incomprehensible in terms of the ideas it seeks to displace. So it was with Einstein's paper, whose publication was totally ignored by the world's professional physicists – all thousand or so of them – and by all other scientists too. You can hardly blame them: his ideas made no sense at all in terms of Maxwell's theory, which had passed every experimental test it had been put through for forty years. Even the equation $E = hf$ for the energy of the quantum was bizarre as it linked the unlinkable – the energy of a *quantum* to the radiation's frequency, a concept that made sense only if the radiation was a *wave*. And who *was* this Albert Einstein, anyway?

Einstein was not downhearted. By January 1906 the papers he had written the previous year had been published without quibble from Planck or Wien and he had become 'Herr Doktor Einstein'. Although he was happy working his eight-hour day at the patent office, he was getting itchy feet and seriously considered taking up a full-time career in school teaching. Meanwhile, he was thinking about the relationship between his light quantum idea and Planck's earlier work. Einstein had originally thought that the two theories were complementary, but he had now come to realize that Planck had used the light quantum idea, albeit implicitly. Einstein also realized that in Planck's theory the energy of each of the atoms in the cavity wall was also quantized – something Planck certainly had not appreciated. If each atom vibrates at a fixed number of times each second – that is, with a fixed frequency – then the energy of the vibrating atom can come only in whole-number multiples of Planck's constant multiplied by the frequency. So the smallest energy the vibrating atom can have is $E = hf$, and it could also have the energy values $2hf$, $3hf$, $4hf$ etc. Einstein was saying that the equation $E = hf$ applies to *every* atom in a solid: it was not a mathematical subdivision of the total energy of *all the atoms*, as Planck had proposed.

In November 1906 Einstein showed how to test this idea by considering how solids absorb heat, something that classical physics was at a loss to explain. He pictured an idealized crystalline solid, its regularly spaced atoms jiggling about with the same frequency, independently of each other in a three-dimensional array. Einstein knew that this picture was not completely realistic because the atoms don't move independently, but he hoped that it was a simplification he could get away with. By also assuming that each atom's vibrational energy is quantized, he predicted that the average energy of the atoms in a solid falls slowly with temperature, until it is zero. His predictions agreed well with previously puzzling measurements taken twenty-five years before. This successful comparison, the first of only three instances in which Einstein published a graph to compare his theoretical predictions with experiment, immediately gave credence to the emerging quantum theory. Because radiation was not involved, physicists could be impressed without having to wring their hands over the troubling contradictions with Maxwell's theory.

Mainly as a result of his relativity theory, Einstein's reputation had spread rapidly among physicists, many of them amazed to discover that the author of this great theory was working eight hours a day in a patent office. The first to recognize Einstein's talent was one of the theoreticians he most admired, Max Planck. The men first met in September 1909 at a conference in Salzburg, where Planck invited Einstein to take a few days off work at the office to give what was to be his inaugural lecture to the luminaries of theoretical physics. To their surprise, Einstein chose to speak not on his acclaimed and profoundly fashionable relativity theory, but on 'The nature and constitution of radiation'.

His talk was a tour de force. In keeping with the tradition of Clausius, he concerned himself neither with experimental odds and ends nor with mathematical detail, but with issues of high principle. His audience must have been startled to hear this new member of their fraternity: he did not give the diffident presentation of the tyro, rather he took up the cudgels and set out a new research manifesto on the nature of light. He took forward his already controversial ideas about radiation quanta when he argued that, like an electron, every quantum travels in a specific direction – in technical language, it has momentum. For the first time, Einstein suggested publicly that radiation consists of particles. He also pointed out that because relativity theory had rendered the ether superfluous, it was no longer necessary to think of radiation as existing in anything, but 'as something existing independently, just like matter'. So important was the problem of understanding radiation, in Einstein's view, that 'everyone should work on it'.

Einstein's audience must have been as nonplussed as Schoenberg's were at the premiere of his second quartet nine months before in nearby Vienna. Planck, speaking for almost all his colleagues, noted respectfully that it was too early to give up Maxwell's equations and he was reluctant 'to assume the light waves themselves to be atomistically constituted'. Planck had, however, softened his opposition to the quantum idea and had come to accept reluctantly that when radiation interacts with matter, energy is transferred to the constituent atoms in discrete quanta. But, like virtually all other physicists, he certainly did not accept that the energy of radiation itself comes in quanta.

In July 1909 Einstein handed in his notice at the patent office in order to take up his first academic post as an extraordinary associate professor of theoretical physics at Zurich University. He spent most of his time thinking about his favourite problem – understanding light quanta. He tried many different approaches and even tried tinkering with Maxwell's equations, but to no avail. The language he used to describe radiation quanta was invariably circumspect and he never went the whole hog and stated unequivocally that the quanta exist; rather he talked of radiation behaving 'as if' its energy was quantized. No wonder many of his colleagues thought his commitment to the quanta was half-hearted.

Among those who had been struck by Einstein's ambivalence towards light quanta was the American Robert Millikan, an enormously energetic experimenter with a penchant for tackling the most pressing questions of the day. In 1912, during a six-month sabbatical from the University of Chicago, he visited Berlin colleagues and spent a good deal of time with Planck, who shared his interest in measurements of fundamental constants. The two also discussed radiation quanta and Planck made clear how vehemently he disagreed with Einstein's ideas. Like many visitors, Millikan was invited to the Plancks' home for one of their musical soirées. Planck accompanied a recital of German lieder by his wife, and he improvised skilfully on the piano, Millikan recalled forty years later.

Millikan was quick to see how important it was to do experiments that would shed light on how radiation ejects electrons from matter, the photoelectric effect. Although he didn't for a moment believe that Einstein's theory of the effect was correct, he appreciated that it made eminently testable predictions that contrasted sharply with ones made by the competing theories. Here was an opportunity, if he could overcome the difficulties of doing the experiments, to shed light on what leading physicists regarded as the most pressing problem of the day – understanding radiation. As soon as he returned to Chicago, Millikan began his photoelectric experiments, which were to take him three years to complete.

Meanwhile, Einstein's light quanta had attracted little interest and few advocates. In June 1913 Planck confidentially put his reservations in writing when he and three of his Berlin colleagues proposed Einstein for the prestigious membership of the Prussian Academy of Sciences. In what was otherwise a cloying tribute to Einstein's achievements, Planck apologized for his younger colleague's having 'gone overboard in his speculations' with this work, and he asked that this 'should not be held against him too much, for without occasional venture or risk no genuine innovation can be accomplished even in the most exact sciences'. Nor was Einstein wholly complimentary about Planck, whom he described as 'stubbornly attached to preconceived opinions that are undoubtedly false'.

Both of them, however, were intrigued by news from Copenhagen that Denmark's brightest young physicist had successfully applied quantum ideas to the structure of atoms. Niels Bohr, then a young man of blazing ambition, had elaborated on a popular picture of the atom as a tiny nucleus surrounded by orbiting electrons. In a trio of papers published in 1913, Bohr decreed that the electrons in every atom could have only certain allowed orbits, corresponding to the atom's characteristic energy values.[17] Although it made no sense in terms of classical physics, the theory explained at a stroke why every atom gives out and absorbs light of certain characteristic wavelengths – they corresponded to quantum jumps between the atom's energy values. Better still, the theory made predictions that agreed with experiment: Bohr's picture of the atom accounted wonderfully well for the wavelengths of the light given out and absorbed by the lightest atom of all, the hydrogen atom. This success quickly persuaded quantum theorists to focus their efforts on understanding atoms. Einstein thought carefully about this problem and later showed how Bohr's model of the atom explained Planck's cavity-radiation law, using a picture in which the radiation quanta, each with the energy $E = hf$, were given out and absorbed by the atoms in the cavity.

Four months before the outbreak of World War I, Planck wooed Einstein to Berlin from Prague, where he had been working for a year. Planck was rejoicing in the orgy of patriotism that preceded the war and later publicly supported the German cause, while the pacifist Einstein deplored the conflict, which he regarded as an international eruption of insanity. The two men put aside their political differences, however, and they became close. Within a few weeks of Einstein's arrival, Planck invited him to one of his musical evenings, where they played Beethoven's D major piano trio with a professional cellist, Einstein a little ragged in the soaring violin part, maestro Planck revelling in the operatic second movement. The two great physicists enjoyed making music together; for the next eighteen years, while they were fatally undermining the foundations of classical physics by day, they liked nothing more than to explore the classical chamber canon by night. Not for them the musical revisionism and dissonance of Berg's *Wozzeck*, Stravinsky's *Oedipus Rex* and Strauss's *Salome*.

By the spring of 1915, Millikan had completed his photoelectric experiments and had been amazed to find that Einstein's photoelectric law was correct. But that did not necessarily imply that his quantum picture of radiation was right, Millikan pointed out.[18] In the very first sentence of the paper in which he published his results, he wrote that Einstein's law 'cannot in my judgement be looked upon at present as resting upon any sort of satisfactory theoretical foundation'. Millikan's results nonetheless gave additional credence to the $E = hf$ equation, as Einstein had used it when he wrote down his photoelectric law.

The congenitally sceptical scientific community was not going to be persuaded by a single piece of experimental evidence in favour of an equation that appeared flagrantly at odds with dozens of other experiments. Some physicists disputed Millikan's results and several groups spent years checking and extending his results, although none of them found any serious disagreement with Einstein's predictions. But that certainly did not prove that radiation consisted of particles: the photoelectric-effect theory was about the radiation's energy – it assumed nothing about the directions in which the quanta were travelling. To prove the particle model of radiation, experimenters would need to demonstrate that each radiation quantum speeds through space in a specific direction. Millikan's results, together with an increasing confidence in the underlying theory, convinced Einstein by 1917 that particles of radiation really did exist. 'I do not doubt any more the *reality* of radiation quanta,' he wrote to a friend a few months later, 'although I stand quite alone in this conviction', forgivably forgetting the handful of lesser scientists who were broadly in agreement with him.

So matters stood when Einstein became a front-rank international celebrity in early November 1919, when British astronomers announced that their results appeared to confirm the superiority of Einstein's theory of gravity over Newton's. Exhausted and demoralized by the war, the world was ready for the consolations of hero-worship.

Haydn once remarked that although his friends often flattered his talent, he always knew that his younger colleague Mozart was far above him. Planck

believed the same thing about his relationship with Einstein whom he called
'the new Copernicus'. Planck was therefore among those least surprised when
Einstein became that great rarity – a celebrity scientist. His work on relativity – never his work on quanta – became the butt of jokes in nightclubs all
over Europe and, no doubt, a theme of modish chatter at Jay Gatsby's parties.
Many people thought that Einstein was preoccupied by this work but the
truth is otherwise: he later wrote, 'I have thought a hundred times as much
about quantum theory as I have about general relativity theory.'

'Einstein fever' gripped New York in April 1920, when the newly famous
scientist sailed into the city to be greeted by thousands of cheering admirers lining the streets. Over in St Louis, the arguments over his particle
description of light were about to be settled in a series of experiments by
Arthur Compton, one of the most talented young physicists in America.[19]
Compton was thinking about the scattering of radiation waves by electrons,
and he pictured it much like water waves bouncing off buoys in a harbour.
If this picture were correct, the wave's frequency would be the same after
the scattering as it was before. Yet Compton found that the two frequencies
were different: the radiation before the scattering had a higher frequency
than it had afterwards. It was as if the mere act of scattering had turned blue
light into red. After years of trying, Compton had not been able to explain
this strange discovery, either by revising widely held assumptions about the
size and shape of the electron or by adapting the quantum picture of radiation. Then, in November 1922, the fog lifted.

The key to the problem, Compton realized, is that radiation quanta each
have momentum, though he was unaware that Einstein had proposed this
several years before. So the radiation as well as the electron could be pictured as particles and their collision was like a microscopic version of
billiards. No slouch as a theorist, Compton quickly worked out the consequences of this and found that his new theory accounted beautifully for the
results of his experiments. The explanation hinged on the $E = hf$ equation:
the energy of the scattered X-ray photon is lower than the energy of the
original photon because some of the energy of the original photon is transferred to the electron. So it follows that the frequency of the scattered
radiation should also be lower. Compton promptly announced his results
three weeks before Christmas at a physicists' meeting in freezing Chicago.
The sensational news that radiation quanta really do behave as particles –
each with both energy and momentum – spread like a tsunami across the
usually calm waters of the international physics community.

Like every truly remarkable observation, Compton's was doubted by the cautious and the disbelieving. Two years of checking followed, but by the end of 1924 the consensus was that Compton's results were correct and that he was right to say that his discovery entailed 'a revolutionary change in our ideas regarding the process of scattering of electromagnetic waves'. The hegemony of Maxwell's wave picture of light was over. Unaware of Einstein's earlier thinking about the particle nature of radiation, Compton did not mention Einstein when he wrote up his discovery for publication. But Einstein was pleased, as well he might have been – he had become the first person successfully to predict the existence of a fundamental particle.

Einsteinons would have been a plausible name for the new particles but, mercifully, no one suggested it. But in 1926 the American chemist Gilbert Lewis came up with an appellation that immediately proved popular. He coined it in a now discredited paper in which he introduced the idea that radiation comes in atoms. He called these 'atoms' by a name now universally used as a synonym for particles of radiation – photons.

IV

Meanwhile, a few scientists outside the Planck–Einstein circle had used the $E = hf$ equation to open another rich seam of the submicroscopic world. While the cream of the world's physicists had been trying to understand how – or whether – the equation applied to radiation, an obscure physicist working in France made the astonishing suggestion in 1923 that it described not just radiation but matter as well. It was this crucial insight that led swiftly to modern quantum mechanics.[20]

The little-known physicist was Louis de Broglie, a thirty-year-old graduate student who had begun his scientific career in earnest only two years before, after he had been demobilized from military service in World War I. De Broglie, a taciturn and reflective man, was no ordinary scholar. He was a bona fide prince and a member of one of France's most illustrious families (his left-leaning great-great-grandfather Charles welcomed the French Revolution but was nevertheless condemned to the guillotine a month before the fall of Robespierre). De Broglie graduated in history and law at the Sorbonne in 1913, but decided to take up a career in physics, thanks to the persuasiveness of his brother Duc Maurice de Broglie, a distinguished experimenter who was wealthy enough to fund a laboratory in his own house. Louis's plans were interrupted after only a few months, when he was called up for military service. He later spent most of the war working without special distinction as a radio engineer at the Eiffel Tower, after it had begun to be used as a post for wireless telegraphy.

De Broglie began his doctorate on quantum theory by working in his brother's private laboratory in his multi-storey town house, two minutes' walk from the Arc de Triomphe and a stone's throw from the Champs-Elysées. While Compton was puzzling over his data in St Louis, Maurice de Broglie was inducting his brother into the latest experimental techniques and drawing his attention to the need for both wave and particle models of radiation. Louis later wrote that 'long conversations with my brother about the properties of X-rays . . . led me to profound meditations on the need of always associating the aspects of waves with that of particles.' It was this line of thought that led him to ask a profound question whose answer was to change the course of science: if electromagnetic waves can behave like particles, can particles such as electrons behave like waves?

Einstein had touched upon this theme a year or so before, but Louis de Broglie's insight was deeper and more vivid. De Broglie later recalled having a Eureka moment, much rarer in scientific reality than in folklore: 'After long reflection in solitude and meditation, I suddenly had the idea, during [August] 1923, that the discovery made by Einstein in 1905 should be generalized to all material particles, notably to electrons.' De Broglie had hit on the idea that the equation $E = hf$ applied to matter as well as to radiation. When Einstein used the equation to describe radiation, he had to explain how the energy of a quantum could be linked to a property of a wave, its frequency. De Broglie had the opposite problem: everyone was familiar with electrons and the idea of other particles of matter, but what on earth was the wave associated with it? According to de Broglie, every particle has some sort of associated matter wave. Using an analogy to the familiar formula $E = hf$, he wrote down a simple formula for the wavelength of the wave associated with a free particle, one with no net force acting on it.[21] As the particle's momentum increases, its wavelength shortens, de Broglie argued, and the size of the wavelength depended on the value of Planck's constant h.

'It looks crazy, but it's a completely sound idea,' Einstein told a colleague in December after he had read de Broglie's PhD thesis. It was all very well to have the master's imprimatur, but what did nature think of it? De Broglie had quickly seen that if matter has wave properties, then a beam of electrons should be spread out – diffracted – like a beam of light. It was, however, a student who saw most clearly how the theory could be checked. Walter

Elsasser, a young physicist at Göttingen University in Germany, pointed out that if de Broglie was right, a simple crystal should diffract a beam of electrons shone on it. Elsasser calculated that if the electrons are accelerated by 150 volts, they should have a wavelength of a tenth of a billionth of a metre, just less than the spacing of atoms in a typical metal. These are just the right conditions for diffraction, so if de Broglie was right, experimenters should be able to detect peaks and troughs in the number of electrons that spread out from a crystal at different angles. Like water and light waves, electrons should also be diffracted.

'Young man, you are sitting on a gold mine,' Einstein told Elsasser. However, the student was to be no more than a catalytic bystander when two experimenters brought his idea to fruition. In August 1926 they had the good fortune to attend a meeting in Oxford, where the theoretical ideas of de Broglie and Elsasser were in the air. The meeting was the annual conference of the British Association for the Advancement of Science, where scientists mingle with each other and the public. Although the two experimenters did not meet at the conference, they both left thinking about how they might confirm de Broglie's ideas. Soon afterwards, they had closed the final chapter of the $E = hf$ story.

The first of the experimenters was Clinton Davisson, a slight and rather frail man who had slowly built his reputation as an electron-scattering expert through work he had carried out at the Bell Telephone Laboratories on West Street in south Manhattan, a few blocks from the meat markets.[22] For five years he had been pursuing a programme of routine but increasingly accomplished experiments to find out what happens when a beam of electrons strikes a metal target. To smash things together like this may seem a crude way of doing science, but experiments like these have proved especially fruitful – witness the success of the post-World War II atom-smashers. Davisson's results were nothing special until April 1925, when a liquid-air bottle in his laboratory exploded, one of the most fortunate laboratory accidents in the history of science. Before the incident, Davisson's nickel target consisted of a random patchwork of tiny crystals, but after he had treated it to put right the effects of the explosion, he inadvertently changed the nickel sample into virtually a single crystal. As soon as he used this new sample to scatter his electron beam, his results changed completely, but electron diffraction was still not in evidence.

Davisson had taken his results to the Oxford conference and had sailed home with an inkling that quantum theory might explain his data. On returning to New York, he used de Broglie's theory to predict the location of the diffraction peaks, but they were nowhere to be found. Undeterred, he and a colleague set out a detailed research programme designed to check once and for all whether electrons really were diffracted. His reward duly arrived in early January 1927, when he detected several pikestaff-plain diffraction peaks. A few simple calculations demonstrated that they occurred precisely where de Broglie's formula predicted, inspired by an analogy with the $E = hf$ equation. Here was proof of something that barely five years before no one had even contemplated – that particles of matter could be diffracted.

Three thousand miles away, in the dour granitic city of Aberdeen, the English physicist George Paget Thomson was meanwhile doing similar experiments with electron beams, but with much higher energies than Davisson was using.[23] After Thomson had returned from the Oxford conference, he began working with a student to try to detect electron diffraction, not from a single crystal but from specially prepared thin films. The experiments were strikingly successful: using a celluloid film, Thomson observed a diffraction pattern whose shape was exactly as de Broglie's formula predicted. This observation completed a unique double for the Thomson family: George Paget Thomson discovered that the electron was a wave, only twenty-eight years after his father 'J.J.' had discovered that it was a particle. And, as in so many father–son disagreements, they were both right: the electron does behave like a particle when experimenters probe its interactions, whereas it behaves like a wave when they study its propagation.

The upshot of all this is that both light and matter can behave like particles and like waves. So ends the $E = hf$ story. Today's scientists routinely use the equation with scarcely a thought for the twenty-six years of work it took to tease out its meaning. The equation is best known for giving the energies of photons, not so much for the crucial role it played in stimulating debate about the duality of matter. As de Broglie was the first to see, radiation and matter are both Janus-faced: they show their particle face in their interactions, and their wave face in their propagation. All science students are confused when they come across this duality, and you can hardly blame them – it perplexed the world's best scientists for years. The problem's resolution came in the late 1920s, when physicists developed quantum theory into what became known as quantum field theory,[24] which made possible a unified description of radiation and matter. Among all the formidable mathematical symbols of quantum field theory, the simple $E = hf$ equation remains, a pristine relic of quantum history.

V

'There are scientific revolutions, but there are no revolutionaries,' suggests the science historian Simon Schaffer. For him, it is undeniable that there are occasionally major shifts in the way scientists think about nature and study it, but he does not believe in the usual picture of the scientific innovator as a lone hero who changes the way science is done after a flash of inspiration. 'No one spearheads a scientific revolution on their own. Not Planck, not even Einstein.'

Does the story of the energy quantum bear out Schaffer's view? There is no doubt that this was a revolution – the acceptance of quantum theory entailed a complete break from classical tradition. Planck and Einstein certainly deserve their reputations as great scientists, but can we attribute the origins of the quantum revolution solely to either of them? Scientists traditionally regard Planck as the father of quantum theory but it is far from clear whether he initially understood the extent of the changes in our thinking about energy that his first quantum theory entailed, as Thomas Kuhn has stressed. For Kuhn, the quantum revolution began with Einstein. But is it fair to regard him as a lone revolutionary? He undoubtedly drew heavily on Planck's work, and for some fourteen years Einstein's advocacy of radiation quanta was decidedly tentative. Even in 1924, when he was enthusing about the importance of Compton's proof of the particle picture of radiation, Einstein did not go the whole hog and say that the particles were real. He wrote in a newspaper article, '[The Compton experiment] proves that radiation behaves as if it consists of discrete energy projectiles . . .' Note: 'as if'.

In my view, it is best to think of the quantum revolution as the work not of a single revolutionary but of a disorganized junta. It had no manifesto, no badge of loyalty, and some of those who belonged to it would have found life a good deal more congenial under the *ancien régime*. It is not even clear when the junta formed and who were its bona fide members. In addition to Planck and Einstein, it would have to include de Broglie and a bevy of experimenters: the Reichsanstalt cavity-radiation experts who supplied Planck with the crucial data; Robert Millikan, who verified Einstein's photoelectric law; Arthur Compton, who first demonstrated the existence of photons; and certainly Clinton Davisson and George Thomson, who proved that electrons can behave as waves.

This junta had its share of dissidents, too. Millikan and Compton, for example, did much of their great experimental work while refusing to believe the quantum explanations they later espoused. No one fought harder than Planck to preserve the heritage of classical physics, and it is a moot point whether he ever fully signed up to all the implications of the quantum revolution. In 1927, two years after virtually the entire science community had accepted that photons exist, Planck wrote a short article for the Franklin Institute in Philadelphia explaining why he still was not fully prepared to believe in them. Nor did he ever change his mind, so far as I know.

Planck's friendship with Einstein had always been the closest relationship in the junta, but the amity came to an end when Planck made an accomodation with Hitler. They first met in 1933, shortly after Hitler became chancellor, and Planck made numerous concessions to the Nazi regime – he regularly gave the Hitler salute, lectured in halls decorated with swastikas, and made speeches in which he pointedly omitted the names of leading Jewish physicists, including Einstein. Appalled as he was to see the Führer's corruption of the apparatus of state, Planck's devotion to the institutions of German physics, his patriotism and, arguably, his gullibility led him to make increasingly more humiliating compromises. In many ways he behaved like a character from an Ishiguro novel, as someone whose way of looking at the world we see gradually exposed as being more and more unequal to events.

Towards the end of the war, on the night of 15 February 1944, an air raid destroyed the Grünewald suburb where Planck had lived for some fifty years. His library, in which he probably first wrote down the $E = hf$ equation, was destroyed along with the rooms in which he had made music with Einstein. Planck's stoicism was to be tested almost to breaking point: he had lost his elder son in the trenches of World War I, and both daughters in childbirth, and his younger son was murdered by the Gestapo in February 1945.

Death finally came to Planck, almost as a redemption, in October 1947, a few months before what would have been his ninetieth birthday. At his commemoration ceremony, on a bright spring day in Göttingen, chamber music by Bach, Brahms and Beethoven interspersed the eulogies. It is a safe bet that the mourning physicists all thought of the photon as a particle like any other, yet Planck had never wholeheartedly been able to join them. The explanation may lie in words that he had written shortly before, in his autobiographical notes: 'A new scientific truth does not triumph by convincing its opponents and making them see the light, but rather because its opponents eventually die, and a new generation grows up that is familiar with it.'

Einstein was one of the dignitaries who sent a message to the mourners. Although he never forgave Planck for working with the Nazis, Einstein wrote a touching tribute to his former friend, beginning, 'A man to whom it has been given to bless the world with a great creative idea needs not the praise of posterity.' Einstein composed those words at the Institute of Advanced Study in Princeton, where he had lived since 1933, having fled Germany the previous December, seven weeks before the Nazis came to power. He never set foot in the country again.

In the eyes of most of the world, Einstein was an authority capable of giving God a boost just by giving him a mention, as Saul Bellow later wrote. But for most of Einstein's fellow scientists he was by then an embarrassingly semi-detached figure, obstinately refusing to accept the fully fledged quantum theory that physicists had formulated in the 1920s. Einstein was, as always, happy to work on his own, to be an intellectual loner. Music continued to be a passion. Although he had given up playing the violin shortly after World War II, the Juilliard Quartet did once persuade him to give an impromptu performance when they visited him in his home in the autumn of 1952. Asked to select a piece, Einstein unhesitatingly chose Mozart's soulful G minor quintet and played second violin, scarcely referring to the score.

If Planck was the Moses of quantum science, Einstein was its Joshua. From his Mount Pisgah in Berlin, Planck saw the Promised Land and he led his followers to it, but he could never make the journey himself. It was Einstein who took them there, though for him the journey was not yet over, as is plain from his final words on radiation quanta, written in December 1951: 'All these fifty years of brooding have brought me no nearer the answer to the question "What are light quanta?" Nowadays, every Tom, Dick and Harry thinks he knows it, but he is mistaken.' To understand the photon properly, Einstein believed, quantum theory was not enough.

Fifty years later, the theory has yet to be troubled by conflict with experiment, let alone to be superseded. If, however, such a counter-revolution were to come, Einstein would be seen as its junta's founding member, posthumously vindicated. And his celebrity would be given another boost in a way that would appropriately underline his bravest contribution to science.

Acknowledgements

Many friends and colleagues have given me valuable advice during the preparation of this essay. It is a pleasure to give special thanks to David Cahan, Stuart and Corinne Freake, John Heilbron, Dieter Hoffmann, Russell McCormmach, Simon Schaffer, Chuck Schwager and Andrew Warwick.

The Sextant Equation

$$E = mc^2$$

Peter Galison

On 17 November 1945, John Wheeler, Princeton physicist, Manhattan Project veteran and herald of a new age of physics, stood before a symposium audience to survey the state of his science. He began by recalling that first moment of the nuclear age, early in the war, at the University of Chicago. A key figure in the war effort had telephoned Washington to tell the president of Harvard and head of the National Scientific Research Board, James Conant, about the events that the refugee physicist Enrico Fermi had just directed: 'The Italian navigator has discovered America.' 'Splendid,' Conant replied, 'and is the new country safe to enter?' The report: 'Yes and Columbus finds the natives are friendly.' That was 2 December 1942, and the coded discussion told Conant and those responsible for the American scientific war effort that the world's first nuclear reactor had safely begun a self-sustaining chain reaction. Physicists had landed on the continent of applied nuclear fission where they could begin to imagine producing power or detonations from energy buried in the heart of the uranium atom. Over the following thirty-two months the scientists of the atomic-bomb project drove relentlessly towards the delivery of nuclear arms, ending, or rather pausing, in the cataclysmic blasts over Hiroshima and Nagasaki in August 1945.

Now, as Wheeler was speaking, just three months had passed since the war's end. Physics, not long before a relatively obscure academic redoubt, now lay front and centre in the nation's attention. Surveying physics and the society around it, Wheeler had a vision of the 'formation of the new world' augured by nuclear physics, 'the great continent which lies beyond [fission] and which represents the last untraversed portion of knowledge of the physical universe'. Mathematicians at the time of Columbus, Wheeler commented, could delude themselves about just how far the explorer had reached in his quest to circumnavigate the globe. Physicists of the mid-1940s, by contrast, could not deceive themselves about what remained to be found. For scientists now held in their hands a sextant of a simplicity that left no room for self-deception. Such a theoretical instrument, such a measure of scientific progress, would at any given moment tell the human race just how far it had progressed towards the total annihilation of matter into energy. Powerful as it was, uranium fission took humankind but a thousandth of the way towards the goal of total energy conversion, for only a thousandth of the mass of a uranium atom blew into pure energy when the uranium nucleus split. A pure swap of matter into energy would, by contrast, provide the final limit to energy production, the ultimately efficient production of energy that could be used to construct a new industrial world. Or to provide a weapon of unequalled power. And the sextant of modern science giving the measure of success, showing humankind its precise location on the scale of total conversion, was Albert Einstein's $E = mc^2$, the most famous equation in the history of science.

What this means is that if a mass of m grams is lost in the splitting of a uranium atom (the parts weigh less than the whole), then the amount of energy released in that fission process is E (in ergs), where E is given by the mass times the speed of light in a vacuum (30 billion centimetres per second) squared. Surprisingly enough, in Einstein's first paper he did not use E for *Energie* or Energia in German and Greek respectively, and c for *celeritas* (swift in Latin), but rather L for energy (surely after *lebendige Kraft* 'living' or kinetic energy) and V for the velocity of light. Although the particular symbols of $E = mc^2$ feel inevitable to us now that we have grown used to those particular symbols, Einstein settled on the E and c only in 1912. Energy can be released in various forms – in the simplest possible version of nuclear fission, a uranium atom divides into two smaller nuclei flying away from each other at a furious rate. The energy released by the splitting of a single uranium atom would be enough to budge a grain of sand visibly off a table; releasing the fission energy contained in the million billion billion atoms of a kilogram of uranium would – and did – destroy several square miles of city.

By the end of 1945, fission, the physics of nuclear reactors and atomic bombs, still held open questions, but it was, in large measure, an understood science. Beyond the cascading neutrons of the fission chain reaction (neutrons splitting nuclei in such a way that more neutrons emerged to split other nuclei that in turn made more neutrons . . .), however, still lay a panoply of problems entirely outside the physicist's command. How do neutrons and protons produce new particles through collisions? Startling new information about these novel processes arrived every month from experimenters' observations of the cosmic rays, mostly protons, that rained down on the Earth's upper atmosphere from deep space. Again Wheeler: 'The possibility of the complete conversion of matter to energy is suggested by present incomplete information on the production of particles of lower mass by or from protons in the upper atmosphere of the Earth.' Wheeler dreamed of a process that would convert *all* of a piece of matter into energy.

Grasping the nature of these particle transformations fascinated Wheeler and his contemporaries. Soon he was launching teams of physicists on quests high in the atmosphere using bombers just back from the war front; Wheeler joined captured German scientists at the White Sands Proving Grounds where they fired unmanned V-2 missiles laden with instruments over a hundred miles into the threshold of space. Glimpses of high-energy particles from deep space beckoned – there was physics to be had up there, but the particles were too rare to be the basis for a full-scale campaign of physical research. What was needed was a consistent and copious source of energetic particles – in this respect, deep space could not compete with the campaign to build larger and more powerful particle accelerators. Needed too would be observations that would record the changes induced in bits of matter when struck by high-energy particles. And finally, physicists would have to produce a new, consistent theory that would capture the relations between elementary particles and the forces that governed their interaction.

The sextant equation $E = mc^2$, according to Wheeler, would guide physicists as they manipulated accelerators, cosmic rays and theories towards the creation of a new field of science: elementary-particle physics. So it has – over the next decades particle accelerators pounded stationary targets with ever-faster projectiles and then shifted towards ramming particles into their antiparticles. Electrons slammed into positrons and protons into antiprotons, each cutting-edge accelerator upping the amount of energy produced and pushing further into the physics of the very small. From the late 1940s into the first decade of the twenty-first century, that burgeoning domain of accelerator-based physics used energy-mass conversion to call into observable existence the basic constituents of matter. Starting with the proton, the neutron, the electron and the positron, the particle zoo's population proliferated as physicists used the energy produced in collisions to create new and different kinds of particles. Already in 1932, the positron, antiparticle to the electron, appeared in an experimentalist's vessel, showing dramatically that matter and antimatter annihilate each other to produce pure energy, and conversely that pure energy could produce a particle and its antiparticle twin.

In the decades after World War II, it became possible to produce and then to manipulate particles like the pion that were of intermediate mass between the proton and the electron. Heavier, excited versions of protons and neutrons issued from the collisions of protons and mesons on nuclei – and the menagerie grew. When electrons and anti-electrons, pions and antipions, or protons and antiprotons could be skilfully battered against one another, their annihilation was complete and the totality of their conjoint energy became available for the production of new subatomic entities. Over the 1960s and 1970s, these pairs of particles were joined by subnuclear quark-antiquark pairs in their various combinations, along with heavier versions of the electron and new force-carrying particles to form the 'standard model' of particle physics. It is the equation $E = mc^2$ that lies behind the enormous accelerators that for three decades have driven particles into their antiparticles. Directly out of these colliding-beam facilities came the canonical formulation of particle physics of the 1970s. It has since remained essentially intact.

In those months towards the end of World War II, the interconvertibility of energy and mass held limitless promise and threat. In June 1945, Wheeler mused, 'Discovery of how to release the untapped energy on a reasonable scale might completely alter our economy and the basis of our military security. For this reason we owe special attention to the branches of ultra-nucleonics [physics beyond the by then rather well-understood physics of nucleons, that is neutrons and protons].' That more distant field would embrace new physics not seen in the wartime laboratories: cosmic-ray phenomena, meson-physics field theory, energy production in supernovae, and particle-transformation physics. Abstract inquiry into physics that probed below the scale of the nuclear, according to Wheeler, would clearly fuse with the 'country's war power'. For Wheeler knew perfectly well that among the tasks of 'ultra-nucleonics' lay the possibility of a more complete use of the energy proclaimed by $E = mc^2$ than that tiny thousandth part liberated by nuclear fission.

Fission's only partial release of energy meant that Hiroshima had been destroyed by the conversion of mass weighing considerably less than the eraser at the top of a pencil. Such thoughts had led Wheeler – and many other physicists – to wonder whether the sextant equation might point the way towards a much more complete release of energy.

Before the site of the Los Alamos weapons laboratory was anything but a country boys' school, a small group of other nuclear illuminati gathered at Berkeley to discuss nuclear weapons. J. Robert Oppenheimer was there as America's foremost quantum theorist. So was Hans Bethe, the physicist who, before he fled 1930s Germany, had figured out the nuclear physics that explained why the sun shines. They were joined by a stellar group, including the Hungarian refugee Edward Teller, later known as the 'father of the H-bomb'. In the brash hothouse environment of those early days, fission weapons seemed trivial to them: ram enough fissile uranium together and it would detonate. They assigned the problem to a young Berkeley physicist, Robert Serber, arrogating for themselves an infinitely more subtle and challenging problem: the hydrogen bomb, or H-bomb. The H-bomb would work by forcing together low-mass nuclei such as those of hydrogen, rather than prying apart heavy nuclei like uranium. But as the high-mesa laboratory of Los Alamos began to take shape, it became ever more obvious that building an A-bomb was anything but trivial. Project leaders, including Oppenheimer and the German refugee Hans Bethe, shunted the H-bomb aside in order to produce a usable weapon by war's end. Edward Teller, however, held tenaciously to the idea and, over the course of the war, moved determinedly away from the mainline fission work towards the defence and development of the weapon that gripped his imagination.

On 12 August 1945 Wheeler, then on the Pacific island of Tinian, which served as a staging area for the nuclear strikes, penned a letter to Teller: 'Dear Edward, with the conclusion of the war today my work here will soon reach its conclusion . . . What I can now do most effectively is, I believe, fundamental research. But I do not feel quite at ease to do so over the next five-year period.' He recalled Teller's previous invitation to work on the fusion weapon, and his own conviction that the H-bomb was a weapon destined for the next war, not the present campaign against the Axis. With the Japanese surrender, contemplation of that next conflict had for Wheeler become inevitable – he expected war with the Soviets to occur in the very near future. And that would be a conflict in which fusion, converting a higher proportion of matter into energy (according to $E = mc^2$), would be vital. For security reasons Wheeler went metaphorical:

Here is a group of men absolutely isolated on an island. They have got into a fight. Two groups of men with quite different ways of doing things have teamed together to try to put down the troublemakers. Our group has learned to put together a bow and arrow. By that means we have put an end to the fighting. Our ally is observant. Now that the fight is over he has gone back and is spending part of the time behind his wall. We know that some of his men would get delight out of building a bow and arrow of their own. We suspect that some of his men would not hesitate to use that bow and arrow on us if someday we happened to get into a disagreement on who is to get the pears from that fine-looking pear tree over there. For some reason or other the two former allies don't seem to be able to get together to turn the bow and arrow weapon over to a custodian whom both can trust . . . Some people in our group say, 'So what' and are making plans to go fishing. I'm one of the people who feels that if we're going to get into an armament race, we'd better start now, and we'd better try to build the best weapon we know how to build – a machine gun which will outmode the bow and arrow.

Wheeler ended by saying that he thought he had better begin thinking about the 'machine gun' if conflict might break out over the next five to ten years. He did, launching a major Princeton-based H-bomb design effort known as 'Matterhorn B' alongside his more pacific inquiries into the transformations of mass into energy. In fact, just next door to Wheeler's part of the Matterhorn project stood another – with the aim of producing energy for civilian use through nuclear fusion. But with nuclear fusion, a thousand times more energy would be released per nuclear collision than in fission. Suddenly one could imagine that bombs the physical size of the ones used against Hiroshima and Nagasaki might deliver the explosive equivalent of 10–20 million tons of TNT, not the 10,000–20,000 tons of TNT equivalent to the World War II atomic bombs. And in principle one could picture making bombs of unlimited destructive capability – within a few years people began to discuss the production of gigaton hydrogen bombs that would blow a hole all the way through the atmosphere. Throughout, Wheeler saw the sextant equation as a compass that would give coordinates on a map leading towards the total conversion of matter into energy, for weapons and for wisdom. For despite its awful destructive force, even the hydrogen bomb still left much of the original mass unconverted to energy.

One peaceable exploration led Wheeler to imagine a new kind of atom: an electron and a positron orbiting one another. After a mere ten-billionth of a second, the new 'positronium' atom would decay as the two partners fell into each other in mutual annihilation, releasing their energy as two photons. Here was a beautiful, pure example of $E = mc^2$: if the positron and electron each had mass m, then the energy released would be $2\ mc^2$ and the two photons would each have a frequency f given by the equation $hf = mc^2$ (since, as Einstein had shown back in 1905, the photon's energy is $E = hf$.) One could look for those energetic photons that flew off back to back. Not too long afterwards physicists at MIT found them in an experiment, bang on the frequency where Wheeler – using Einstein's equation – said they would be.

From there a myriad of other transformations seemed to beckon. Nucleons smacked into each other in cloud chambers (vats of water vapour that made visible the tracks of particles) and yielded a host of new entities. How did these 'nuclear explosions' increase in probability with the energy of the incoming particles? What characterized the kind and number of explosion products? As far as Wheeler and many of his colleagues were concerned, answering questions like these would take the world of physics ever closer to a full understanding of their marvellous sextant $E = mc^2$.

Fission, fusion, positronium, accelerators, cosmic rays, black-hole dynamics – so much of late-twentieth-century physics ties back to that simply stated equation. But its origins lay far from the big physics of laboratories like Fermilab outside Chicago or CERN on the Swiss-French border, far from the weapons laboratories of Los Alamos, Livermore – or Arzamas-16. Young Einstein could scarcely have foreseen these developments when he first wrote down the equation.

We must go back to Einstein's world, the world surrounding him as he stood, as a clerk, in the patent office of 1905. This was a world in which electrification was a central pillar of modernization. Construction crews were ripping streets apart to build electric tramways, electricians were tearing gas lamps from ceilings and walls to make room for electric lighting. Crisscrossing their way across the United States, Europe and Russia, industrial power companies spun a great web of power lines, electric generators and measuring devices so they could deliver power to factories, cities and dwellings. Einstein's own family – his father and uncle – ran a fairly typical small electrotechnical business where they manufactured clocklike devices to measure power and other electrical quantities. Maxwell's equations, now taught in every elementary physics class, were in the 1870s still new enough to be only incompletely taught even in advanced schools, and Einstein clearly found the new theories and devices to be as fascinating as anything in science. The Bern patent office hired the twenty-three-year-old Einstein specifically to handle electrotechnical innovations – it was his job to assess their degree of novelty, to isolate and articulate the principles by which they worked.

It was from that Bern patent office, in his *annus mirabilis* of 1905, that Einstein published five extraordinary papers. Einstein's first article arrived at the journal *Annalen der Physik* on 18 March, displaying his theory of the light quantum; it was, in many ways, the paper that launched quantum physics. Six weeks later the young physicist submitted his doctoral dissertation, where he showed how to estimate the size of molecules by reasoning about the way in which big molecules like sugar contributed to the viscosity of sugar water. On 11 May Einstein submitted his account of Brownian motion, demonstrating the effect of physically real atoms and molecules pounding on small suspended particles – think of smoke dust diffusing through air. It was a powerful intervention for the reality of atoms – atoms, in Einstein's reckoning, were not merely helpful fictions to be used in reckoning chemical processes. They were physical objects statistically slamming into the suspended particle, bit by bit driving it around the liquid.

Our concern here is with Einstein's fourth and fifth papers, submitted at the end of June and September. For it was in those two papers, no doubt his most famous ones, that Einstein introduced the special theory of relativity and derived as a consequence the famous equation of energy and mass. The relativity piece itself, 'On the electrodynamics of moving bodies', built on two simply stated starting principles and moved towards predictions from there. Avoiding detailed assumptions about this or that feature of the way particular objects were built or interacted, Einstein's theory hardly resembled the work of senior physicists of the time. Instead, it had an outsider's style – or perhaps a return to an older form of clarity.

As in his ideal physical theory, thermodynamics, Einstein wanted above all to start with *principles*. In thermodynamics, all rests on the twin pillars of the conservation of energy and the ever-increasing entropy of the world. In relativity, Einstein had in mind two other founding principles. First, Einstein asserted, the old starting point of classical physics would hold good for electricity and magnetism as well. That is, physicists since Galileo had accepted the proposition that one could not use mechanical means to tell whether one was 'really' moving if one was in a constantly moving enclosed box. (Galileo imagined the observer to be below deck in a sailing boat cutting evenly across the open sea; Einstein, not surprisingly, chose the train, sliding along smooth steel tracks, as the site of his thought experiments.) Einstein's insistent message was that Galileo still spoke to us. In the windowless hold of an evenly moving ship you couldn't watch a fish swim in a fish tank or drop a ball or conduct any mechanical experiment that would reveal your 'true' motion. So, Einstein added, would it be impossible to conduct any experiment in a smoothly running train with electricity, magnetism or light that would reveal that one was 'truly at rest'. This is the relativity principle.

Einstein's second starting point was, he confessed, at first quite surprising: within an inertial reference frame (one not accelerated), light travels at the same speed independently of the velocity of its source. Sit in a railroad station and measure the speed of light from a lantern fixed on the front of a stationary train engine – its light emerges at 186,000 miles per second. Now imagine a train hurtling through the station at half the speed of light, 93,000 miles per second. In ordinary classical physics a ball thrown from the moving train (in the direction that the train is moving) would sail by the station at the speed of the train *plus* the speed the thrower gave the ball. Astonishingly, Einstein says, this is not so for light. Sitting in the station, you would see the lantern light from the front of the high-speed train travel by you at 186,000 miles per second – and not a bit faster. Moreover, applying the first (relativity) principle, if you ran after a light beam shining away from you, you would never even begin to catch up. Regardless of inertial reference frame, regardless of the speed of the source, light always will be measured to be travelling at the same speed that we abbreviate by c. This is the second principle: the absolute speed of light.

From these two simply stated propositions, the physical equivalence of inertial reference frames and the absolute nature of the speed of light, Einstein changed physics for ever. In the process he overturned notions of space and time that had been the foundation of physical understanding since the time of Newton. Completing this work in May 1905, he began shortly afterwards to reflect on some consequences of the new physics. From Bern, he wrote to his friend Conrad Habicht on a summer Friday of 1905:

> I would love to have you here. You would soon become your old mischievous self again. – The value of my time does not weigh heavily these days; there aren't always subjects that are ripe for rumination. At least none that are really exciting . . . A consequence of the study of

electrodynamics did cross my mind. Namely, the relativity principle, in association with Maxwell's fundamental equations, requires that the mass be a direct measure of the energy contained in a body; light carries mass with it. A noticeable reduction of mass would have to take place in the case of radium. The consideration is amusing and seductive; but for all I know God Almighty might be laughing at the whole matter and might have been leading me around by the nose.

Evidently persuaded that he was not causing God to laugh, Einstein wrote up his three-page $E = mc^2$ paper, 'Does the inertia of a body depend upon its energy content?', in September 1905; *Annalen der Physik* received it on the 27th of the month.

Before Einstein, there was already a great deal of discussion of how electromagnetic energy might be related to mass. In fact, some of the leading physicists of the day aimed to explain the existence of all inertial mass (the resistance of matter to being set in motion) as nothing other than the fact that charged particles, reacting to their own electric and magnetic fields, were hard to accelerate. Einstein himself never subscribed to such a reductionist programme; that is, one that aimed to show that everything, even inertia, was at root nothing but charge and electric and magnetic fields. It was also well established that a container of electromagnetic energy (a mirrored box filled with light, for example) would have a mass that rose in proportion to the electromagnetic energy that it held.

But Einstein was after far bigger fish – not content with an analysis of light, he was arguing that *any* form of energy had inertial mass associated with it. Not surprisingly, his $E = mc^2$ paper triggered debate. One of Einstein's allies, Max Planck – one of the leaders of German theoretical physics – lost little time in pointing out that a transfer of heat also adds mass. So, it seemed, a hot frying pan would weigh more than an otherwise identical cold one. This was new: nothing in Newtonian physics led one to expect that mass could possibly vary as a result of energy alone.

When Johannes Stark, a well-known senior physicist who later became an ardent Nazi, saw Planck's and Einstein's results, he attributed the discovery of the equivalence to Planck. That was too much for the young Einstein (who had not yet developed his Delphic style): 'I find it somewhat strange that you do not recognize my priority regarding the connection between inertial mass and energy.' Stark backed down quickly: 'You are greatly mistaken, esteemed colleague, if you think that I have not been doing sufficient justice to your papers. I champion you wherever I can, and it is my wish to be given the opportunity to propose you for a theoretical professorship in Germany quite soon.' To which a mollified Einstein replied, with regrets, that he had let 'a petty impulse goad me into making that remark about priority. . . . People who have been granted the privilege of contributing to the progress of science should not let their pleasure in the fruits of joint labour be spoiled by such things.'

Over the years following 1905, Einstein worked hard to generalize the result – to show that the equivalence of energy and mass was truly complete. Always pressed to come back to the equation, he offered three ways of deriving his best-known result. In the first, the original paper of 1905, Einstein imagined a body that emitted an equal burst of light back to back. Then he recalled how from the special theory of relativity he could look at the same situation from a different, unaccelerated reference system. Combining the two results he could deduce $E = mc^2$ but to show this properly one has to see precisely how energy transforms from one frame to another. Some twenty-nine years later, in a Pittsburgh lecture, Einstein presented a different argument for $E = mc^2$, this time using the fact that energy and momentum should be conserved in all inertial frames of reference. But it was his third, single-page argument that was the simplest: in 1946 he produced an $E = mc^2$ argument for the *Technion Journal* that required nothing from relativity theory but the basic assumptions. Let's take that last method and pause to consider Einstein's reasoning.

Suppose, as Einstein suggests, that one accepts four principles:

1) That the principle of special relativity holds good: that is, all reference frames that are not accelerated are equivalent. No one frame is 'truly' at rest, for example, only *relative* motions can be discussed in a physically meaningful way.
2) That momentum is conserved – after all, a fundamental article of faith even in classical physics. For ordinary matter, momentum is equal to mass multiplied by velocity. Conservation of momentum is the principle that if one adds up all the momentum, for example, for all the billiard balls on a table before they collide with one another, the same amount of momentum will be present after the collision.
3) That radiation has momentum – an experimentally tested result long accepted. (It is, for example, sunlight that pushes the tails of comets away from the Sun.)
4) That a moving observer sees a source of light as undergoing a change in its apparent angle ('stellar aberration'). In other words, it had

long been known that an observer on Earth, for example, sees starlight as coming from a position shifted by a small angle α from the star's true place in the heavens. That angle depended on the velocity of the Earth, v, and was generally accepted to be, for velocities that were small compared with the speed of light, c, approximately $\alpha = v/c$. (This effect is easy to understand. If rain is falling straight down towards the ground and you run through it, you experience the rain as driving towards you at a certain angle. The faster you run, the bigger the 'aberration' of the rain from straight down; the angle would depend on the ratio of your speed to the speed of the rain. If, as you ran, you had a 'telescope' consisting of a long cardboard tube, you would have to angle it away from the vertical to have the raindrops fall straight through the tube. Similarly, because of the Earth's motion, optical telescopes need to be angled from a 'true' star position to see that star's light.

Suppose too, Einstein added, that we have one reference frame, the 'rest frame', which we might anachronistically identify as the frame of a space shuttle that is floating, engines-off, in deep space far from any objects like stars or planets, that would exert significant gravitational forces on it (Figure 1). In this frame a book hovers without moving in the middle of the shuttle, before two flashlights spaced equally on opposite sides each simultaneously flash a burst of light of energy $E/2$ directly towards the book. The energy of both flashes is then absorbed by the book, so its energy increases by E. In the 'rest' reference frame of the space shuttle, the book doesn't go anywhere because it has been hit with equal impact by the light flashes coming from opposite directions.

Now, Einstein continued, let's look at exactly the same process from a different, 'moving' frame of reference (a Russian spaceship, say) moving

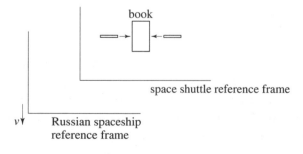

Figure 1

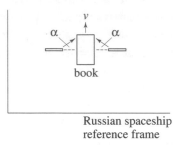

Russian spaceship
reference frame

Figure 2

steadily downwards with velocity v. Viewed from this frame the scene looks like slightly different. As observed from the Russian spaceship, before our worthy book is hit by the twin flashes of light, the book will be moving *upwards* with velocity v (Figure 2). This means that in the spaceship frame, before the light beams hit the book, of mass M, the book's momentum is just Mv. The classical theory of light tells us that the momentum of a light burst of energy $E/2$ is just $E/2c$. Now in the Russian spaceship frame, the flashes appear not to travel horizontally but (because of the aberration effect) to arrive at a small angle, $\alpha = v/c$ to the horizontal.

In the Russian spaceship frame, the book's momentum *after* it is hit by the flashes is the sum of the original upward momentum of the book (Mv) and the momentum the book gets from the two light flashes, which in the Russian spaceship frame hit it at this 'aberration' angle.[1] Consequently, the light beams contribute a momentum Ev/c^2 to the book, which already had a momentum Mv: so the total momentum of the book in Russian spaceship frame after the absorption is $Mv + Ev/c^2$.

Although the book's momentum has increased, its final upward velocity is still v, the opposite of the Russian spaceship's velocity. (The book's velocity must remain v in the Russian spaceship frame: in the space shuttle frame, the light flashes hit the book in opposite directions and so leave it stationary; therefore, even after absorption, the book is still moving at v in the Russian spaceship frame.) So, as Einstein realized, the energy absorption must have increased the mass of the book – because the book's velocity does not increase, this is the *only* way of accounting for the increase in its momentum. If we denote the final mass of the book by M', then, in the frame of reference of the Russian spaceship:

Final momentum of the book = $Mv + Ev/c^2 = M'v$.

Dividing v from this equation and then subtracting M from both sides yields: $M' - M = E/c^2$, which is just another way of saying $E = (M' - M)c^2$. Now $M' - M$, the difference of the mass of the book before and after the arrival of the light flashes, can be abbreviated by the mass gained, m, giving the object of our desire,

$$E = mc^2$$

Now since one form of energy can always be converted into another, this result is not simply about light beams. It means that any form of energy adds to the inertial mass: a hot billiard ball is more massive than a cold one, and a spinning planet has more mass than a still one. In fact, if mass is allowed to turn into energy, it will. What might stand in its way? Conservation laws can – a conservation law is a statement that certain quantities don't change in a closed system. For example, you can't create electrical charge out of nothing. Or momentum – the tendency of a body to stay in straight motion once it is in motion – remains the same unless you apply a force to it. Because of these conservation laws, a single electron can't, in relativity theory, simply vanish into pure energy – that would strike electrical charge from the universe. Now if an electron hits an *anti*-electron (which has the opposite charge from that of an electron), the story is very different. Then the sum of their charges is zero (plus one added to minus one) so it *is* possible for the mass of the electron and positron to be converted completely to energy. Conversely, if the conservation laws are obeyed, pure energy can turn into mass – such as a positron and an electron.

Over the decades following 1905 $E = mc^2$ came to the laboratory. In 1932, two physicists from the famous Cavendish Laboratory in Cambridge, England, the experimentalists John Cockroft and Ernest Walton, showed that they could accelerate protons to bust apart a lithium nucleus. The resulting fragments of the lithium nucleus, it turned out, weighed less than the original lithium nucleus. At first it seemed as if mass had simply vanished. But by measuring the total energy of the flying fragments, the Cantabrigians, using $E = mc^2$, could show that the energy 'lost' in the mass change as the lithium nucleus broke precisely accounted for the energy contained in the fast-moving pieces that shot out from the broken nucleus. Einstein's formula had struck again.

But the world-changing use of $E = mc^2$ came with the discovery that neutrons could cause nuclear fission in uranium. For years the physicist Lise Meitner had been working with the chemist Otto Hahn in the Kaiser

Wilhelm Institute for Chemistry.[2] There in the leafy Berlin suburb of Dahlem, the physicist and the chemist bombarded nuclei with neutrons, using chemistry to sort out the products. For several years not only they, but also others, including Enrico Fermi's group in Rome, had concluded that the reaction products they were seeing after bombardment were actually new elements beyond uranium on the periodic table. Such 'transuranics', as they were called, appeared sensational, perhaps the greatest discovery of the new radioalchemy. In the Berlin collaboration the two kinds of skills they brought to the laboratory complemented each other: Meitner was the physicist of the outfit, Hahn the chemist. But compatibility in the lab meant nothing as the Nazis closed in and Meitner, who was Jewish, found her fate hanging by a thread. Finally, having been smuggled out of Germany by train on 13 July 1938, Meitner set up a rather threadbare scientific life in Sweden, where she anxiously awaited news from her collaborators, while the world hovered on the edge of war.

In Berlin the lab results only grew more confusing to Hahn, who continued the experiments. He and Meitner had long ago grown used to seeing products from the collision that in some reactions behaved like elements much lighter than uranium. But that, Hahn and everyone else believed, was merely a chemical illusion, an impossibility – the elements must be near uranium on the periodic table. 'Breaking' a nucleus into much smaller parts was simply impossible. One could chip off a proton or an alpha particle (two protons bound together with two neutrons). But breaking a nucleus squarely into two seemed, as one physicist later put it, like exploding a house by throwing a ball through a window. If a reaction product looked like barium, for example, it was probably the chemically related radium. Then things got really odd, and late one December night in 1938, Hahn wrote Meitner:

> 19.12.38 Monday eve in the lab. Dear Lise! . . . It is now just 11 p.m.; at 11.45 Strassmann [their other collaborator] is coming back so that I can eventually go home. Actually there is something about the 'radium isotopes' that is so remarkable that for now we are telling only you. . . . Our Ra[dium] isotopes act like Ba[rium].

'So please,' Hahn implored, 'think about whether there is any possibility' that there might be a variety of barium that was much heavier than usual.

Hahn sent his paper to the publisher three days after writing to Meitner, concluding his article with the conflicted sentiment that his and

Strassmann's chemical and physical souls were at war. They saw what looked like familiar light elements, but this just could not be: 'As chemists . . . we should substitute the symbols [of light elements] for [the heavy elements we have been discussing]. As "nuclear chemists" fairly close to physics we cannot yet bring ourselves to take this step which contradicts all previous experience in nuclear physics.'

When the letter of 19 December reached her, Meitner and her nephew the physicist Otto Robert Frisch, who had also fled, set out in the snow for a walk and began to pry apart the puzzling epistle. What would happen, they began to wonder, if the uranium nucleus, when hit by a neutron, began to oscillate like a fat water droplet? Thinking of the nucleus as such a drop had been current for some years. Suppose, they continued, that the whole droplet was normally in a rather delicate equilibrium, its 92 protons repelling each other furiously, and yet the whole held together by short-range but powerful nuclear attraction of the 238 or so protons and neutrons for one another. Then it might be that the droplet would distend itself as it oscillated, perhaps to the point where it would resemble a viscous barbell with globes at each end joined by a slender nuclear bar. At such a point of distension, the mutual repulsion of the protons located in the two globes could be more than the short-range nuclear bonds could counteract. Suddenly, driven by the electric repulsion of the globes, the nucleus could cleave into two parts, its two globes hurtling away from one another with the repelling force of two roughly equal bags of 46 protons. Meitner calculated. Two lighter nuclei would weigh less than they would together. And that mass difference, converted into energy according to $E = mc^2$, would be enormous. She and her cousin knew what no one else in the world suspected: back in Dahlem there was nuclear fission.

Events moved fast. The Danish physicist Niels Bohr, considered by many of his colleagues to have been the father of quantum theory, hearing Meitner and Frisch's interpretation, immediately understood where all his previous reasoning had gone wrong. Wheeler, who took the boat to America with Bohr in 1939, joined him in the mid-Atlantic composition of a comprehensive theoretical analysis of fission. One question led to another, as the physics sensation of the split atom jumped from laboratory to headline. And the next, immediate issue was vital to an unstable world: would the miscellaneous neutrons splattered about when the nucleus divided cause additional fissions? Could the fission of uranium cause a chain reaction? If it could, the enormous energy released by fission would multiply geometrically. Within months several physicists began to suspect that the fission process might, in the not too distant future, lead to the construction of nuclear bombs. Several pleaded with Einstein to write that fateful letter of 2 August 1939 to President Roosevelt:

> In the course of the last four months it has been made probable – through the work of Joliot in France as well as Fermi and Szilard in America – that it may become possible to set up nuclear chain reactions in a large mass of uranium, by which vast amounts of power and large quantities of new radium-like elements would be generated. Now it appears almost certain that this could be achieved in the immediate future. At stake, however, was more than abstract energy production. This new phenomenon would also lead to the construction of bombs, and it is conceivable – though much less certain – that extremely powerful bombs of a new type may thus be constructed. A single bomb of this type, carried by boat or exploded in a port, might very well destroy the whole port together with some of the surrounding territory.

Contact, Einstein insisted, would be needed between the administration and physicists. Ominously, Germany had stopped the sale of uranium. An intermediary representing the scientists' position saw President Roosevelt on 1 October 1939, and the atomic advocates pursued their concerns with a more technical memorandum by the Hungarian refugee Leo Szilard, the discoverer of the nuclear chain reaction. By then the Nazis had invaded Poland, and the snowball began its crashing descent. Fears of a German nuclear bomb rose; Pearl Harbor was attacked; and not long after the initiative the British prepared a seed project for a nuclear weapon. US committees evolved into laboratories and laboratories into the largest factories – of any kind – that the world had ever seen. Some years later, thinking back on those days, Einstein mused on the morality of what he had helped set in motion, first with the speculative scribbles of a young patent officer, and then later as the most famous scientist in the world:

> I made one mistake in my life – when I signed that letter to President Roosevelt advocating that the atomic bomb should be built. But perhaps I can be forgiven for that because we all felt that there was a high probability that the Germans were working on this problem and they might succeed and use the atomic bomb to become the master race.

Indeed, when Einstein was pressed to explain why people could discover atoms but not the means to control them, he replied, 'That's simple, my friend: because politics is more difficult than physics.'

By the end of the war, when John Wheeler delivered the address with which I began, $E = mc^2$ was, for physicists, a sign of the atomic times – celebrated for ending a war forced on them and regretted for precipitating an arms race. It was at one and the same time a guide to the future and a memorial of what had gone wrong.

After World War II, $E = mc^2$ was everywhere: it had long since left the physicists' control. One small marketing firm named itself after the equation: "You have to work smarter, not harder,' their self-description added. 'With images of Albert Einstein throughout the office as the "mascot" for the business, it would be difficult to do otherwise.'[3] $E = mc^2$ is also the name of a soft drink, a teenage science camp in Texas and the banner for a consortium of school districts in New Jersey that aims to improve science teaching. It is the title of a French best-seller by Patrick Cauvin ($E = mc^2$, *mon amour*), a love story about two eleven-year old geniuses who take flight to Venice. Not surprisingly, you can order up a two-by-three-foot poster of Einstein himself emblazoned with this equation.

You might not expect the equation to make a particularly good piece of music, but Big Audio Dynamite did and, as best I can tell, nearly a dozen other rock groups have titled songs after the equation. A film – distributed on video – also bears the equation as its title: 'An Oxford physics professor tries to take Einstein one step further while balancing the demands of his wife and girlfriend – all the makings of nuclear fission!' There are $E = mc^2$ Japanese graphics companies and French Internet systems, Arizonian study groups and art installations from several countries. It is everywhere: symbol of genius, sign of power, harbinger of destruction.

Perhaps we should not be surprised. Unlike any other equation of physics, $E = mc^2$ binds to the broader culture in four ways. First, the equation itself is compact, easy to write, and dramatic in its implications for the laboratory and for the world. Einstein's equation governing the gravitational field, on the other hand, is more or less unpronounceable to the average person: $R_{ab} - \frac{1}{2} R\, g_{ab} = -8\pi G\, T_{ab}$, *mon amour* doesn't quite have that commercial snap, and is, I would wager, rather harder to write into a rock-and-roll hit – though physicists might rightly complain that the equation governing general relativity merits greater veneration than the mass-energy equivalence.

Second, the equation $E = mc^2$ captured, at least partially, the extraordinary fascination the broader culture of the arts and humanities has had with relativity's modification of ideas of space and time. Even before relativity, the painter Claude Monet was already fascinated by issues of simultaneity, speed, time and the alteration of space. When physics offered a world of non-Euclidean space-time and a fusing of temporality and spatiality, those notions, or at least metaphorical analogues of them, fell on fertile ground.

Third, after the British astronomer Arthur Eddington's eclipse expedition of 1919 proclaimed that Einstein's theory had correctly predicted the bending of starlight, Einstein became a cult figure standing all at once (at least for his adoring fans) as individual genius, pre-war pacifist, post-war conciliator and moral exemplar. Misunderstood, vilified and then lionized beyond measure, Einstein became a symbol of hope for anyone doing anything against the grain. For his enemies he was, of course, the anti-hero: cosmopolitan, anti-nationalist, Jew, abstract theorist, democrat, cut off from the so-called intuitions of earth, blood and nation. Even before World War II, Einstein, and through him his most famous equation, stood for the mixture of philosophy, physics and modernity that alternately seduced and horrified the world around him.

With the long hot and then cold war stretching from 1939 to 1989, the equation came to stand for something else – nuclear weapons – encapsulating in its sparse symbols both power and knowledge. Here the 'sextant equation' gained a fourth meaning, because these weapons seemed to combine the most esoteric understanding with the most terrible destructiveness. The equation came to signify an almost mystical force, embodying instantaneous and apocalyptic death.

It is in the confluence of these various cultural currents that we find the lines of affect that cluster around this equation. At once philosophy and genial fantasy, practical physics and terrifying weapon, $E = mc^2$ has become metonymic of technical knowledge writ large. Our ambitions for science, our dreams of understanding and our nightmares of destruction find themselves packed into a few scribbles of the pen.

The Rediscovery of Gravity

The Einstein Equation of General Relativity

Roger Penrose

Introduction

Einstein's theory of general relativity provided an extraordinary revolution in our understanding of the physical world. Yet it did not come about through the findings of experimenters' laboratories. It was purely a product of one particular theoretician's insight and imagination. It was thus a revolution that stood in stark contrast with the conventional picture of how a scientific revolution should take place. That picture would hold that a previously accepted scientific viewpoint would be overthrown only when there is a sufficiently impressive accumulation of observational data in contradiction with it. The twentieth century indeed saw some extraordinary revolutions in fundamental physics, each of which led to a thorough overhauling of basic principles and a shattering of previous views as to the nature of physical reality. In the main, they were in accordance with such a conventional picture. But we shall be seeing that general relativity was very different.

In a broad sense, there were two quite distinct fundamental revolutions in twentieth-century physics. The first was relativity, concerned with the nature of space and time, and the second was quantum theory, concerned with the nature of matter. But the theory of relativity itself involved what might be called *two* revolutions, going under the respective names of 'special relativity' and 'general relativity'.

Special relativity is concerned with the strange modifications that must be made to Newtonian physics when bodies travel with speeds that approach that of light, whereby space and time coordinates mysteriously transform among one another, leading to the combined notion of *space-time*. This theory essentially grew out of observational conflict with the idea of an all-pervasive 'ether', which would have defined an absolute state of rest. The most famous conflict with this notion of an ether came from the Michelson–Morley experiment (1887), which attempted to measure the speed of the Earth through the ether; it produced a null result. That experiment, among others, made it increasingly difficult to hold to a Newtonian view of space and time. The revolution that was special relativity came somewhat relentlessly through the work of several scientists: George Fitzgerald, Joseph Larmor, Hendrik Lorentz, Henri Poincaré, Albert Einstein and Hermann Minkowski. I believe that it should, accordingly, be viewed as an example of a revolution of the 'conventional' type, where it was experiments, in the main, that drove the theorists to move away from the Newtonian scheme of things (even though Einstein's own route to the special theory was not particularly experiment-based).

Quantum theory, also, was very experiment-driven. In fact this was true to a far greater degree than in the case of special relativity. Physicists were forced to introduce this new theory to cope with the behaviour of very small-scale matter when they were faced with a vast body of observational data that was in gross conflict with ordinary Newtonian ideas.

The *general* theory of relativity, on the other hand, with its description of gravity as an effect of the 'curvature of space-time' rather than as Newton's gravitational *force*, seemed to have been pulled out of the blue by Einstein, with no apparent need at all for such a revolutionary new viewpoint. At the turn of the twentieth century Newton's beautiful picture of universal gravitation, acting according to an inverse square law of force between particles, was in wonderful accord with observation, to an accuracy of something like one part in ten million. There were still a few minor anomalies, but these all eventually turned out to result from errors of observation or calculation, or from the fact that some disturbing influence had not been taken into account. Well, not quite all – for there was still something not completely accounted for in the tiny details of the motion of the planet Mercury. This was not unduly troubling to astronomers at the time, however, and it was believed that a more careful analysis of the situation would also resolve this apparently minor problem within Newton's scheme of things. Observationally, so it seemed, there was no real expectation that Newton's theory would not suffice.

But Einstein had found himself to be guided to a very different perception of gravitation from that of Newton. It was not observational data that influenced Einstein. Perhaps this is not quite fair. There *was* one piece of observational data that he relied upon, but it was not of the twentieth century nor of the nineteenth, nor even of the eighteenth or the seventeenth. What troubled Einstein had been well established by Galileo in the late sixteenth century (and had been noticed by others even earlier), and was a familiar part of accepted gravitational physics. For more than four centuries the true significance of Galileo's observation had lain dormant. But Einstein saw it with new eyes, and only he perceived its hidden meaning. It led him to the extraordinary view that gravitation is a feature of *curved space-time geometry*, and he produced an equation – now known as Einstein's equation – of unprecedented elegance and geometric simplicity. Yet, to calculate its implications would present enormous technical difficulty, though the results would be almost invariably indistinguishable from those of Newton. Occasionally they would not be, however, and remarkable new effects would come out of Einstein's theory. In one case the precision of Einstein's theory could be seen to advance beyond that of Newton's by another factor of about ten million!

What is this paradigm of a beautiful equation, the *Einstein equation* that governs general relativity? It is commonly written

$$R_{ab} - \tfrac{1}{2}R\, g_{ab} = -8\pi G\, T_{ab},$$

but what does this mean? Why should this conglomeration of symbols be regarded as beautiful? Clearly, without the meaning that lies behind these symbols, there is neither beauty nor physical significance. We shall come to some real understanding of what this equation means shortly, but for the moment we must settle for a brief interpretation. The quantities on the left-hand side of this equation refer to certain measures of this mysterious 'space-time curvature'; those on the right, to the energy density of matter. Einstein's $E = mc^2$ tells us that energy is essentially equivalent to mass, so the right-hand terms refer equally to the *mass* density. Recall, also, that mass is the source of gravity. Einstein's field equation[1] thus tells us how space-time curvature (left-hand side) is directly related to the distribution of mass in the universe (right-hand side).

Before we begin, a few words about reading mathematical equations may be helpful, as there are indeed some equations in what follows. If you find these things intimidating, then I recommend a procedure that I normally adopt myself when I come across such an offending line. This is, more or less, to ignore that line and skip over to the next line of actual text. Well, perhaps one should spare the equation a glance, and then press onwards. After a while, armed with new confidence, one may return to that neglected equation and try to pick out its salient features. The text itself should be helpful in telling us what is important about it and what can be safely ignored. If not, then do not be afraid to leave an equation behind altogether.

The Principle of Equivalence

Let's see if we can appreciate what Einstein was striving to achieve in putting forward his general theory of relativity. Why did he feel that there was a physical need to go beyond Newton's highly successful theory? Why did Einstein introduce the notion of space-time curvature? What, indeed, *is* space-time curvature?

The central principle that Einstein believed must be incorporated into gravitational theory in a fundamental way was what he referred to as the *principle of equivalence*. The essential ingredient of this principle was, in effect, known to Galileo at the end of the sixteenth century (and before him by Simon Stevin in 1586, and by others going back to Ioannes Philiponos in the fifth or sixth century). Imagine that a large and a small rock, each of whatever material composition may be chosen, are dropped simultaneously from (say) the top of the Leaning Tower of Pisa. If we may ignore the effects of air resistance, then the rocks will fall at the same rate and reach the ground together. Let us picture a video camera placed on the large rock, aimed at the small one. Since the two rocks fall exactly together, the image that the video camera records is of a small rock just hovering, seeming to be stationary and therefore apparently unaffected by gravity. To the rocks (until they hit the ground), the Earth's gravity seems to have completely vanished!

This observation contains the essence of the principle of equivalence. By falling freely in under gravity, one can eliminate its local effects, so that apparently the gravitational force has disappeared. Conversely, it is possible to produce effects indistinguishable from those of gravity by referring things to an accelerating reference frame. This apparent gravity due to acceleration is a familiar feature of modern high-speed transport. As a car accelerates forwards, the occupants are pressed to the backs of their seats as though a new gravitational force had suddenly appeared, pulling the occupants to the rear. Similarly, if the driver suddenly applies the brakes, then the occupants seem to be pulled forwards, as though there is a gravitational force pulling them to the front of the car. If the car swings to the right, then there would appear to be a gravitational force pulling the occupants to the left, and so on. These effects are particularly manifest on an aeroplane, as it is often difficult to tell which direction is actually 'down' – i.e. towards the Earth's centre – owing to the confusion of effects from the plane's acceleration and the Earth's actual gravitation. The principle of equivalence tells us that this confusion is a fundamental property of gravity. The physical laws that appear to be operating if measurements are taken with respect to an accelerating reference frame are just the same as those that operate if the reference frame is considered to be *un*accelerating but where an appropriate gravitational field of force is introduced, in addition to those forces already present.

It should be remarked that this 'equivalence' property is something that holds only for the *gravitational* field, and not for any other type of force. It certainly does *not* hold if we take an electric field in place of a gravitational one. Consider, for example, a corresponding situation to that outlined above, where rocks are imagined to be dropped from the Leaning Tower, but now with electric forces replacing the gravitational ones. The acceleration rate at which a body 'falls' in a background electric field is by no means independent of its compositional nature. This acceleration depends upon what is referred to as the body's charge-to-mass ratio. To take an extreme case, we could imagine that the two bodies have equal mass but their charge values are opposite (so that one is positively charged and the other negatively). Then the bodies would accelerate in the background electric field in opposite directions! A video camera placed on one body would certainly not register the other as being unaccelerated.

The issue with regard to the charged bodies in a background electric field, as opposed to the massive bodies in a background gravitational field, is that the force on the charged body is proportional to its *charge*, whereas its resistance to motion – i.e. its *inertia* – is proportional to its *mass*. What is special about the gravitational case is that the force on the body and its resistance to motion are *both* proportional to its mass. From the perspective of Newtonian theory, this fact seems entirely fortuitous. The equality between *gravitational mass* (controlling the strength of the gravitational force on a body) and *inertial mass* (controlling resistance to change of motion) is by no means an essential requirement for a dynamical theory of the Newtonian type, but this equality in the case of gravity makes things a little simpler, since one does not have two kinds of mass to worry about.

Although these matters were known for a long time – basically since Galileo's early considerations and certainly appreciated by Newton – it was Einstein who first realized the profound *physical* importance of the principle of equivalence. What importance was this? Let us first recall Einstein's development of *special* relativity. He had then taken the 'principle of special relativity' to be a fundamental principle. According to this principle, the laws of physics are the same with respect to any uniformly moving (unaccelerated) observer. Although Larmor, Lorentz and Poincaré before him had had the basic transformation laws of special relativity, none of them had adopted Einstein's viewpoint that this *relativity principle* should be fundamental and therefore respected by all the forces of nature. Einstein's fundamentally 'relativistic' attitude on this matter had led him to ponder upon whether there is really anything particular about the restriction to *uniform* motion in the statement of the relativity principle. What about the way in which physical laws are perceived by an *accelerating* observer?

At first sight, it would appear that accelerating observers simply perceive laws that are *different* from those perceived by uniformly moving observers. In Newtonian language, one needs to introduce 'fictional forces' (i.e. 'unreal' forces) to cope with the effects of acceleration. Here is where the principle of equivalence comes in. According to Einstein, such fictional forces are no less real (and no more real) than the gravitational force that we all seem to feel pulling us downwards to the centre of the Earth. For the force of the Earth's pull can appear to be eliminated if we fall freely with it. Recall our imagined video camera attached to one of Galileo's falling rocks. In the accelerating frame of the video camera, the Earth's field seems to have disappeared. It seems to have been rendered 'fictional' by the simple procedure of referring things to a reference frame at rest with respect to the video camera.

With Einstein's viewpoint, an accelerating observer perceives the same laws as those of the unaccelerated one provided that an appropriate new *gravitational field of force*, arising from the acceleration, is introduced in addition to all the other forces involved. In the case of the falling video camera, this additional field would be a gravitational field directed upwards which just cancels the Earth's downward field. In the video camera's reference frame, therefore, the gravitational field has been reduced to zero.

In a speech Einstein gave in Japan in 1922, he recalled the moment at which he happened on this idea, which occurred to him late in 1907:

> I was sitting in my chair in the patent office when all of a sudden a
> thought occurred to me: 'if a person falls freely he will not feel his
> own weight'. I was startled. This simple thought made a deep impres-
> sion on me. It impelled me toward a theory of gravitation.

Elsewhere, Einstein referred to this realization as 'the happiest thought of
my life'. For it contained the seeds of his wonderful general theory of
relativity.

Yet the reader may be forgiven for worrying that Einstein seems to have
eliminated gravity altogether with this point of view. Surely there *is* an
effect that we call gravity! The planets surely *do* move in ways that are
beautifully accounted for by Newtonian theory. And there surely *does* seem
to be something that holds us to our chairs! The Einsteinian view would
appear to be telling us that there is no such thing as gravity, since we can
always eliminate the gravitational force by simply choosing a frame of reference that is in free fall. Where has gravity gone in this Einsteinian view?
In fact it has not gone away, but is concealed in some subtleties that I have
glossed over. In the next section, we shall see where the gravitational field
is indeed hiding.

Tidal Forces

The considerations of the previous section are essentially local. I have
ignored how Newton's gravitational field of force might be varying from
place to place. The direction 'down' is not quite the same here in Oxford
as it is in London, owing to our differing locations on the globe. If I try
to eliminate the gravitational field where I sit here at my desk, by considering my descriptions relative to a rigid reference frame that falls
freely to the ground here in Oxford, then this frame will not quite do the
job for someone in London. Thus, the 'elimination' of the gravitational
field by adopting a freely falling frame is not really a straightforward
matter.

To make the situation a little more specific, let us imagine an astronaut
called Albert – but we shall refer to him simply as '*A*' for short – who falls
freely in the vicinity of the Earth. We could imagine that *A* simply drops
towards the ground, but this might be considered to a little inhumane. We
are concerned just with accelerations and not velocities directly, so it is just
as good to suppose that Albert is safely in free orbit about the Earth. Let us

suppose that A is surrounded by a small sphere of particles, initially at rest
with respect to A. Each particle will have an acceleration towards the
Earth's centre C, and this will be in accordance with Newton's inverse
square law. The two particles P_1 and P_2 that lie on the straight line CA will
have accelerations in the direction of C, but the acceleration of the lower
point P_1 will be a little greater than that at A, and the acceleration at the
higher point P_2 a little less than that at A. Thus, *relative* to Albert, P_1 will
be accelerating slowly down towards the Earth's centre C but P_2 will be
accelerating up away from C. Both P_1 and P_2 will appear, to A, to be
accelerating away from A. On the other hand, any particle P_3 on the horizontal circle of particles centred at A will accelerate slightly inwards, as
it is pulled towards the Earth's centre C, since C is a definite finite distance from A, with a slightly different 'down' direction. Relative to A, the
acceleration of such a point P_3 will appear to be inwards towards A. The
entire sphere of particles will become distorted into a prolate (cigar like)
ellipsoidal shape, moving inwards towards A in horizontal directions relative to A, and moving outwards along the line from A to the centre C.
(See Figure 1.)

This distortion effect is referred to as the *tidal effect* of gravity. The
reason for the description 'tidal' is that it is precisely this same effect that
is responsible for the tides of the Earth's oceans, as governed by the location of the Moon. To see this, let us now imagine that A represents the
centre of the Earth and that the sphere of particles represents the surface of
the Earth's oceans. Let C now represent the location of the Moon. Again
there will be slightly differing accelerations towards the Moon's centre C
at all the points on the ocean's surface. The resulting effect, relative to the
Earth's centre A, will be to cause a prolate ellipsoidal distortion of the
ocean surface, which bulges in a direction towards the Moon (C) and also
in the opposite direction. This is precisely the main effect that gives rise to
the tides. (Subsidiary influences are the Sun's similar, but smaller, tidal
effect and the frictional and inertial influences on the actual motion of the
water in the oceans.)

It is a particular (defining) feature of Newton's *inverse square law* that
the *volume* of the sphere of particles remains initially constant in its momen-
tary distortion into an ellipsoid. (What this amounts to saying is that the
outward acceleration at P_1 and P_2 is twice the inward acceleration at the
horizontal points like P_3.) This fact depends upon there being no mass density within the sphere itself. If there were a significant amount of massive
material *within* the sphere, then there would be an additional inward

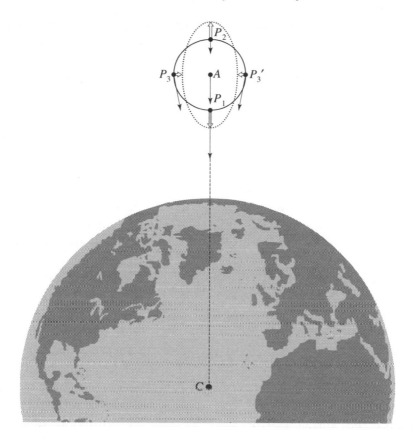

Figure 1 The tidal effect. Open arrows show relative acceleration.

acceleration that would serve to *reduce* the volume of the sphere in its initial motion. The amount of this (initial) volume reduction is, quite generally, proportional to the total mass surrounded by the sphere. In fact, Newton's magnificent gravitational theory is effectively encompassed within the simple facts that I have just described.

A particular example of this volume reduction would occur if we consider our sphere of particles to surround the Earth completely, in the vicinity of the Earth's surface, where we are now concerned with the Earth's gravitational field *itself*, rather than the small corrections due to the Moon which are (mainly) responsible for the tides. The distortion of our sphere is now of the pure volume-reducing type. This is an inward acceleration all around the Earth, and it supplies us with the familiar gravitational field that indeed holds us to our chairs.

Space-Time Curvature

Although the idea of space-time has not yet featured in these considerations, and we shall be coming to this more fully in the following section, it is useful to get some feeling for why the above way of looking at Newtonian gravity is actually telling us that Einstein's perspective on gravitational theory, where the principle of equivalence is regarded as fundamental, leads naturally to the notion that gravitation is manifested in a form of *space-time curvature*. Let us try to imagine that the history of the universe is laid out before us as a *four-*dimensional continuum. We are not, for the moment, trying to depart from Newtonian physics; we are merely looking at the Newtonian universe in an unusual way – as a piece of four-dimensional geometry! In addition to having three spatial coordinates, say x, y and z, we shall also introduce the time coordinate t describing a fourth dimension. Of course the visualization of the full four dimensions creates difficulties, but such a complete visualization is not really necessary. Let us temporarily 'forget' the space coordinate y, so that we now have a three-dimensional space-time coordinatized by x, z and t. Figure 2 gives us some idea of what is involved. An individual point particle is now represented as a *curve* in the space-time; this curve, describing the particle's history, is called the *world-line* of the particle.

We shall try to understand the history of the sphere of particles surrounding Albert (from Figure 1) and see what this has to do with the notion of space-time curvature. At the right-hand side of Figure 2, I have tried to

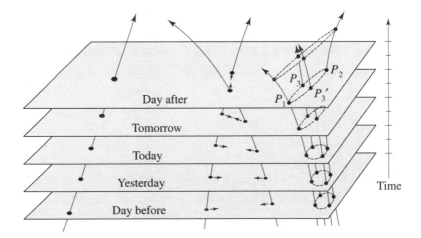

Figure 2 Space-time (Newtonian case). Geodesic deviation (tidal effect) is illustrated on the right.

depict the history of the evolution of this sphere, with one of the spatial dimensions (namely the horizontal dimension coordinatized by y) suppressed. The sphere (in this reduced dimensionality) now appears as a circle, and as time evolves it gets distorted into an ellipse. Note that there is a bending outwards of the world-lines of the vertically displaced particles P_1 and P_2 (major axis of the ellipse), this bending being outwards away from Albert's central world-line. On the other hand there is an inward bending for the world-lines of the horizontally displaced particles P_3 and P_3' (minor axis of the ellipse).

We are to compare this 'bending effect' with the behaviour of geodesics on a curved surface. A geodesic is a curve of minimal length on such a curved surface. We may think of a piece of string stretched taut over the surface. It will describe a geodesic on that surface. If the surface has what is called *positive curvature* (like the curvature of an ordinary spherical surface), then slightly displaced geodesics that start out parallel to each other will begin to curve in towards each other. If the surface has what is called *negative curvature* (like the surface of a saddle), then slightly displaced geodesics starting out parallel will begin to diverge away from each other. (See Figure 3.) This manifestation of curvature is referred to as *geodesic deviation*.

In our space-time picture of the tidal distortion, as illustrated at the right in Figure 2, we see a combination of these two kinds of curvature. There is positive curvature (inward-bending) for the horizontally displaced world-lines of P_3 and P_3', whereas for the vertically displaced world-lines of P_1 and P_2 we have negative curvature (outward-bending). This interpretation of the distortion of world-lines that occurs in the *tidal effect* as a geodesic deviation of some kind becomes justified when we are able to think of the world-lines of particles, freely moving under gravity, as geodesics in space-time. For this

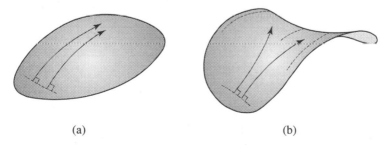

(a)   (b)

Figure 3 (a) Positive curvature causes convergence of geodesics – like the
surface of an orange
(b) Negative curvature causes divergence of geodesics – like a saddle.

we shall need to have an appropriate notion of 'distance' in space-time. We shall be coming to that in the next two sections. We shall be seeing that the tidal effect is indeed an instance of geodesic deviation, and it is thus a direct measure of space-time curvature.

We observe that the notion of curvature in higher dimensions is a more complicated thing than it is in the two-dimensional case. In two dimensions, we find that the curvature at any point is given simply by a *single number*,[2] which would be a *positive* number in the sphere-like case of positive curvature and a negative number for the saddle-like case of negative curvature. In more than two dimensions the curvature is described by *several* numbers, called *components* of the curvature, these basically measuring the two-dimensional type of curvature in various different directions. In the example just considered we have seen, in effect, a positive curvature component referring to the horizontal direction from A to P_3 and $P_3{}'$, and a negative curvature component referring to the vertical direction from A to P_1 and P_2. In fact, in the four dimensions of space-time, there are *twenty* independent components to the curvature, and these can be collected together to describe a mathematical entity referred to as the 'Riemann curvature tensor'. I shall defer discussion of the notion of a tensor until a later section, but it is worth pointing out here that the Einstein equation is itself a tensor equation, and the little indices (such as the a and b on R_{ab}) simply provide a labelling for such components in different directions.

So far, we have not really been doing general relativity, but merely Newtonian gravitational theory from the Einsteinian perspective.[3] In order to move forward to full general relativity, we shall have to understand a little more about *special* relativity: why is it really a four-dimensional space-time theory, and what is the appropriate notion of 'distance' in this space-time geometry? Let us come to all this next.

Minkowski's Notion of Space-Time Geometry

Einstein based his 1905 special theory of relativity on two basic principles. The first was already referred to earlier; for all observers in uniform motion the laws of nature are the same. The second was that the speed of light has a fundamental fixed value, not dependent upon the speed of the source. A few years earlier, the great French mathematician Henri Poincaré had a similar scheme (and others, such as the Dutch physicist Hendrik Lorentz, had moved some way towards this picture). But Einstein had the clearer vision that the underlying principles of relativity must apply to *all* forces of nature.

Historians still argue about whether or not Poincaré fully appreciated special relativity before Einstein entered the scene. My own point of view would be that whereas this may be true, special relativity was not *fully* appreciated (*either* by Poincaré *or* by Einstein) until Hermann Minkowski presented, in 1908, the four-dimensional space-time picture. He gave a now famous lecture at the University of Göttingen in which he proclaimed, 'Henceforth space by itself, and time by itself are doomed to fade away into mere shadows, and only a kind of union of the two will preserve an independent reality.'

Einstein seems not to have appreciated the significance of Minkowski's contribution initially, and for about two years he did not take it seriously. But subsequently he came to realize the full power of Minkowski's point of view. It formed the essential background for Einstein's extraordinary later development of *general* relativity, in which Minkowski's four-dimensional space-time geometry becomes *curved*.

The physical interpretation of this curvature is basically that which has already been given, but there is still an essential missing ingredient, namely the interpretation of the world-lines of particles moving freely under gravity as *geodesics* in space-time geometry. Examples of such geodesics would be the world-lines of our astronaut *A* and of the surrounding sphere of particles. To understand this interpretation, it will be important first to appreciate the *flat* four-dimensional mathematical structure that Minkowski actually introduced in order to describe special relativity.

For this it is helpful to start by considering familiar three-dimensional Euclidean geometry. It is convenient to introduce Cartesian coordinates x, y, z to label points in Euclidean 3-space. Then the *distance l* from the origin (with coordinates $x = y = z = 0$) to the point (X, Y, Z) (i.e. coordinates $x = X, y = Y, z = Z$) is given by the Pythagorean relation

$$l^2 = X^2 + Y^2 + Z^2.$$

(The reader will recall the Pythagorean theorem, which states that the squared length of the hypotenuse of a right-angled triangle is equal to the sum of the squared lengths of the other two sides. This would be the two-dimensional formula $l^2 = X^2 + Y^2$, since the distance between two points in the plane is the hypotenuse l of a triangle whose other two sides' lengths are X and Y. The extension to three dimensions is a two-step consequence of this.) We can also use the above formula to express the Euclidean distance between *any* two points, where now X represents the *difference* between the *x*-coordinate values of the two points, and similarly for Y and Z.

It is easy to generalize the formula to four dimensions and obtain the squared distance from the origin to the point $w = W$, $x = X$, $y = Y$, $z = Z$ in Euclidean *4-space* as

$$l^2 = W^2 + X^2 + Y^2 + Z^2.$$

However, Minkowski's space-time geometry differs subtly but importantly from this. Although space and time coordinates indeed get mixed up with one another in relativity theory according to a kind of rotation (a 'Lorentz transformation'), the way that an ordinary Euclidean rotation mixes up the (w, x, y, z)-coordinates does not give us quite the correct prescription. There is a qualitative distinction between the space and time coordinates in Minkowski's description, which shows up as a *sign* difference in the above distance formula.

In place of the fourth spatial coordinate *w* we introduce a time coordinate *t*. How do we modify the above formula so as to obtain the correct Minkowskian measure of 'distance' τ? In fact, in order to arrive at the most directly physical such measure, it is appropriate to reverse the signs of *all* the spatial contributions, leaving the one temporal coordinate $t = T$ to contribute with a positive sign:

$$\tau^2 = T^2 - X^2 - Y^2 - Z^2.$$

Here I am using units of distance and time so that the speed of light comes out as *unity*. Thus, if we were to use the year as the time unit, then we should have to use the light year as the unit of spatial measure; if we use the second as the time unit, then we must use the light second as the unit of spatial measure (about 186,000 miles).

What kind of 'distance' is then defined by the quantity τ? It is better to think of τ as a measure of *time*. It is what is called the *proper time*. If the space-time point P with coordinates $t = T$, $x = X$, $y = Y$, $z = Z$ is such that the quantity on the right-hand side of the above expression is *positive*, then P is *timelike separated* from the origin O, which means, physically, that it is theoretically possible for the world-line of a particle to pass from O to P (if T is positive) or from P to O (if T is negative). If this particle moves uniformly in a straight line from O to P, then the quantity τ (taken with the positive sign) is the *time* (proper time) actually experienced by the particle between O and P as measured by an ideal clock situated on the particle. (The fact that this time is not simply the Newtonian t, but involves the spatial coordinate differences also, is an expression of the 'relativity of time' that occurs with special relativity.) As with Euclidean geometry above, these considerations apply also when the origin O is replaced by some arbitrary point P', but where now the quantities T, X, Y, Z refer to the *differences* between the respective t, x, y, z coordinates of the two space-time points P and P', and where t is the time experienced by the particle moving inertially from P to P'.

Minkowskian geometry has the curious property that the 'distance' between two points P and P' can sometimes be zero even though P and P' do not coincide. This happens when a light ray can contain both P and P' (which we think of as a 'particle of light', or photon, travelling with the speed of light). Thus, noting the above interpretation of 'Minkowski distance' as proper time, we find that a photon would not experience any passage of time at all (if photons could actually experience anything!). For fixed P, the locus of such points P' constitutes the (future) *light cone* of P. The light cones are important because they determine the *causality properties* of Minkowski space, but I shall not be much concerned with them here. The one essential point that will be needed is that the world-line of a particle with mass must lie *within* the light cone at each of its points. This simply expresses the fact that the particle does not exceed the speed of light anywhere. Such a world-line is referred to as a *timelike* curve. Any massive particle's world-line must be a timelike curve.

Now any timelike curve (i.e. allowable particle world-line) has a Minkowski 'length' whether or not the curve is straight. A curved world-line describes an *accelerating* particle. This 'length' is, physically, simply the (proper) time that is experienced by the particle. To obtain this length mathematically, we just do the same thing that we would do in ordinary Euclidean geometry, except that we must take into account the sign differences, noted above, that are involved in passing from Euclidean to Minkowskian geometry. To do this explicitly, we need the *infinitesimal* expression for the length that measures the 'distance' between two infinitesimally separated points. We then 'add up' (technically: *integrate*) all these infinitesimal separations along the curve to get the total length. In Euclidean three-dimensional geometry, this infinitesimal separation 'dl' is related to standard Cartesian coordinates x, y, z by the formula

$$dl^2 = dx^2 + dy^2 + dz^2.$$

In the Minkowski case, we must modify this to

$$d\tau^2 = dt^2 - dx^2 - dy^2 - dz^2,$$

but the interpretation is completely analogous. (Those unfamiliar with the relevant calculus notations can imagine dt to stand for $t'-t$ and dx to stand for $x'-x$, etc. where P' lies infinitesimally close to P within the light cone of P.) The total lapse of time (proper time) between two points on a world-line, as measured by an ideal clock, is the total 'length' of the world-line between these points.

An important feature of length in Euclidean geometry is that among all the curves joining two points, the length is *minimum* when the curve is straight. ('The shortest distance between two points is a straight line.') There is a closely analogous property in Minkowskian geometry, except that things are the other way around. If we select a pair of timelike-separated points, then among all timelike curves joining them, the proper time is a *maximum* when the curve is straight. Physically, this provides us with what is sometimes referred to as the 'clock paradox' (or 'twin paradox'), whereby a traveller to a distant star and back ages less (because of a shorter 'Minkowski distance') than his twin sister whom he leaves behind on Earth. The Earthbound twin has a straight world-line, and therefore she experiences a greater duration of time than does her space-travelling brother, whose world-line is curved because of the acceleration. It is very misleading, however, to think of this as a paradox. Admittedly it takes some getting used to, but it is not actually paradoxical, and many experiments have now confirmed this effect to great accuracy. Minkowski geometry makes the time difference between the two twins seem almost 'ordinary'.

Why was Einstein led to modify Minkowski's beautiful space-time geometry and introduce curved spacetime? We have seen that in special relativity, particles moving freely in the absence of forces – i.e. *inertially* moving particles – have straight world-lines in Minkowski space. Einstein's desire to incorporate the principle of equivalence into physical theory led him to the view that a *new* concept of 'inertial motion' was required. Since the gravitational force can be locally eliminated by use of a freely falling reference frame, we are not to consider the gravitational force as 'real' according to Einstein's viewpoint. So Einstein found that he needed to introduce a different notion of inertial motion, namely *free fall under gravity*, with no other forces acting. Because of the tidal effects that we encountered above, we cannot think of the 'inertial' particles (in Einstein's sense) as having straight (i.e. geodesic) world-lines in Minkowski's geometry. For this reason we need to generalize this geometry so that it becomes *curved*. Einstein found that, indeed, the world-lines of his inertial particles could now be *geodesics* in this curved geometry – locally maximizing the 'length' rather than minimizing it, in accordance with the above – and the tidal distortion is indeed an instance of geodesic deviation, providing a direct measure of space-time curvature. Let us try to understand this curvature a little more fully.

Curved Space-Time Geometry

In the nineteenth century two great German mathematicians, Carl Friedrich Gauss and Bernhard Riemann, introduced the general notion of 'curved geometry'. To get a feeling for this kind of geometry, think of the surface of a tennis ball divided in half. It can be flexed in various ways, but what is called its *intrinsic* geometry remains unchanged under such deformations. Intrinsic geometry is concerned with distances measured *along* the surface. It is not concerned with the space (here our ordinary Euclidean three-space) in which the surface may be pictured as embedded. Distances measured directly across from one point to another taken *outside* the surface are not the concern of intrinsic geometry. The length of a curve drawn *on* the half tennis ball is unchanged by the flexing, however, and such lengths are the basic concern of intrinsic geometry.

Gauss introduced this idea of intrinsic geometry in 1827 in the two-dimensional case, like our tennis-ball surface just considered. He showed that there is a notion of curvature in this geometry that is entirely intrinsic, so that it is completely unaffected by changes in the way that such a surface might be embedded. This curvature can be calculated from the length measures along the surface, where we think of the lengths of curves on the surface as obtained by integrating an *infinitesimal* measure of length dl along the curve, just as above. In practice, one introduces some convenient system of coordinates on the surface, say u, v, and we find an expression for dl in the form

$$dl^2 = A\,du^2 + 2B\,du\,dv + C\,dv^2$$

where A, B and C are functions of u and v (this expression being locally the same as the infinitesimal 'Pythagorean' expression for distance $dl^2 = dx^2 + dy^2$ that we had earlier but now written in terms of the general coordinates u, v).

In 1854 Riemann showed how to generalize Gauss's intrinsic geometry of surfaces to higher dimensions. The reader might be puzzled about the motivations here. Why might mathematicians be interested in higher-dimensional intrinsic geometry? Ordinary space has just three dimensions and it is hard to see how to make sense of 'flexing' a three-dimensional 'surface' – let alone a higher-dimensional surface – within it. The first point to make is that this picture was helpful only for getting us started in understanding the notion of 'intrinsic geometry'. We should really be thinking of the intrinsic geometry of our surface as being something that stands on its own, without the need for an embedding space at all. Indeed, one of Riemann's original motivations was that the physical three-space within which we actually find ourselves might have a curved intrinsic geometry, without it having to 'reside' within some higher-dimensional space.

But Riemann also considered n-dimensional intrinsic geometries, and one might question the motivations for that. Two considerations are relevant here. It turns out that the mathematical formalism that has been developed for handling curved three-spaces is basically the same as that for handling curved n-spaces in general, so there is nothing to be gained by restricting attention to the case $n = 3$. The other consideration is that curved (intrinsic) n-geometry is important in many contexts where the n does not refer to the number of dimensions of ordinary space, but to the number of degrees of freedom of some system. There are abstract mathematical spaces known as 'configuration spaces', a single point of which represents the entire arrangement of parts of some physical structure. The dimension n of such a space can be very large indeed, when the system has many parts, and Riemann's higher-dimensional geometry can be of great relevance to these spaces.

The notions of 'metric' and 'curvature' in the n-dimensional case are natural generalizations of those introduced by Gauss for ordinary two-dimensional surfaces, but because of the large number of components involved, we need a suitable notation in order to handle them all. In place of the three 'metric components' A, B and C that occur in the above expression for dl^2 in the two-dimensional case, we need *six* such quantities for three dimensions. These are the components of the *metric tensor*, generally denoted g_{ab}. This quantity serves to define the appropriate notion of a 'distance' between neighbouring points, frequently denoted by ds.[4]

In Riemann's geometry, we obtain the *length of a curve* in the space by *integrating* ds along the curve in just the same way as in the flat-space case already discussed. A geodesic in a Riemannian manifold is a curve that (locally) minimizes length (so it describes 'the shortest distance between points', in an appropriate sense). The *curvature* of the Riemannian space is the quantity that describes the amount of geodesic deviation in all the various possible directions in the space (as indicated above). Not surprisingly, there are many components to the curvature, there being lots of possible directions in which this geodesic deviation may be measured. In fact, all this information can be collected together in the quantity called the *Riemann tensor*. The Riemann tensor (or its collection of components) is commonly written R_{abcd}, where those little indices refer to all the different possible ways in which the geodesic deviation might be measured.[5]

Einstein's general relativity is formulated in terms of a concept of curved four-dimensional space-time which bears the same relation to Minkowski's flat space-time as Ricmann's concept of curved geometry bears to flat Euclidean geometry. The metric g_{ab} can be used to define curve lengths, but as in Minkowski's flat space-time geometry, this 'length' is best thought of as defining the *time*, as mcasurcd by a particle along its world line. Those world-lines that locally maximize this time measure are the geodesics in space-time and are considered to be the world-lines of inertially moving particles (where 'inertial' is taken in Einstein's sense of 'freely moving under gravity', as already described).

Now we recall that the geodesic deviation in space-time that is caused by gravitation (in Newtonian theory) has the property that in vacuum there is initially no volume change, whereas when there is matter present in the vicinity of the deviating geodesics, the *volume reduction* is proportional to the total mass that is surrounded by the geodesics. This volume reduction is an *average* of the geodesic deviation in all directions surrounding the central geodesic (this central one being the astronaut A's world-line). Thus, we need an appropriate entity that measures such curvature averages. Indeed, there is such an entity, referred to as the *Ricci tensor*, constructed from R_{abcd}. Its collection of components is normally written R_{ab}. There is also an *overall* average single quantity R, referred to as the *scalar curvature*.[6] We recall that R_{ab} and R, together with g_{ab}, are precisely the things that appear on the left-hand side of Einstein's equation.

The quantities g_{ab}, R_{abcd}, and R_{ab} are (sets of components of) entities called tensors, and tensors are fundamentally important in the study of Riemannian geometry. The reason for this has to do with the fact that in this subject, one is not really interested in the specific choice of coordinates that happen to be used for a description of the manifold. (This is an implication of a strict adherence to the principle of equivalence.) One set of coordinates may be used, or another set may be used equally well. It is just a matter of personal convenience. The *tensor calculus* was a marvellous technical achievement, developed in the late nineteenth century by several mathematicians as a means of extracting *invariant information* about the manifold, its metric and its curvature, where 'invariant' essentially means 'independent of any particular choice of coordinates'.

In Einstein's deliberations about how to incorporate the principle of equivalence fully into a physical theory of gravitation, he eventually realized that he needed a formulation that is 'invariant' in the sense referred to above. He called this requirement the *principle of general covariance*. The space-time coordinates that refer to two differently accelerating frames of reference can be related to each other in some (often complicated) way, and neither is 'preferred' over the other. Einstein had to enlist the help of his colleague Marcel Grossmann to teach him what he needed to know of the 'Ricci calculus' (as the tensor calculus was then called). The only essential difference between the curved space-time geometry that he required and the Riemannian geometry that the Ricci calculus was designed for (in the four-dimensional case) was the change in 'signature' that was needed in passing from the locally Euclidean structure of Riemannian spaces to the locally Minkowskian structure needed for a relativistic space-time.

Full General Relativity

Let us return to Albert, our astronaut A surrounded by a sphere of particles. All these particles, as well as A, are moving inertially, in Einstein's sense (i.e. freely under gravity), and he postulated that inertially moving particles should have world-lines that are geodesics in space-time.[7] We recall that the initial volume reduction of this sphere is proportional to the mass enclosed, in Newtonian theory, and that it is the Ricci tensor that measures this volume change. Accordingly, we may expect that the appropriate relativistic generalization of Newton's theory would be one in which there is an equation relating the Ricci tensor of space-time to a tensor quantity that appropriately measures the mass density of matter. The latter quantity is what is referred to as the *energy–momentum tensor*, and its family of components is normally written T_{ab}. One of these components measures the mass-energy density; the others measure momentum densities, stresses and pressures in the material.

There is a factor of proportionality, in Newton's theory, between the inward acceleration and the mass density, that is Newton's gravitational constant G. This led Einstein to anticipate something like the equation

$$R_{ab} = -4\pi G \, T_{ab}.$$

The 4π comes from the fact that we are dealing with densities rather than individual particles, the minus sign coming from the fact that the acceleration is inwards, where my own conventions for the sign of the Ricci tensor are such that outward acceleration counts positively – but there are innumerable different conventions about signs etc. in this subject.

This equation is indeed what Einstein first suggested, but he subsequently came to realize that it is not really consistent with a certain equation,[8] necessarily satisfied by T_{ab}, which expresses a fundamental *energy conservation* law for the matter sources. This forced him, after several years of vacillation and uncertainty, to replace the quantity R_{ab} on the left by the slightly different quantity $R_{ab} \quad \frac{1}{2} R g_{ab}$, which, for purely mathematical reasons, rather miraculously *also* satisfies the same equation as T_{ab}! By this replacement, Einstein restored the necessary consistency of the resulting equation, which is his now justly famous and very remarkable *Einstein equation*[9]:

$$R_{ab} - \tfrac{1}{2} R \, g_{ab} = -8\pi G \, T_{ab}.$$

This 'volume reduction' in the geodesic deviation that this equation gives rise to is just slightly different from what we expect from Newtonian theory, because of the additional term '$-\frac{1}{2} R g_{ab}$' that now occurs in the left-hand side of the above equation. Instead of the 'source of gravity' (i.e. source of volume reduction) being simply $4\pi G$ multiplied by the *mass* density (in the sense of the mass-energy term in T_{ab}), it now turns out to be $4\pi G$ multiplied by the mass density *plus the sum of the pressures* in the material, in three mutually perpendicular directions (coming from other components of T_{ab}). For ordinary materials, like that composing ordinary stars and planets, the pressures are very small as compared with the mass densities (because the constituent particles of such bodies move slowly in comparison with the speed of light), so agreement with Newtonian theory is very precise. There are, however, certain circumstances (such as with the instability of supermassive stars, as they collapse to become black holes) in which this difference actually has important effects.

Classical Tests of General Relativity

It might seem from the preceding discussion that Einstein's general relativity is simply a technical modification of Newtonian theory, the latter having been rephrased so that it is in accordance with relativity and the principle of equivalence. Indeed, this could be said to be the case, although the way in which I have presented the comparison with Newtonian theory is not the way in which this was originally done. By concentrating on the tidal force of Newtonian gravity as something that cannot be eliminated in free fall, we have been able to see more directly its relationship to space-time curvature and, therefore, to the framework of Einstein's general relativity.

In fact, it is remarkably hard to find clear-cut observational differences between the two theories. Originally, there were the so-called 'three tests' of general relativity. The most impressive of these three was the explanation of the perihelion advance of the planet Mercury in its orbit around the Sun. It had been known from work in the nineteenth century that there was a curious discrepancy with Newtonian theory in Mercury's motion. When the perturbing effects of all the other known planets are taken into account, there is still a slight extra component to Mercury's motion, amounting to a swing in the axis of its orbital ellipse of 43 seconds of arc per century. This amount is so tiny that it would take about 3 million years for the ellipse of Mercury's orbit to swing completely around owing to this effect alone. Astronomers had tried various explanations, including the prediction of another planet within Mercury's orbit, which they had christened Vulcan. None of these ideas worked, but Einstein's theory exactly accounted for the discrepancy, and provided a rather impressive test of the theory.[10] The other two tests concerned the slowing of ideal clocks in a gravitational field and the bending of light by the Sun's field. The clock-slowing effect was convincingly confirmed by an experiment by Pound and Rebka in 1960, although it was recognized that this was a rather weak test of general relativity, being a direct consequence of energy conservation and the equation $E = hf$ for the energy of a photon.

The light-bending effect has a more interesting history. Before he had found the full general relativity, Einstein had used considerations from the principle of equivalence to predict, in 1911, that the Sun would bend light by an amount that is only one half of what the full theory actually predicts. This effect should be observable during a favourable solar eclipse and it was proposed to make an expedition to the Crimea in 1914 to test Einstein's 1911 version of his theory. From Einstein's point of view, it was fortuitous that World War I prevented the expedition from taking place. By the time Arthur Eddington led a corresponding expedition to the Island of Principe to view light bending during the eclipse of 1919, Einstein fortunately had found, in 1915, the correct theory, and the observations were hailed as a triumph for that theory. In the light of modern analysis, these observations may be regarded as less convincing than they were thought to be at the time, when they were taken as a resounding success for Einstein's theory. Nevertheless, modern observations of this effect, and of a related time-delay effect noted by Shapiro, supply convincing support for Einstein's prediction.

Einstein's light bending is now so well established that it is used as a very impressive tool for observational astronomy and cosmology. Distant galaxies provide complicated lensing influences on even more distant light sources. This can give important information, not reliably obtainable in any other way, concerning the distribution of mass in the universe. Einstein's prediction has been turned around to provide a superb probe of matter in the distant universe.

Gravitational Waves

One of the most striking predictions of Einstein's theory is the existence of *gravitational waves*.[11] Maxwell's theory of electromagnetism had led to the prediction that waves of oscillating electric and magnetic field should be able to propagate through space at the speed of light, and Maxwell had postulated, in 1865, that light itself is an effect of this nature. Maxwell's prediction is now thoroughly confirmed in many experimental situations. Einstein's theory of gravity has many similarities with Maxwell's theory of electromagnetism, one being the existence of corresponding gravitational waves, these being distortions of space-time that propagate with the speed of light. Such waves would be emitted by gravitating bodies in orbit about one another, but the effect is generally very small. In our solar system, the largest emission of energy in the form of gravitational waves comes from the motion of Jupiter about the Sun. The amount of this energy loss is only about that in the light of a 40-watt light bulb!

In fact (perhaps partly owing to the influence of his colleague the esteemed Polish physicist Leopold Infeld) Einstein seems to have wavered in his belief that a freely gravitating system might actually lose energy in the form of gravitational waves. In the early 1960s, when I was first becoming actively interested in Einstein's theory, there was a debate raging concerning this issue. At about this time, some important advances were beginning to be made in general relativity. For many years earlier, even stretching back to the time of the theory's conception, little interest was shown by serious physicists, and the subject was thought of as rather a playground for pure mathematicians. But in the early 1960s something of a renaissance of interest in general relativity occurred. In particular, the work of several theoreticians provided what to me was a convincing demonstration of the existence and generation of gravitational waves as a *real* physical phenomenon, the energy loss due to these waves being in accord with a formula that Einstein had put forward much earlier, in 1918.

In more modern times Einstein's theory has acquired an extraordinary boost from the observations (and theoretical analysis) of Joseph Taylor and Russell Hulse. In 1974 they first observed pulsar signals from the double neutron-star system PSR 1913+16. The variations in these signals give detailed information about the masses of the stars and their orbits, and one can cross-check this information with what general relativity predicts. There is an extraordinary overall agreement between theory and observation. Over the twenty-five-year period during which this system has been observed, there is a precision in the timing of the signals to roughly one part in 10^{14}; that is, one part in one hundred million million. To a first approximation, this gives a check on the Newtonian orbits of the stars. To a second, there is detailed confirmation of the general-relativistic corrections to the orbits (of the nature of that which occurs with the perihelion advance of Mercury). Finally, the loss of energy from the system in the form of gravitational waves, which is predicted by Einstein's theory, is seen to be in precise agreement with the theory. In 1993 Hulse and Taylor were awarded the Nobel Prize for Physics for the discovery and analysis of this remarkable system. From its uncertain beginnings, when general relativity had seemed an outlandish and rather flimsily supported theory, it now stands in extraordinary agreement with observation. In this one instance at least, a physical theory appears to be in detailed accord with nature to a precision greater than that which has been ascertained for any other individual physical system.

The existence of gravitational waves seems very well established in the PSR 1913+16 system. But such waves have not yet been convincingly observed *directly* here on Earth. There are several detectors in various stages of construction which should be able to observe such waves in the future. Moreover, the totality of these detectors, at different locations about the globe, should, within a few years, provide us with a remarkable Earth-scale gravitational-wave telescope able to obtain information about cataclysmic events (such as collisions between black holes and the like) occurring in very distant galaxies. This should give a completely new kind of window on the universe in which gravitational waves replace the usual electromagnetic ones. As with the light-bending effect, Einstein's prediction of gravitational waves may thus be turned around to provide a wonderful new observational tool to tell us something important about the distant universe.

Some Difficulties with General Relativity

We have now seen something of the extraordinary successes of general relativity. What about its limitations? The traditional view of the subject has been that its equations are notoriously difficult to solve. Indeed, despite the relatively very simple appearance of the Einstein equation, it hides a very considerable complication that is revealed when the expression $R_{ab} - \frac{1}{2}Rg_{ab}$ is written out explicitly in terms of the components g_{ab} and their first and second partial derivatives with respect to the coordinates. For many years only few solutions of the equations were known explicitly, but more recently numerous mathematical devices have been employed to find hosts of different solutions. Many of these are of mainly mathematical interest and do not directly relate to situations of particular physical relevance. Nevertheless, quite a lot is now known from the nature of exact solutions concerning, in particular, rotating bodies, black holes, gravitational waves and cosmology.

This notwithstanding, it is still hard to find particular exact solutions that describe situations that one may be interested in. Most notorious among these issues is the 'two-body problem': find an exact solution of Einstein's equation describing, say, two stars in orbit about one another. The difficulty here is that owing to the emission of gravitational waves, the two would spiral in towards one another, whence the situation possesses no symmetry. (The presence of symmetry is a great help in solving equations generally.) In fact, the difficulty in finding exact solutions to equations in physics is not now regarded as a particular limitation on a physical theory. With the advent of modern high-speed electronic computers, physicists can often get a much better picture of the evolution of the equations from a numerical simulation than they might obtain from an explicit exact solution. Considerable efforts have been devoted to developing computer techniques in general relativity, and some very good progress has been made.

Some of the main problems involved in solving the Einstein equation are of a rather different kind from sheer complication, however, and arise from an ingredient that is specific to general relativity: the principle of general covariance. Thus, when a solution is found, by computation or by analytical methods, it may not be clear what the solution *means*. Many features of the solution might merely reflect some aspect of the particular choice of coordinates, rather than expressing something of interest concerning the physics of the problem. Techniques have been evolved for answering such questions, but much more needs to be done in this area.

Finally, there is the profound issue of *singularities* in solutions of Einstein's equation. These are places where the solution 'diverges', thereby giving infinite answers rather than something physically sensible. For many years, there was great confusion in the subject because such singularities may turn out to be 'fictional'; that is to say, they may be merely the result of some inappropriate choice of coordinates rather than of some genuinely singular feature of the space-time itself. The most famous example of this kind of confusion occurred with the renowned *Schwarzschild solution* – the most important of all solutions of the Einstein equation. It describes the static gravitational field surrounding a spherically symmetrical star, and was found by Karl Schwarzschild in 1916 as he lay dying from a rare disease contracted on the eastern front in World War I, the same year that Einstein published his first full account of general relativity. At a certain radius, now known as the *Schwarzschild radius*, a singularity appeared in the metric components, and this region of the space-time used to be referred to as the 'Schwarzschild singularity'. People did not tend to worry much about this singularity, however, because the region would normally lie far beneath the surface of the star where, owing to the presence of a matter density (the T_{ab} of Einstein's equation), Schwarzschild's solution would cease to hold. But in the 1960s, the discovery of quasars led astronomers to wonder whether some highly compressed astrophysical bodies, as small as the scale of their Schwarzschild radii, might actually exist.

In fact, as early as 1933, Monseigneur Georges Lemaître had shown that with an appropriate coordinate change, the singularity at the Schwarzschild radius can be seen to be fictitious. Accordingly, this non-singular region is now *not* called a singularity, but is referred to as the Schwarzschild *horizon* – of a black hole. Indeed, any body that is compressed down to smaller than its Schwarzschild radius must collapse inwards towards the centre and a black hole is the result. No information can escape from within the Schwarzschild radius, which is why it is now referred to as a 'horizon'.

Space-Time Singularities

At this point it may be appropriate to relate how I myself became professionally involved with general relativity. In the late 1950s I was a young research fellow at St John's College, Cambridge. My official area of interest was in pure mathematics, but a friend and colleague of mine, Dennis Sciama, had taken it upon himself to acquaint me with many of the exciting things that were going on in physics and astronomy. I had had a significant but amateur interest in general relativity, since that subject was something that could be comprehended, and its beauty appreciated, by someone such as myself, with merely a love of geometry and an appreciation of the relevant physical ideas. Although Dennis had fired my interest in physics, I had not thought of general relativity as a subject into which I would research in a serious way, mainly because I had thought of it as somewhat peripheral to the main concerns of the fundamental quantum physics of the small-scale universe.

Nevertheless, probably some time in 1958, Dennis persuaded me to accompany him to attend a seminar given in London by David Finkelstein. This was on the extension of the Schwarzschild solution through its Schwarzschild radius. I remember being particularly struck by this lecture, but I had been troubled by the fact that although the 'singularity' at the Schwarzschild radius had been eliminated by a change of coordinates, the singularity at the *centre* (zero radius) still remained, and could not be removed in this way. Might it be, I had thought to myself, that there is some underlying principle that prevents the complete elimination of singularities from a broad class of solutions of the Einstein equation, including that of Schwarzschild?

Upon returning to Cambridge, I tried to think about this problem, though I was completely inadequately equipped to tackle it. At the time, I was concerning myself with a formalism known as the 2-spinor calculus, which has application to the study of spinning quantum particles. My pure-mathematical work had led me to study the algebra of tensors in a rather general way, and I had become intrigued by 2-spinors, because they seemed to be, in some sense, the square roots of vectors and tensors. In a certain clear sense, 2-spinors constitute a system that is even more primitive and universal in the description of space-time structures than that provided by tensors. Accordingly, I tried to see whether the employment of spinors might provide novel insights into general relativity, and whether these might be of use for the singularity problem.

Although I did not find that spinors told me much about singularities, I did find that they meshed extraordinarily well with the Einstein equation itself, providing unexpected insights that are not easy to come by, by other means. The elegance of the resulting expressions was striking, and I was hooked! For the ensuing forty-two years, general relativity has been one of my deepest passions, particularly in relation to its affinity to certain unusual mathematical techniques.

In 1964 I became interested in the singularity problem again, largely because John A. Wheeler pointed out that recent observations of those objects now known as quasars indicated that the Schwarzschild radius is being approached by actual astrophysical objects. Could the singularity that arises as a result of the collapse of a body down *through* that radius – the one at the centre that I had worried about in Finkelstein's lecture – actually be avoided? The exact solution of such a collapse (now referred to as a black hole), as found by Oppenheimer and Snyder in 1939, indeed possessed a genuine singularity at the centre. But a crucial assumption of their model was exact spherical symmetry. It could well be imagined that with irregularities present, the infalling matter might *not* simply be focused to an infinite-density singularity at the centre, but might instead pass through a complicated central configuration to be flung outwards again, and no actual singularity might be the result.

My earlier worries that such singularities may be inevitable had led me to doubt such a possibility, and I began to wonder whether some ideas that I had been more recently playing with, involving qualitative topological considerations – rather than the usual direct attempts at exact solution of Einstein's equation – might be able to resolve this issue. In due course, this unorthodox line of thinking led me to the first 'singularity theorem' of physical relevance in general relativity, which showed that, under some very reasonable general assumptions, *any* gravitational collapse to within a region that qualitatively resembles the Schwarzschild radius (but with no special assumptions of symmetry) results in a genuine space-time singularity.

Later work by Stephen Hawking, and by the two of us together, generalized this result, showing that in addition to the black hole situation, such singularities are also inevitable in the Big Bang origin of the universe, irrespective of any symmetry assumptions. The standard cosmological models derive from the original cosmological solutions to Einstein's equation found in 1922 by the Russian Alexander Alexandrovich Friedmann. Here, exact spatial homogeneity and isotropy is assumed, and the solution expands away from the initial Big Bang singularity. What the singularity theorems show is that we cannot eliminate the Big Bang singularity just by dropping the symmetry assumptions of homogeneity and isotropy.

All this is dependent upon the validity of Einstein's equation (and upon some physically reasonable assumptions concerning T_{ab}). Some people regarded these singularity theorems as revealing a profound shortcoming in Einstein's general relativity. My own attitude is somewhat different. We know, in any case, that Einstein's theory cannot be the last word concerning the nature of space-time and gravity. For at some stage an appropriate marriage between Einstein's theory and quantum mechanics needs to come about. What the singularity theorems reveal is an inner strength in Einstein's classical theory, in that it points clearly to its *own* limitations, telling us where we must look for an extension into a quantum world, and telling us also something of what to expect from an eventual quantum/gravitational union. We shall try to glimpse something of this in the next section.

The Beginning and Ends of Time

In the discussion above, we have caught sight of two situations in which space-time singularities arise in Einstein's theory: in gravitational collapse to a black hole and in the Big Bang origin of the universe. It seems clear that Einstein was very unhappy about both of these seeming blemishes to his theory. He appears to have been of the opinion that realistic departures from the high symmetry that is assumed in the standard exact solutions ought to lead to *non*-singular solutions. Unfortunately we shall never know what his reaction to the singularity theorems would have been, but apparently one of his reasons for trying, in his later years, to generalize general relativity to some kind of 'unified field theory' was his attempt to arrive at a singularity-free theory.

Initially he favoured a spatially closed-up universe that is *static* – so it would remain unchanged for all time. He found that he could achieve this only by introducing (in 1917) a *cosmological constant* Λ into his equation, which then becomes

$$R_{ab} - \tfrac{1}{2}R\, g_{ab} + \Lambda\, g_{ab} = -8\pi G\, T_{ab}.$$

Later, he regarded this modification as his 'greatest mistake'. If he had not insisted on a static model, but just let his original equation carry things along so as to obtain the Friedmann picture of a universe expanding away from a 'big bang', then he would probably have predicted the expansion of the universe, which was actually discovered observationally by Edwin Hubble in 1929.

There is much discussion today of whether the observational evidence now actually favours the existence of a (very small) cosmological constant. Some cosmologists (especially the proponents of what is referred to as the 'inflationary scenario') claim that such a constant is necessary in order to fit recent observations. Yet there are some seeming contradictions as things stand, and it will be better to wait until the dust settles before coming to any clear conclusions about this.

According to my own perspective on these issues, while we must be cautious about claims concerning the observational status of the large-scale universe, we must accept that the Big Bang and black-hole singularities are indeed part of nature. Rather than shrinking from them we must try to learn from them something of the 'quantum geometry' that should ultimately replace them. What can we learn? Although little is known in detail about the nature of singularities, some general comments can be made.

The first is that although unstoppable gravitational collapse must sometimes occur (such as with a supermassive star or collection of stars at a galactic centre), we do not know for sure that a black hole would be the result even though the singularity theorems tell us to expect space-time singularities. There is a still unproved assumption, referred to as 'cosmic censorship' (that I pointed out in 1969), which asserts that the resulting singularity cannot be 'naked', which means in effect 'visible from the outside'. If naked singularities do not occur, then a black hole must indeed be the result. (In any case, naked singularities would be, in a clear sense, 'worse' than black holes!) A black hole swallows material in its immediate vicinity and (cosmic censorship being assumed) destroys it all in the singularity at the centre. To the infalling material, this singularity represents the 'end of the universe', and it plays a role like a big bang reversed in time.

Despite this particular unpleasant feature, the *exterior* space-time to a black hole has a large number of very elegant properties. Moreover, large black holes appear to lie at the centres of virtually all galaxies, and the extraordinary physics that sometimes goes on in their immediate neighbourhoods seems to be responsible for the stupendous energy output of quasars, which can easily outshine entire galaxies. They also represent the regions of highest entropy known in the universe, and a famous formula due to Bekenstein and Hawking tells us exactly what that entropy should be in terms of the surface area of the hole's horizon.

In effect, cosmic censorship may be interpreted as telling us that there are just two kinds of space-time singularity in the universe, the *past* type (in the Big Bang) and the *future* type (in black holes). Matter is created at the past-type singularity and it is destroyed at the future-type ones. At first sight, these two types of singularity would appear to be simply time-reverses of each other. However, when we examine this in a little more detail, we find a gross distinction between these two types of singularity. This is related to the enormously large entropy of black holes. In everyday terms, 'entropy' means 'disorder', and the famous second law of thermo-dynamics tells us that the entropy of the universe increases with time. It turns out that the physical origin of the second law can be attributed to this gross asymmetry between past- and future-singularity structure, where the past-type singularities are particularly special and simple, whereas the future-type ones are general and extraordinarily complicated. Using the Bekenstein–Hawking formula for black-hole entropy, one can conclude that the 'specialness' of the Big Bang was quite stupendous, namely to one part in at least $10^{10^{123}}$.

Quantum Gravity?

Where does this gross time-asymmetry in space-time-singularity structure come from? The issue continues to stir up much controversy, but my own view is that there is a clear implication that the 'quantum gravity' that is sup-posed to account for the detailed nature of space-time singularities must be time-asymmetrical. I am continually amazed by the fact that so few workers in the area of quantum gravity seem to have come to the seemingly obvious conclusion that whatever the nature of this still-missing 'quantum gravity' theory may be, it *must* be a fundamentally time-asymmetric scheme. It is true that Einstein's equation is symmetrical under reversal of time, and so also is the Schrödinger equation which governs the evolution of a quantum state. Accordingly, any 'conventional' application of the rules of quantum mechan-ics to Einstein's theory ought to lead to time-*symmetrical* conclusions. In my own opinion, this provides a clear indication that the sought-for 'quantum gravity' must be an *unconventional* quantum theory, according to which the rules of quantum mechanics must themselves be expected to *change*. This is in addition to changes that must in any case be expected to take place in the classical rules of Einstein's general relativity. Thus I agree with Einstein in his belief that quantum mechanics is incomplete.

However, this is not the position of the great majority of those who attempt to combine quantum theory with general relativity. Despite the wealth of unusual and fascinating ideas that have been put forward as candidates for a 'quantum gravity' theory – such as 'space-times' of ten, eleven or twenty-six dimensions and ideas involving supersymmetry, strings, etc. – none of these candidates takes on board the possibility that the very rules of quantum mechanics may have to change. In my own view (and in the view of a sizeable minority of researchers into the foundations of quantum mechanics), changes in the rules of quantum theory are to be expected in any case, because of what is known as the 'measurement problem'.

What is the measurement problem? For this, we need to understand a little of the actual rules of quantum theory. There is a mathematical quantity referred to as the *quantum state* (or wave function), frequently labelled Ψ, which is supposed to contain all the necessary information defining the quantum system under consideration. The time-evolution of the state Ψ is governed by Schrödinger's equation until a measurement is made on the system, whereupon the state *jumps* (randomly) to one of a set of allowed possibilities defined by the specific measurement being performed. This 'jumping' does not take place in accordance with Schrödinger's equation, however, and the measurement problem is to understand how this random jumping comes about, given that the state is supposed actually to evolve by the deterministic Schrödinger equation.

I believe that a strong case can be made that the pure Schrödinger equation does *not* apply rigorously at all scales, and needs modification when gravitational effects begin to become significant. Accordingly, such a modification would necessarily be part of the correct 'quantum gravity' theory. Moreover, the measurement problem would find its resolution within this 'correct quantum gravity' theory. One of the main reasons for believing this comes from strong arguments that point out a fundamental conflict between the principle of general covariance and the basic principles of standard Schrödinger wave function evolution. According to this reasoning, quantum jumping (which I take to be a physically *real* phenomenon rather than the 'illusion' that it is often assumed to be) comes into play as a feature of the resolution of this conflict.[12] Now whatever form this modification of Schrödinger's equation would take, it would have to be *time-asymmetric*, and a gross asymmetry between past and future singularities would be expected, in accordance with the arguments I have given here.

As things stand, no plausible such modification of Schrödinger's equation has yet come to light, so a unification of quantum theory with general relativity along these lines remains as elusive as a unification along any of the more conventional lines that have so far been suggested. Finding the correct unification presents the twenty-first century with one of its greatest challenges. If this challenge is successfully met, then it will have profound implications running far beyond those that we can directly perceive at the moment. It will not be met, however, if the strange and wonderful principles underlying Einstein's beautiful equation are not thoroughly respected.

Erotica, Aesthetics and Schrödinger's Wave Equation

Arthur I. Miller

A good friend of Erwin Schrödinger recalled that 'he did his great work during a late erotic outburst in his life'. The epiphany occurred in the Christmas holidays of 1925, when the thirty-eight-year-old Viennese physicist vacationed with a former girlfriend at the Swiss ski resort of Arosa near Davos. Their passion was the catalyst for a year-long burst of creative activity. Like that of the dark lady who inspired Shakespeare's sonnets, her name remains a mystery, though most likely Schrödinger's wife was not in the dark about her husband's latest infidelity. Perhaps we owe to this unidentified woman the marvellous fact that apparently unconnected strands of research coalesced, and Schrödinger discovered the equation that bears his name.

In its form, at least, Schrödinger's equation was familiar to many scientists and its appearance was almost comforting in the light of the assault on familiar concepts coming from younger quantum physicists. It appeared to be the long-sought-after expression of the quantum theory that had been first articulated by its reluctant discoverer Max Planck in 1900, and had then been further refined by Albert Einstein and Niels Bohr, among others. In essence, it did for the subatomic world what Newton's laws did for the large-scale world some two centuries before – Schrödinger's equation enabled scientists to make detailed predictions about how matter behaves, while being able to visualize the atomic systems under study. Armed with the Schrödinger equation, it was possible for the first time to understand atomic structure in detail, whereas Newton's equations simply did not make sense in the microworld.

A different quantum physics of the atom had been found some six months before Schrödinger's creative outburst. This was accomplished by Werner Heisenberg, a brilliant young German theoretician, working at the University of Göttingen. The twenty-four-year-old Heisenberg had discovered a distinctly different approach to atomic physics, couched in an unfamiliar and difficult mathematics that offered no scope for visualization of atomic processes and no equation analogous to Newton's for classical systems. In fact, one motive for Schrödinger's version of atomic physics was his distaste, amounting to disgust, for Heisenberg's. Schrödinger even proved the mathematical equivalence of the two versions – so which one was better? Schrödinger preferred his own and was adamant on this point. Heisenberg thought otherwise and immediately and forcefully staked his own claim.

But there is a paradox. Although the Schrödinger equation was superficially easy to use, it featured a quantity called the wave function that was extremely difficult to interpret and impossible to observe directly. Heisenberg vehemently disagreed with Schrödinger's interpretation of the wave function as representing an atomic electron's smear of electricity around the nucleus. An enormous and bitter controversy arose that is not fully resolved even today. Schrödinger himself was never happy with the most common interpretation of the wave function's meaning.

In this essay, I want to explore how Heisenberg's *interpretation* came to dominate over Schrödinger's, even though Schrödinger's *method*, suitably reinterpreted, replaced Heisenberg's in almost all areas of physical theory. The issues under intense debate at that time – how we are to visualize atomic behaviour, and whether it is purely understandable in terms of probability – still echo through physics today. From a practical point of view, however, quantum theory has proved enormously successful. It has formed the basis of our understanding of the microworld, enabling technologists to develop increasingly effective transistors, microprocessors, lasers and fibre-optic cables. The theory is mainly implemented through Schrödinger's equation, which is used as a routine research tool by scientists all over the world.

Born in Vienna in 1887, the cultural and political capital of the Austro-Hungarian Empire, Schrödinger attended a gymnasium, or high school, which emphasized the study of Greek and Latin classics. Schrödinger also taught himself English and French.[1] He excelled at school and was recognized as a student of genius calibre. This wide and deep education served to ingrain in him a profound respect for classical tradition. His book *Nature and the Greeks*, published in 1948, is an elegant exposition of ancient physical theories and their relevance. Schrödinger developed a life-long interest in philosophy, which would lead to more than casual readings in Eastern texts such as the Vedanta, which he wrote about in 1925 in an intensely personal account of his beliefs, *Seek for the Road*. It is influenced by Hinduism, and is an argument for the essential oneness of human con-sciousness and for the unity of humanity and nature. This was not published until 1961, the year before his death, as part of the volume called *My View of the World*.

Although Heisenberg also attended a gymnasium and had a flair for music and philosophy, his mind-set differed radically from that of the more conservative Schrödinger, his senior by a decade and a half.[2] Heisenberg revelled in situations that were in flux. It cannot be irrelevant that he came of age in one of the most turbulent periods in German history, amid defeat in World War I, the collapse of the monarchy and revolution spreading across the Reich. Like Schrödinger, Heisenberg came from a cultured family; he was a pianist of near-concert level. Music was central to his life, while Schrödinger had no feeling for it. They were united, however, by their vigour and youthfulness, which both men sustained to an advanced age.

In 1906 Schrödinger entered the University of Vienna, where he had excellent teachers. He flourished in this atmosphere, deepening his under-standing of physics and adding to it an interest in biology, to which, some forty years later, he contributed some profound ideas in his short book *What is Life?* (James Watson, co-discoverer of the structure of DNA, cited it as an inspiration.)

By this time Schrödinger's highly developed erotic instinct had begun to emerge. It differed in his mind from the traditional male-chauvinist goal of female domination. Rather, Schrödinger believed he was exploring the essence of female sensuality. And he kept a logbook with comments, dates and names of encounters, his *Ephemeridae*. Like the then avant-garde artist Gustav Klimt, Schrödinger forever sought 'to capture the feeling of female-ness'. We can imagine that Schrödinger's calculated casualness of dress and appearance, with his high forehead, carefully combed hair and intense gaze, in conjunction with a seemingly inexhaustible well of knowledge, was very attractive to women. Despite his bourgeois demeanour and correctness, there was something Byronic about Schrödinger.

Like such Viennese compatriots as Ludwig Wittgenstein, he played an active role in World War I, serving with distinction on the Italian front in an artillery unit of the Austro-Hungarian army. Schrödinger was cited for his leadership in the face of fierce counter-battery fire in October 1915 in the course of the bloody battles made famous by Ernest Hemingway in *A Farewell to Arms*. Soon after, he was promoted to *Oberleutnant* and finished the war in Vienna in the cushy position of teaching introductory meteorology to army officers, while publishing papers on gas theory and general relativity.

By 1925 Schrödinger differed in every way from the brash younger men bursting onto the scene in quantum physics. Even his dapper style of dress contrasted with Heisenberg's, who is remembered as looking 'like a simple farm boy, with short, fair hair, clear bright eyes, and a charming expression'.[3] Compared with Heisenberg and his colleague and confidant, the hypercritical and acerbic Wolfgang Pauli, Schrödinger was already a senior figure, with a professorship at the University of Zurich.

Heisenberg's undergraduate education was extraordinary. Fortuitously, when he entered the University of Munich, during the winter semester of 1920–21, the famous physicist Arnold Sommerfeld was about to teach the atomic-physics part of his theoretical-physics cycle. In this way Heisenberg was thrown right onto the cutting edge of research. He often recalled that he learned physics backwards, studying atomic physics before Newton's, which is supposed to be the stepping stone to advanced topics.

The atomic theory current at the time was formulated in 1913 by the twenty-seven-year-old Danish physicist Niels Bohr. Bohr's was a frighteningly intense search for clarity, in a lifelong journey that he would share with colleagues and students in deep, critical dialogues. For this reason many of Bohr's scientific papers are almost opaque, having been worked and reworked so many times that omitting a word or a sentence can completely distort its meaning. He thought at his best in conversation, when he had someone against whom to bounce ideas. But in 1913 he was a young man in a hurry and moved with a gracefulness honed by the high level of football that he played. Ten years later he would begin to assume the heavy, gloomy veneer that reflected the weight of the problems he took upon himself in seeking the meaning of a new physics, one that defied all preconceptions of what a theory ought to be.[4]

Bohr's atomic theory of 1913 is best remembered for its hallmark imagery of atoms as minuscule solar systems. It was a magnificent pastiche of Newton's celestial mechanics with adroit insertions from Planck's radiation theory. Bohr's use of Newton's theory permitted its imagery to be transported into the atomic realm. This enabled him to restrict the electrons bound in atoms, that is, atomic electrons, to certain orbits about their central sun, or nucleus. These allowed orbits are called stationary states or energy levels. Consider the hydrogen atom, which is the simplest atom because it comprises a single electron that is bound to a positively charged nucleus. According to Bohr's theory, this electron can exist only in certain orbits. The lowest allowed orbit – the one closest to the nucleus – is referred to as the atom's ground state. An astounding consequence of Bohr's theory is that while in an allowed state the electron just perches, like a bird in a tree, doing nothing but waiting. By contrast, according to the accepted electromagnetic theory of the day, combined with Newton's mechanics, the electron ought to be orbiting the nucleus like a planet around the sun. According to the traditional laws of physics, the orbiting electron would continuously give off radiation energy. Consequently the atomic electron would lose energy and eventually spiral into the nucleus. The result is that matter would be totally unstable. We know this is not the case because, for example, you are sitting here and reading this article, instead of exploding. Explaining atomic stability was considered to be a key problem at the time. Bohr, however, had the great creative insight to realize that it was for the present insoluble and so just accepted it as a fact of life. This was his reason for postulating the existence of a lowest stationary state, or orbit, in which the electron neither drops nor radiates any light – and no further questions, thank you.

By, for example, illuminating the atom with light it is possible to excite the electron into a higher allowed orbit. Once there the electron is again like a perching bird, only now waiting to descend back to the ground state. Eventually it will come down, either directly or perhaps by transitions between states above the ground state. These transitions are not smooth but discontinuous and so are called quantum jumps. It is while making such transitions between stationary states that the electron emits radiation in bursts – that is, discontinuously. An enormous success of Bohr's theory was its ability to account for the wavelengths of the radiation emitted by hydrogen to within 1 per cent of the values experimenters had observed. Moreover, it successfully predicted previously unobserved wavelengths to a similarly high accuracy.

Bohr's theory caused great excitement in the physics community. One eminent and very sober physicist of whom more later – Max Born – said of Bohr's theory that it performed 'a great magic on mankind's mind; indeed its form is rooted in the superstition (which is as old as the history of thought) that the destiny of men could be read from the stars'. Einstein immediately praised the theory as 'an enormous achievement'.

By early 1925, however, the situation in atomic physics had become direly confused. The consensus among physicists was that Bohr's atomic theory was at a dead end. It could not treat with any accuracy anything except simple instances of the hydrogen atom. By 1923 data began to accrue from the interaction of atoms with light, to the effect that atoms did not respond like minuscule solar systems after all.

Physicists quickly cobbled together a hybrid version of Bohr's theory and this served as a stopgap. In this lash-up, no attempt whatever was made to visualize what was going on at a subatomic level but it was assumed that atoms could somehow lose energy by making a transition between one energy level and a comparatively lower one – by making a 'quantum jump'. Likewise, the atom could gain energy by jumping between energy levels to a higher one. In each of these processes, the energy lost or gained is carried by a burst of light corresponding to radiation with a particular wavelength. This explains why atoms emit and absorb radiation with special wave-lengths, known as spectral lines. Another crucial feature of this unsatisfactory theory was the innovative idea that it was not possible to predict exactly when atoms made quantum jumps – it was possible only to quote the probability of such an event's taking place at a particular instant. Bohr imported this 'probabilism', which came to be a central feature of quantum thinking, from a successful theory Einstein had introduced in 1916, in his theory about the interaction of radiation and atoms. These three features of the improved quantum theory of the atom – probabilism, quantum jumps and non-visualizability – were sufficient to make the theory serviceable until the beginning of 1925, when it, too, folded.

Physicists interpreted probabilities as a sign of not truly understanding the mechanisms of individual processes. They believed that eventually the mechanism by which electrons made transitions in atoms would come to be understood, and some as yet unknown version of Newton's mechanics would be formulated. In the end it would be business as usual, and probabilities would be unnecessary. This would turn out not to be the case. Although the modified version of Bohr's theory ultimately failed, it served as Heisenberg's stepping stone to his dramatically new atomic theory which was based on unvisualizable electrons and radical discontinuities. Its foundation was a mathematics that posed extreme difficulties in actually applying it. Heisenberg himself, in his first paper on the quantum mechanics, did not understand how to use it.

He had stumbled across mathematical quantities called matrices. This is because Heisenberg was interested in finding a sort of bookkeeping method for all possible atomic transitions between stationary states. Matrices are a natural way to do this, in addition to providing machinery for calculating the characteristics of spectral lines. To be a bit more precise: matrices are square arrays of numbers, and in quantum mechanics each entry represents a possible atomic transition, either up or down in energy. Through a well-known mathematical method the energies of the atom can be calculated. These are called the matrices' proper values, or eigenvalues, and their calculation is usually arduous. Wolfgang Pauli, one of the strongest calculators of the day, took over forty pages to deduce the energy levels of the simple hydrogen atom from Heisenberg's theory. By the end of 1925, certain long-standing problems had been solved by Heisenberg and his co-workers, which had eluded Bohr's theory. Heisenberg's 'matrices' version of quantum mechanics seemed to promise a great deal.

Heisenberg's hybrid education was undoubtedly one of the sources of his daring, hell-for-leather approach to atomic-physics research. Less than a year after entering university, he wrote his first paper, in which he chose not to work from certain rules for translating results from Newton's physics into quantum physics, as was the accepted method, but from a model already somewhat in agreement with quantum ideas. As one of Heisenberg's colleagues said later, 'A wonderful combination of profound intuition and formal virtuosity inspired Heisenberg to conceptions of striking brilliance.'

At this time Schrödinger was pursuing, as usual, a wide variety of interests. Besides investigations in general relativity, since 1917 he had been studying the perception of colours. Then there was his interest in problems concerning sound and elastic media, which led to his investigation of wave theory, soon to come in handy.

On the personal side, Schrödinger was living in Zurich with his wife of some five years, Annemarie, known affectionately as Anny. They lived in a slightly subdued north Swiss version of the bohemian Weimar culture that so scandalized German conservatives and nationalists: the sexually shocking, ambiguous world we associate with Marlene Dietrich and expressionist art and cinema. The violent reaction to this liberated milieu was personified by Hitler and his Nazi Party; the rise of such thuggish violence would give Schrödinger second thoughts about leaving Zurich for Berlin in 1927 to take Planck's chair. Meanwhile, the marriage was clouded on the one hand by

Anny's never conceiving the child that Erwin so badly wanted and on the other by his compulsive womanizing. They were an odd couple. Anny had limited intellectual interests and worshipped Erwin's looks and brilliance. After their passion waned – about a year into their marriage – they both sought sex elsewhere, and yet they remained married and cared for each other as friends. As Anny commented some years later, 'You know it would be easier to live with a canary than a racehorse, but I prefer the racehorse.' Schrödinger never had a close male friend in his entire life. His flair for dress and his romantic intensity towards women were fuelled by his love for the theatrical.

From considerations based on relativity theory, in 1923 Louis de Broglie suggested that electrons can also be waves, whereas previously everyone thought of them exclusively as particles. At once Einstein recognized the importance of de Broglie's observations and elaborated on them in research on gas theory. Einstein was enthusiastic and wrote to a colleague that de Broglie 'has lifted a corner of the great veil'. But de Broglie and Einstein were only part of the impetus behind Schrödinger's orgy of creativity, as he explains in the third of the papers he published in the spring of 1926:

> My Theory was inspired by L. De Broglie, *Ann. de Physique* (10) 3, p. 22, 1925 (Thèses, Paris, 1924) and by short but incomplete remarks by A. Einstein, *Berl. Ber.* (1925) pp. 9 ff. No genetic relationship whatever with Heisenberg is known to me. I knew of his theory, of course, but felt discouraged, not to say repelled, by the methods of transcendental algebra, which appeared very difficult to me and by the lack of visualizability.[5]

The sense of aesthetics that Schrödinger alludes to here is his preference for a mathematics that is more familiar and also not as ugly as Heisenberg's 'transcendental algebra' (or matrices), but which also permits visualizability of atomic processes. This becomes clearer in what follows.

In a more objective tone, one of Schrödinger's principal criticisms of Heisenberg's quantum mechanics is that it appeared to him 'extraordinarily difficult' to approach such processes as collision phenomena from the viewpoint of a 'theory of knowledge' in which we 'suppress intuition and operate only with abstract concepts such as transition probabilities, energy levels, and the like'. Indeed, in Heisenberg's formulation during 1925–26 it was possible only to calculate atomic energy levels; that is, to deal only with electrons bound in atoms. On the other hand, the concept of what is abstract is relative: Bohr, Heisenberg and Pauli considered energy levels 'and the like' to be perfectly concrete. Schrödinger admitted in 1926 that there may exist 'things' that cannot be comprehended by our 'forms of thought', and hence do not have a familiar Newtonian space and time description, but 'from the philosophic point of view' he was sure that 'the structure of the atom' does not belong to this set of things.[6]

But Schrödinger did realize, along with Heisenberg and other physicists of the time, that visual imagery taken wholesale from the world of sense perception would not suffice. In order to avoid it altogether, Heisenberg based his quantum mechanics on unvisualizable particles. Schrödinger sought a means of visualizing electrons that was different from the way in which scientists had become accustomed to thinking of them; that is, as particles. He realized that such an approach to the electron had been made available by de Broglie and Schrödinger and set out to exploit it. This may have been an aesthetic preference, but it was one on which theories could be based. Building on de Broglie's daring idea that electrons can be waves as well as particles, Schrödinger applied this hypothesis to electrons bound in atoms.

Schrödinger's basic idea was to formulate a theory for electrons bound in atoms in which they are analogous to a vibrating string fixed at both ends. How the string vibrates is an indicator of the electron's energy. This sort of wave theory also avoids quantum jumps. The reason is that atomic transitions were understood as occurring in the manner of waves representing the electron's charge density, surrounding the nucleus and decreasing their radius in order to pass between allowed states.

Let me describe how Schrödinger applied his ideas to the simplest atom of all, a single electron orbiting a nucleus – a hydrogen atom. As a thought experiment, consider the electron as a string fixed at both ends; that is, bound in hydrogen atom. When the string is vibrating at its lowest energy as a standing wave, there is exactly one half-wavelength between the ends. At the next highest energy, there are two half-wavelengths between the ends; then, at the next highest energy, three half-wavelengths, and so on. The point is that each configuration of the vibrating string corresponds to a particular energy, or eigenvalue, of the string.

Schrödinger's equation, when applied to the hydrogen atom, yielded much the same relationship between energy levels and allowed wave functions. The equation predicts the possible energy values that the electron can have (its energy levels, each denoted by E) along with the so-called wave functions that describe its behaviour (each of them is denoted mathematically by the Greek letter ψ, psi). The equation says

$$\hat{H}\psi = E\psi$$

The letter $\hat{H}$ stands for the mathematical expression (known technically as an operator) that represents the total energy of the atom. After the mathematics has been done, one ends up with a set of energy levels, each with at least one corresponding wave function.[7, 8]

The amazing thing was that this simple mathematical operation predicted exactly the right energy levels for the hydrogen atom, reproducing the success of Bohr's planetary model. But how should we picture orbiting atomicelectrons in Schrödinger's picture? That's where it gets difficult. Schrödinger visualized the atomic electrons as a distribution of electronic charge whose distribution in space is related to the electrons' wave function.

Great though Schrödinger's achievement was in writing down his equation, it is at odds with the special theory of relativity. Basically the equation is inconsistent with guidelines set down by Einstein's principle of relativity according to which an equation should have the proper mathematical form so as to be able to include measurements made on systems moving at high speeds close to that of light. But this was intentionally so because Schrodinger initially tried a relativistic approach and failed. What he did was to insert de Broglie's results into the relativistic equation connecting energy, momentum and mass. Then he specialized to that benchmark for all quantum theories, the hydrogen atom, in order to calculate its spectrum of energy values. He failed because Schrödinger's relativistic equation did not include the electron's spin, a property that was only beginning to be understood at the time. On the other hand, Schrödinger found that a non-relativistic version gave results in agreement with observation. This problem would be overcome in 1928 when the English theoretician Paul Dirac brilliantly proposed a quantum equation for the behaviour of the electron that was consistent with special relativity. This equation naturally explained why the electron has spin.[9] Looking back on this episode, Dirac wrote that Schrödinger should have pursued the relativistic equation because, in his view, 'It is more important to have beauty in one's equations than to have them fit experiment.'[10]

How did Schrödinger derive his equation? The derivations that Schrödinger presented in his published papers for the Schrödinger equation are in fact not derivations at all, but plausibility arguments: he knew beforehand what he wanted. Actually the Schrödinger equation should be considered axiomatic, that is, underivable: its validity comes from the correct solutions it gives to certain problems such as the hydrogen atom spectrum which Schrödinger disposed of in a few pages, compared with Pauli's mathematical gymnastics using Heisenberg's quantum mechanics.

Schrödinger went on to prove the mathematical equivalence of the wave and quantum mechanics and pushed this result to support his disdain for quantum mechanics: when discussing atomic theories he 'could properly use the singular'.[11] For Schrödinger, *sic transit* quantum mechanics. But what sort of picture did Schrödinger offer? He maintains that no picture at all is preferable to the miniature solar-system atom, and in this sense the quantum mechanics is preferable because of 'its complete lack of visualisation'; however, this conflicts with Schrödinger's philosophical viewpoint. Schrödinger argued that the wave function for, say, the electron in the hydrogen atom is related to the electron's distribution of electricity around the nucleus. However, Schrödinger's proof of the localization of the waves representing the electron turned out to be incorrect, as Heisenberg showed in 1927: the waves representing the electron do not in general remain localized, that is, stay together.[12] But Schrödinger was intellectually honest in emphasizing that his claimed visual representation is unsuitable for systems containing more than one electron. The reason is that the wave function representing a single electron can be visualized as a wave in three dimensions because it depends on the electron's position in a three-dimensional space. The wave function for a system of two electrons depends on both of their spatial positions, and so is three plus three or six-dimensional, whereas our visual perception is restricted to three dimensions.

The state of quantum mechanics in the first half of 1926 can be summarized as follows. No adequate atomic theory had existed as of mid-June 1925, but by mid-1926 there were two seemingly dissimilar theories. Although a particle-based theory, Heisenberg's renounced any visualization of the bound particle itself, its mathematical apparatus was unfamiliar to physicists and difficult to apply, and it was based specifically on discontinuities. But discontinuity is anathema to Newtonian physics as well as to the pre-quantum version of electromagnetism in which all processes occur continuously and are visualized as waves. On the other hand, Schrödinger's wave mechanics focused upon *matter* as waves, offered a visual representation of atomic phenomena (albeit restricted to a single electron), and served to account for discrete spectral lines without quantum jumps. The

Schrödinger theory's more familiar mathematical apparatus of differential equations set the stage for a calculational breakthrough, supported by Schrödinger's proof of the mathematical equivalence of the two theories.[13] Wave mechanics delighted the portion of the physics community that resisted discontinuity being built into physics and preferred a version of atomic physics based on a theory similar to Newton's. Although conclusive evidence of the wave-particle duality for electrons would not appear until 1927, experiments performed as early as 1923 agreed with de Broglie's hypothesis. Consequently many physicists tended to accept it. As Einstein wrote to Schrödinger on 26 April 1926, 'I am convinced that you have made a decisive advance . . . just as I am equally convinced that the Heisenberg . . . route is off the track.'

Heisenberg's first recorded comment on Schrödinger's wave mechanics is in a letter of 8 June 1926 to his friend and colleague Pauli, and he was enraged: 'The more I reflect on the physical portion of Schrödinger's theory the more disgusting I find it. What Schrödinger writes on the visualizability of his theory is probably not quite right. In other words it's crap.'

During this troubled period in Heisenberg's professional life, he was most candid with Pauli, then at the University of Hamburg. Pauli's interests were always wide-ranging, and included such esoterica as numerology and the Kabbala. Nor was he adverse to dipping into the Hamburg underworld of drugs and sex. In the early 1930s Pauli became a devotee of the Zurich psychoanalyst Carl Jung. The two went on to co-author a book in which Pauli wrote a memorable Jungian analysis of the great astronomer Johannes Kepler, a man not unlike Pauli in his extra-scientific interests.

In addition to sharp comments in letters to Pauli, Heisenberg responded quickly in print as well, although in a more piano tone. In a paper of June 1926 he wrote that although the physical interpretations of the two theories differ, their mathematical equivalence allows this difference to be put aside; for 'expediency' in calculations he will utilize Schrödinger's wave functions, with the caveat that one must not impose upon the quantum theory Schrödinger's 'intuitive pictures'.[14]

Schrödinger and Heisenberg first encountered each other in July 1926 in Munich, where Arnold Sommerfeld had invited Schrödinger to deliver two lectures on his new theory. There was standing room only. Barely keeping himself in check, after the second lecture Heisenberg rose to deliver what was essentially an impromptu monologue attacking Schrödinger's wave mechanics because it was apparently unable to explain how radiation interacts with matter through quantum jumps. Amid shouts of disagreement from the audience, the angry chairman, an eminent Munich physicist, motioned to Heisenberg to sit down and be quiet. Later on, he told Heisenberg that his physics 'and with it all such nonsense as quantum jumps [is] finished'. Heisenberg was despondent because he seemed to be unable to convince anyone of his views. But he continued to argue his case, and by August 1926 colleagues began to write worried letters to Schrödinger asking how indeed he could explain certain quantum effects without discontinuities. Schrödinger himself began to feel uncertain.

Tension between quantum mechanics and wave mechanics increased with the publication of results by Heisenberg's mentor at the University of Göttingen, Max Born, in July 1926. (Max Born later became a footnote in pop music as the grandfather of Olivia Newton-John.) The forty-five-year-old Born was a rather shy, withdrawn character who directed one of the three institutes where Heisenberg studied. (The other two were Sommerfeld's at the University of Munich and Bohr's in Copenhagen.) Heisenberg had discovered his quantum mechanics while spending a period of time away from Munich at Göttingen. At Born's institute, physicists were interested in exploring the nature of electrons as particles by arranging for them to hit and scatter off atoms. This is a very different sort of physics problem from dealing with electrons bound in atoms. Born was interested in 'free' electrons, that is, electrons that have no net force acting on them. But at this time neither quantum mechanics nor wave mechanics could deal with free electrons, as they move through space.

Originally trained as a mathematician, Born had quickly grasped the subtleties in the Heisenberg and Schrödinger formulations in addition to their physical content. So everyone listened when Born wrote of the deficiencies of both men's theories in accounting for scattering experiments. And so Born decided that 'new concepts' were needed and wave mechanics would be his vehicle, because at least it presented the possibility of some sort of visual imagery.

Born made the stunning proposal that Schrödinger's wave function represents neither the electron's visualizable charge distribution as a wave surrounding the atom's nucleus, nor a group of charge waves moving through space. Rather the wave function is a totally abstract quantity in that it is not at all amenable to any visualization. Instead of being able to calculate from it a density of electricity, one calculates something that acts like a density – a probability density for the electron to be present in some region of space. This dramatic assumption transformed Schrödinger's equation into a radically new form, never before contemplated. Whereas Newton's equation of motion yields the spatial position of a system at any time, Schrödinger's produces a wave function from which a probability can easily be calculated. Schrödinger's equation then tells us not the path of a particle, but how the probability of the particle's detection changes with time. Born's aim was nothing less striking than to associate Schrödinger's wave function with the presence of matter.

By the autumn of 1926 Heisenberg had come to hate Schrödinger not only because his equation was so widely used – professional jealousy should never be discounted among creative people – but also for another, no less important reason, one that impacted directly on the very depths of Heisenberg's own research programme. He recalled, 'Schrödinger tried to push us back into a language in which we had to describe nature by "intuitive methods". That I couldn't believe. That is why I was so upset about the Schrödinger development in spite of its enormous successes.[15] After all, Schrödinger's equation was incredibly simpler to use than the mathematics in Heisenberg's quantum mechanics.' Then came Born's paper in which 'he went over to the Schrödinger theory'. Heisenberg described these developments as very disturbing to his 'actual psychological situation at that time'.

In November 1926 Heisenberg published a paper that got very little attention but, he recalled, 'for myself it was a very important paper'.[16] It was written by an angry man, and in it Born's scattering theory is nowhere cited and Schrödinger is sharply criticized. Heisenberg demonstrates that a probabilistic interpretation can be understood only if there are quantum jumps; that is, discontinuities. The gist of Heisenberg's paper is to prove that the presence of probabilities implies discontinuous phenomena, which in turn requires the presence of particles that are, after all, discontinuities in the fabric of nature. And so he comes down firmly in favour only of a particle viewpoint and, by implication, against Schrödinger's wave mechanics.

In subsequent articles during that year Heisenberg emphasized that phenomena occurring in the small volumes of the subatomic contradict our customary intuition. By this he meant that, contrary to Schrödinger, terms derived from day-to-day understanding of the world like 'wave' and 'particle' cannot be glibly extended into the atomic world. Surprises lurk there such as the wave-particle duality of light, first made explicit by Einstein in 1909, and the wave-particle duality for electrons proposed by de Broglie in 1923. This dual mode of existence is totally counter-intuitive and unimaginable. How can something be continuous and discontinuous at the same time? For this reason, physicists were slow to accept Einstein's light quantum. Their principal reason, stated forthrightly by Planck in 1910, was that when light was shone on alternating strips of opaque and transparent material (known as a diffraction grating) it behaves like water waves, producing a smoothly varying pattern of light that cannot be explained by assuming that light behaves as particles. This profound problem was solved only in 1927 when Born put forward his interpretation of the wave function, which accounted for these diffraction patterns in terms of myriad tiny impacts by individual particles of light. For many physicists, however, the mixture of wave and particle elements in the explanation remained deeply puzzling.

Just as a particle representation for light seemed out of place, so, at first, did a wave representation for the electron, as it was proposed in 1923 by de Broglie. Physicists were eventually persuaded to accept a wave-particle duality for the electron, because experimental data in that year lent some support to de Broglie's hypothesis. Conclusive results were achieved in 1927. Yet evidence for the existence of light quanta appeared in experiments of 1923. But even the person who performed these experiments, Arthur Compton, couldn't believe the results. His principal objection lay in the relation between the energy of the light quantum (which is, after all, a particle, and is therefore localized) and its wavelength (which is not localized). How can such entirely different quantities be related at all? Is this not like trying to link fishes and rocks? Clearly the wave nature of electrons, which would by 1927 be accepted as conclusive, did not upset physicists as much as disturbing the centuries-old sacred representation of light as a wave.

To Heisenberg, as to Schrödinger, the fundamental issue in quantum theory became, as Heisenberg put it, to explore the 'kind of reality' that existed in the atomic world. Physics had become a branch of metaphysics because nothing less was at stake than understanding the nature of physical reality. Heisenberg took on this problem in his classic paper of 1927, 'On the intuitive content of the quantum-theoretical kinematics and mechanics', the so-called 'uncertainty principle' paper.[17] The term 'intuitive' in the title signals that this absolutely fundamental concept has to be redefined in the atomic world. Straightaway Heisenberg makes it clear that the basic issue facing quantum mechanics is the meaning of certain terms when they are extrapolated into the atomic realm: 'The present paper sets up exact definitions of the words: position, velocity, energy, etc. (e.g. of an electron).' Heisenberg insists that it is the *interpretation* of quantum mechanics that is in question: 'Heretofore, the intuitive interpretation of the quantum mechanics is full of internal contradictions that become apparent in the struggle of the opinions concerning discontinuum- and continuum-theory, wave and particles.' He reasoned that a new intuitive interpretation, replete with visual imagery, of the new atomic theory should follow from its equations and be grounded in the 'uncertainty principle'. What this means is that, unlike in classical physics, in the atomic domain the measurement uncertainties in position and momentum cannot be simultaneously reduced to zero. Rather, the product of these uncertainties is an extremely small but nonzero quantity. In concrete terms: the more precisely the particle's position is measured, the less precisely can its momentum be ascertained, and vice versa.

Heisenberg was able to give his ideas precise mathematical form. It involved the uncertainty, or rather 'indeterminacy' or 'imprecision in knowledge', of simultaneous measurements of position and momentum (for the situations he was considering, momentum p = mass × velocity). Denoting the uncertainty in position as Δx (delta x) and the uncertainty in momentum as Δp (delta p), Heisenberg's uncertainty relation is that the product $\Delta x \Delta p$ is at least $h/(2\pi)$, where h is Planck's constant (6.6×10^{-34} joule-seconds).[18] Its perhaps unfamiliar units aside, although Planck's constant is an extremely small quantity, it is not zero. This is why, according to the uncertainty principle, the more precisely we can measure a particle's position, the less we know about its momentum at the same time. This completely contradicts the common sense or intuitive idea in Newton's physics that there is no reason at all why, at any given moment, we cannot know to any desired accuracy both where a particle is and how fast its moving. For example, according to Newton, the accuracy with which you know the position of a falling apple should, in principle, have nothing to do with how accurately you know its speed at the same time.

Having demonstrated that discontinuities and a particle representation were essential to any new atomic theory, and that Schrödinger's suggested visual imagery drawn from familiar phenomena was insufficient, at this point Heisenberg chose to deal with Schrödinger's *ad hominem* comments in his third communication of 1926. He did so in a footnote, almost as an afterthought. He recalled Schrödinger writing of the matrix version of quantum mechanics as a theory that is 'frightening, indeed repulsive in its counter-intuitivity and abstractness'. Heisenberg continued with a double-sided compliment to Schrödinger as having formulated a theory that could not be esteemed highly enough because it permitted 'mathematical penetration of the quantum-mechanical laws'. However, Heisenberg continues, in his 'opinion' its 'popular intuitivity' led scientists astray from the 'direct path' for the consideration of physical problems.

By this time it was clear that Schrödinger had no intention of fighting back in print. But privately Schrödinger persisted in his view of the possibility of a visual imagery of waves for elementary particles, of no probabilities entering the picture, and of no quantum jumps. On 4 October 1927 Schrödinger arrived at Bohr's institute in Copenhagen to lecture on his theory. Heisenberg recalled what happened:

> Bohr's discussions with Schrödinger began at the railway station and were continued daily from early morning until late at night. Schrödinger stayed in Bohr's house so that nothing would interrupt the conversations. And, although Bohr was normally most considerate and friendly in his dealing with people, he now struck me as an almost remorseless fanatic, one who was not prepared to make the least concession or grant that he could ever be mistaken. It is hardly possible to convey just how passionate the discussions were, just how deeply rooted the convictions of each, a fact that marked their every utterance.[19]

Discussing various ways in which the electron could make atomic transitions, Schrödinger concluded, 'The whole idea of quantum jumps is sheer fantasy.' Bohr's reply was simply: 'Yes, in what you say, you are completely right. But that doesn't prove there are no quantum jumps. It only proves that we can't visualize them.'[20] One of Schrödinger's final retorts at Bohr was that 'if all this damned quantum jumping were really here to stay, I should be sorry I ever got involved with quantum theory'.[21] By this time the strain had made Schrödinger ill with fever and he had taken to bed. Bohr's wife took meticulous care of him. But Bohr was relentless – sitting on the edge of Schrödinger's bed, he continued to press his argument; 'But you must surely admit that . . .'[22] Schrödinger refused to capitulate. He continued to believe that atomic processes could be visualized with the old imagery, suitably redefined. But Bohr thought otherwise, and had become increasingly interested in Heisenberg's uncertainty principle, which indicated that the equations of quantum mechanics would point the way to an entirely new visual imagery. Physics had come full circle back to the view of Plato, some 2,000 years earlier, in which mathematics would be the guide to what constitutes physical reality.

The Schrödinger equation turned out to have an enormously wide range of applications. This became clear immediately for chemistry when a new branch of research emerged, quantum chemistry, which studies the bonding of atoms and such complex situations as molecular bonding and chemical reactivity. The earliest triumph of Schrödinger's equation in this area is Walther Heitler and Fritz London's description in 1927 of the bonding of the hydrogen molecule. This sort of problem was, of course, impossible even to approach in the old Bohr theory of the atom. It was based on another of Heisenberg's dazzling discoveries. In 1926 he had deduced the helium atom's spectrum, a problem that had defeated everyone in the old Bohr theory. The dazzling aspect of the discovery is that in quantum theory particles can attract one another by exchanging places extremely rapidly. This exchange phenomenon is at the basis of Heitler and London's theory and would also be central to the first theory of the force that holds the nucleus together, formulated by Heisenberg in 1932.

The Schrödinger equation can also be used to study how chemicals react at a molecular level, the details of which are usually extremely difficult if not impossible to observe experimentally. The wave function of every molecule is very complicated: it has to take into account both the relative positions and the interactions of all the constituent particles. To compute these wave functions from the Schrödinger equation by hand is a virtual impossibility – computers are essential. For this reason, the computation of these wave functions – and the chemists' understanding of chemical processes at a molecular level – has burgeoned since the development of increasingly high-speed computers in the late 1970s. The consequence has been advances in almost all areas of chemistry, from the production of new drugs to the study of the Earth's atmosphere.

The province of the Schrödinger equation is not restricted to the atomic and subatomic domains. It is also needed to explain some extraordinary effects that we see in the large-scale world, notably superconductivity and superfluidity. Superconductors are special materials whose electrical resistance drops suddenly to zero when the temperature falls to below a critical value that is usually below –250 celsius, extremely cold by everyday standards. Such materials have many extraordinary attributes, not least that they all completely expel magnetic fields when they are superconducting. The phenomenon of superfluidity is similarly puzzling. It occurs only in liquid helium at extremely low temperatures, when very strange things happen – it flows practically without viscosity and can even climb up and over the walls of vessels that contain the liquid. The remarkable thing is that both superconductivity and superfluidity can be tackled theoretically by using the Schrödinger equation, applied to the matter's constituent atoms and molecules.

Besides playing an integral role in physics and chemistry, the Schrödinger equation has become an active topic in philosophy. Consider the so-called measurement problem. Whereas in classical physics the interaction between the measurement apparatus with the system under investigation can be ignored, this is not so in quantum theory. For example, consider the following experiment. I want to measure the position of a falling marble, which I can accomplish by, say, photographing it. This process entails that the marble be illuminated and that light be reflected from the falling marble onto a photographic plate. The fact that the marble is being bombarded with light quanta makes pretty much no difference at all to the outcome. In practice the marble's position and its velocity (and so its momentum, too) can be determined simultaneously to any desired degree of accuracy.

But what if the marble is an electron? According to wave mechanics, the falling electron can be anywhere because its wave function is spread out over all of space. The marble, on the other hand, is localized right from the start.[23] Clearly the question 'What is the electron's position?' *really* has no meaning until an actual measurement is carried out, in this case by photographing it. Photographing the electron means illuminating it with at least one light quantum, which becomes part of the measurement system. The interaction of this single light quantum with the electron locates the electron at that moment. This is known as 'collapse of the wave function' because the interaction between the measurement system (light quantum) and system in question (electron) reduces the electron's previously spread-out wave function to a certain well-defined region of space. In other words, of all the possible positions that the electron can have as a wave spread out over all of space, a single one is selected by the measurement process. Therefore, the state of the electron is irreversibly changed from being potentially everywhere to being definitely somewhere. The uncertainty principle informs us that the cost is an enormous uncertainty in the electron's momentum. One of the enduring puzzles of quantum theory concerns what happens to the wave function of an electron (or any other quantum) during a measurement. Before the measurement is made, the electron is in a combination of several quantum states, but the very act of measurement is believed – according to standard quantum lore – to put it in one particular state. What on earth is the underlying mechanism behind this? On this fundamental question, the Schrödinger equation and the other fundamental equations of quantum theory are silent.[24]

There is an interesting photograph of the Nobel Prize winners for 1933 taken at the Stockholm train station. Dirac is to Heisenberg's right and

Schrödinger to his left. Dirac and Heisenberg are in formal suits and over-coats. In most photos Heisenberg is either smiling or in some sort of dignified, serious pose, but here he has turned away from Schrödinger with a look almost like disgust. Schrödinger, alone of the three, has a big grin and seems to be having the time of his life. He is in the flamboyant attire of the day: calf-length trousers with bottoms bloused over elastic ends and high socks, casual coat with large fur collar, and his signature bow tie. Another memorable photograph in which the two adversaries are both present is also telling of their bitterly divergent views. It is at the annual summit gathering of physicists – the 1933 Solvay Conference in Brussels. As is the style in these photos, the elder conferees sit while the younger ones stand. In time the younger ones begin to move into the seats. Schrödinger sits and Heisenberg stands almost but not quite directly behind him.

Although many physicists consider quantum theory to be a closed book, there are still fundamental issues that remain unsettled, and most of them are rooted in the Schrödinger equation. Schrödinger wrote on 23 March 1936 to Einstein of his recent meeting with Bohr in London, 'I found it good that they strive in such a friendly way to bring one over to the Bohr–Heisenberg point of view . . . I told Bohr that I'd be happy if he could convince me that everything is in order, and I'd be much more peaceful.'[25] Bohr never could.[26] Instead, he isolated Schrödinger.

The battle lines were quickly and clearly drawn in the struggle between the waves and particles. Things seemed to be going well for a while for Schrödinger's cause. Until, that is, the winter of 1926, when Bohr summoned Heisenberg to Copenhagen to hammer out the meaning of quantum physics. Their deliberations went on for much of the following year. During this time they worked out the so-called Copenhagen interpretation with its emphasis on probabilities, discontinuities and wave-function collapse, all of which were anathema to Schrödinger. But he was no match for them. Schrödinger did not fight either in print or at the famous 1927 Solvay Conference, leaving it to no lesser a figure than Einstein to fly the flag. But Einstein also got nowhere with Bohr and company, despite some ingenious counter-proposals. The 'war' lasted a year. Whereas Schrödinger never made another great discovery before or after the equation that bears his name, Heisenberg had several notable successes before June 1925 and would go on throughout the mid-1930s to do more great work. He would always remain a force to be reckoned with. In the pantheon of twentieth-century physics, Heisenberg is second only to Einstein.

Ironically, although Heisenberg won the battle, and felt he had won the war, Schrödinger's equation is more widely used than Heisenberg's version of atomic physics. This is the case despite the incompatibility of Schrödinger's equation with relativity, which is unimportant for just about every practical application, notably because most of these applications involve quanta that travel at nowhere near the speed of light. On the other hand, Heisenberg's matrix formalism found its role in deeply theoretical areas such as the quantum field theory of fundamental particle physics.

What I have always found so intriguing about the Heisenberg–Schrödinger dispute is that it was fundamentally one of aesthetic choice. Both versions of atomic physics could account, in principle, for all known experimental data about the hydrogen atom and were fundamentally equivalent, in that they gave the same explanations about, for example, the helium atom. Each man defended his view of nature passionately. Bohr's great realization here is that neither man took serious account of the wave–particle duality of light and matter. And here Bohr made a key point: there is a third aesthetic, in which waves and particles are taken together, within a suitable interpretation of Schrödinger's wave function, which was already to hand – namely Born's.

That there are two versions of atomic physics should come as no surprise, because in our world of perceptions things come in pairs, such as particles and waves, yin and yang, black and white, yes and no, love and hate, light and darkness – there are no intrinsic maybes as there are in the atomic world. Yet through abstraction, through emphasis on conception rather than perception, we can move onto a higher plane and appreciate the power of ambiguity. This is generally uncomfortable in our personal lives, in which we strive to resolve ambiguous situations through decisiveness – once again into an 'either/or' mode. As Einstein and Picasso demonstrated in the first decade of the twentieth century, ambiguity is the key to discovering representations of nature that are beyond mere superficial appearances. Direct viewing can deceive, as Einstein discovered in physics and Picasso discovered in art. In Einstein's relativity theory of 1905, time and space are relative, and are interpreted according to how different observers view them. For example, two events that occur at the same time to one observer will not be simultaneous for another observer in relative motion. In Picasso's great work of 1907 *Les Demoiselles d'Avignon*, from which the cubism of Georges Braque and Picasso developed, the painter discovered a way of representing figures so that many possible perspectives appear on the canvas all at once.[27] In their own ways, Schrödinger and Heisenberg carried this adventure of abstraction into the atomic world.

The literary critic William Empson has argued eloquently that the insights of quantum theory could illuminate literature as well.[28] Before switching to literature in 1928 while a student at Cambridge, Empson had read mathematics and was well versed in physics. He developed new interpretations of Shakespeare's works, seeing fit to 'attach the notion of probability to the natural object rather than to the infallibility of the human mind'.[29] Empson advocated renewing the study of literature through the lens of a reality altered by quantum theory. By this he meant that Shakespeare ought not to be analysed in an 'either/or' mode, but the focus should be on ambiguities, that is, a 'both/and' mode, which can bring out hitherto hidden textual meanings. It is possible for a text to have two contradictory meanings at once, as in the wave/particle duality. One of Empson's examples is how to interpret a character as complex as Falstaff. One must accept him as the sum total of apparent opposites, 'as the supreme expression of the cult of mockery as strength and the comic idealization of freedom, yet as both villainous and tragically ill-used'.[30] In Empson's view, the reader ought to 'hold in mind a variety of things [Shakespeare] may have meant, and weigh them . . . according to their probabilities',[31] just as the physicist represents the state of an atom with wave functions.

The concepts of quantum theory, with its deep abstractions, now permeate every aspect of our life. They have required us to rethink a wide range of subjects, transforming our intuitive understanding of nature. Quantum theory is used daily by almost every physicist, yet few of them have ever paused to think about its interpretive subtleties. Like a great work of literature, quantum theory is open to many different interpretations. Most physicists are unaware of this and assume that what they read in the texts of quantum theory texts is catechism. So ingrained has this attitude become that authors no longer state that they are presenting the Copenhagen interpretation, set down during 1926–27 by Bohr and Heisenberg. It has been my experience in teaching the history and philosophy of physics that the more thoughtful physics students are taken by complete surprise and are troubled, having come to expect certainty in textual exposition, instead of ambiguity in interpretation. As the physicist who did more to delve into the foundations of quantum theory than anyone else since Bohr, Einstein and Heisenberg, John Bell, once put it, 'for all practical purposes' quantum physics works well.[32] He forcefully reminded us, however, that we still do not fully understand the Schrödinger equation. As the great intuitive physicist Richard Feynman wrote in his usual pungent style, 'I think I can safely say that nobody understands quantum mechanics.'[33]

A Piece of Magic

The Dirac Equation

Frank Wilczek

> One cannot escape the feeling that these mathematical formulae have an independent existence and an intelligence of their own, that they are wiser than we are, wiser even than their discoverers, that we get more out of them than was originally put into them.
>
> Heinrich Hertz, on Maxwell's equations of electromagnetism

> A great deal of my work is just playing with equations and seeing what they give.
>
> Paul Dirac

> It gave just the properties one needed for an electron. That was really an unexpected bonus for me, completely unexpected.
>
> Paul Dirac, on the Dirac equation

The power of equations can seem magical. Like the brooms created by the Sorcerer's Apprentice, they can take on a power and life of their own, giving birth to consequences that their creator did not expect, cannot control, and may even find repugnant. When Einstein discovered $E = mc^2$, as the culmination of special relativity theory's consolidation of the foundations of classical physics, neither weapons of mass destruction nor generators of inexhaustible energy entered his ken.

Of all the equations of physics, perhaps the most 'magical' is the Dirac equation. It is the most freely invented, the least conditioned by experiment, the one with the strangest and most startling consequences.

In early 1928 (the receipt date on the original paper is 2 January), Paul Adrien Maurice Dirac (1902–84), a twenty-five-year-old recent convert from electrical engineering to theoretical physics, produced a remarkable equation, for ever to be known as the Dirac equation. Dirac's goal was quite concrete, and quite topical. He wanted to produce an equation that would describe the behaviour of electrons more accurately than previous equations. Those equations had either incorporated special relativity or quantum mechanics, but not both. Several other, more prominent and experienced physicists were working on the same problem.

Unlike these other physicists, and unlike the great classics of physics, Newton and Maxwell, Dirac did not proceed from a minute study of experimental facts. Instead he guided his search using a few basic facts and perceived theoretical imperatives (some of which we now know to be wrong). Dirac sought to embody these principles in an economical, mathematically consistent scheme. By 'playing with equations', as he put it, he hit upon a uniquely simple, elegant solution. This is, of course, the Dirac equation.

Some consequences of Dirac's equation could be compared with existing experimental observations. They worked quite well, and explained results that were otherwise mysterious. Specifically, as I'll describe below, Dirac's equation successfully predicts that electrons are always spinning and that they act as little bar magnets, and the rate of the spin and the strength of the magnetism. But other consequences appeared utterly inconsistent with obvious facts. Notably, Dirac's equation contains solutions that appear to describe a way for ordinary atoms to wink out into bursts of light, spontaneously, in a fraction of a second.

For several years Dirac and other physicists struggled with an extraordinary paradox. How can an equation be 'obviously right' since it accounts accurately for many precise experimental results, and achingly beautiful to boot – and yet manifestly, catastrophically wrong?

The Dirac equation became the fulcrum on which fundamental physics pivoted. While keeping faith in its mathematical form, physicists were forced to re-examine the meaning of the symbols it contains. It was in this confused, intellectually painful re-examination – during which Werner Heisenberg wrote to his friend Wolfgang Pauli, 'The saddest chapter of modern physics is and remains the Dirac theory' and 'In order not to be irritated with Dirac I have decided to do something else for a change . . .' – that truly modern physics began.

A spectacular result was the prediction of *antimatter* – more precisely, that there should be a new particle with the same mass as the electron but the opposite electric charge, and capable of annihilating an electron into pure energy. Particles of just this type were promptly identified, through painstaking scrutiny of cosmic-ray tracks, by Carl Anderson in 1932.

The more profound, encompassing result was a complete reworking of the foundations of our description of matter. In this new physics, particles are mere ephemera. They are freely created and destroyed; indeed, their fleeting existence and exchange is the source of all interactions. The truly fundamental objects are universal, transformative ethers: quantum fields. These are the concepts that underlie our modern, wonderfully successful theory of matter (usually called, quite inadequately, the Standard Model). And the Dirac equation itself, drastically reinterpreted and vastly general-ized, but never abandoned, remains a central pillar in our understanding of nature.

We'll want to look behind the magic, and ultimately to expose its illusion. To begin, here's a look at the equation itself:

$$\left[\gamma^\mu\left(i\frac{\partial}{\partial x^\mu} - eA_\mu(x)\right) + m\right]\psi(x) \;=\; 0.$$

These hieroglyphics are unpacked in the Appendix to this essay. But now let's focus on the equation, to appreciate its many facets.

Dirac's Problem and the Unity of Nature

The immediate occasion for Dirac's discovery, and the way he himself thought about it, was the need to reconcile two successful advanced theories of physics that had got slightly out of synch. By 1928 Einstein's special theory of relativity was already over two decades old, well digested and fully established. (The general theory, which describes gravitation, is not part of our story here. Gravity is negligibly weak on atomic scales.) On the other hand, the new quantum mechanics of Heisenberg and Schrödinger, although quite a young theory, had given brilliant insights into the structure of atoms, and successfully explained a host of previously mysterious phe-nomena. Clearly, it captured essential features of the dynamics of electrons in atoms. The difficulty was that the equations developed by Heisenberg and Schrödinger did not take off from Einstein's relativistic mechanics, but from the old mechanics of Newton. Newtonian mechanics can be an excellent approximation for systems in which all velocities are much smaller than the speed of light, and this includes many cases of interest in atomic physics and chemistry. But the experimental data on atomic spectra, which one could address with the new quantum theory, were so accurate that small deviations from the Heisenberg–Schrödinger predictions could be observed. So there was a strong 'practical' motivation to search for a more accurate electron equation, based on relativistic mechanics. Not only young Dirac but also several other major physicists were after such an equation.

In hindsight we can discern that much more ancient and fundamental dichotomies were in play: the dichotomy of light versus matter; the dichotomy of continuous versus discrete. These dichotomies present tremendous barriers to the goal of achieving a unified description of nature. Of the theories Dirac and his contemporaries sought to reconcile, relativity was the child of light and the continuum, and quantum theory the child of matter and the discrete. After Dirac's revolution had run its course, all were reconciled, in the mind-stretching conceptual amalgam we call a quantum field.

The dichotomies light/matter and continuous/discrete were surely felt by the earliest sentient protohumans. They were articulated clearly and debated inconclusively by the ancient Greeks. Aristotle distinguished fire and earth as primary elements, and argued against the atomists in favour of a fundamental plenum ('Nature abhors a vacuum').

These dichotomies were not relieved by the triumphs of classical physics; indeed, they sharpened.

Newton's mechanics was best adapted to describing the motion of rigid bodies through empty space. Although Newton himself in various places speculated on both sides of both dichotomies, Newton's followers emphasized his 'hard, massy, impenetrable' atoms as the fundamental building blocks of nature. Even light was modelled in terms of particles.

But early in the nineteenth century a very different picture of light, according to which it consists of waves, scored brilliant successes. Physicists accepted that there must be a continuous, space-filling ether to support these waves. The discoveries of Faraday and Maxwell, assimilating light to the play of electric and magnetic fields, which are themselves continuous entities filling all space, refined and reinforced this idea.

Yet Maxwell himself, and Ludwig Boltzmann, succeeded in showing that the observed properties of gases, including many surprising details, could be explained if the gases were composed of many small, discrete, well-separated atoms moving through otherwise empty space. Furthermore J. J. Thomson experimentally, and Hendrik Lorentz theoretically, established the existence of electrons as building blocks of matter. Electrons appear to be indestructible particles, of the sort that Newton would have appreciated.

Thus as the twentieth century opened, physics featured two quite different sorts of theories, living together in uneasy peace. Maxwell's electrodynamics is a continuum theory of electric and magnetic fields, and of light, that makes no mention of mass. Newton's mechanics is a theory of discrete particles, whose *only* mandatory properties are mass and electric charge.[1]

Early quantum theory developed along two main branches, following the fork of our dichotomies. One branch, beginning with Planck's work on radiation theory and reaching a climax in Einstein's theory of photons, dealt with light. Its central result is that light comes in indivisible minimal units, photons, with energy and momentum proportional to the frequency of the light. This, of course, established a particle-like aspect of light.

The second branch, beginning with Bohr's atomic theory and reaching a climax in Schrödinger's wave equation, dealt with electrons. It established that the stable configurations of electrons around atomic nuclei were associated with regular patterns of wave vibrations. This established a wave-like property of matter.

Thus the fundamental dichotomies had softened. Light is a bit like particles, and electrons are a bit like waves. But sharp contrasts remained. Two differences, in particular, appeared to distinguish light from matter sharply.

First, if light is to be made of particles, they must be very peculiar particles, with internal structure, for light can be polarized. To do justice to this property, one needs to have a corresponding property for the particles of light. We cannot achieve an adequate description of a light beam by saying it is composed of so-and-so many photons with such-and-such energies; those facts will tell us how bright it is and what colours it contains, but not how it is polarized. To get a complete description, one must also be able to say which way the beam is polarized, and this means that its photons must somehow carry around arrows that allow them to keep a record of the light's polarity. This would seem to take us away from the traditional ideal of elementary particles. If there's an arrow, what is *it* made of? And why can't it be separated from the particle?

Second, and more profound, photons are evanescent. Light can be radiated, as when you turn on a flashlight, or absorbed, as when you cover it with your hand. Therefore particles of light can be created or destroyed.

This basic, familiar property of light and photons takes us far away from the traditional ideal of elementary particles. The stability of matter would seem to require indestructible building-blocks, with properties fundamentally different from evanescent photons.

Watch now as these last differences fade away, and the unity of nature stands fully revealed!

The Early Payoff: Spin

Dirac was working to reconcile the quantum mechanics of electrons with special relativity. He thought – mistakenly, we now know – that quantum theory required equations of a particularly simple kind, the kind mathematicians call first-order. Never mind why he thought so, or precisely what first-order means; the point is that he wanted an equation that is, in a certain very precise sense, of the simplest possible kind. Tension arises because it is not easy to find an equation that is both simple in this sense and also consistent with the requirements of special relativity. To construct such an equation, Dirac had to expand the terms of the discussion. He found he could not get by with a single first-order equation – he needed a system of four intricately related ones, and it is actually this system we refer to as 'the' Dirac equation.

Two equations were quite welcome. Four, initially, were a big problem.

First, the good news.

Although the Bohr theory gave a good rough account of atomic spectra, there were many discrepant details. Some of the discrepancies concerned the number of electrons that could occupy each orbit; others involved the response of atoms to magnetic fields, as manifested in the movement of their spectral lines.

Wolfgang Pauli had shown, through detailed analysis of the experimental evidence, that Bohr's model could only work, even roughly, for complex atoms if there were a tight restriction on how many electrons could occupy any given orbit. This is the origin of the famous Pauli exclusion principle. Today we learn this principle in the form 'only one electron can occupy a given state'. But Pauli's original proposal was not so neat; it came with some disturbing fine print. For the number of electrons that could occupy a given Bohr orbital was not one but two. Pauli spoke obscurely of a 'classically non-describable duplexity', but – needless to say – did not describe any reason for it.

In 1925 two Dutch graduate students, Samuel Goudsmit and George Uhlenbeck, devised a possible explanation of the magnetic-response problems. If electrons were actually tiny magnets, they showed, the discrepancies would disappear. Their model's success required that all electrons must have the same magnetic strength, which they could calculate. They went on to propose a mechanism for the electrons' magnetism. Electrons, of course, are electrically charged particles. Electric charge in circular motion generates magnetic fields. Thus, if for some reason electrons were always rotating about their own axis, their magnetism might be explained. This intrinsic *spin* of electrons would have an additional virtue. If the rate of spin were the minimum allowed by quantum mechanics,[2] then Pauli's 'duplexity' would be explained. For the spin would have no possibility to vary in magnitude, but only the possibility to point either up or down.

Many eminent physicists were quite sceptical of Goudsmit and Uhlenbeck. Pauli even tried to dissuade them from publishing their work. For one thing, their model seemed to require the electron to rotate at an extraordinarily rapid rate, at its surface probably faster than the speed of light. For another, they gave no account of what holds an electron together. If it is an extended distribution of electric charge, all of the same sign, it will want to fly apart – and rotation, by introducing centrifugal forces, only makes the problem worse. Finally, there was a quantitative mismatch between their requirements for the strength of the electron's magnetism and the amount of its spin. The ratio of these two quantities is governed by a factor called the gyromagnetic ratio, written g. Classical mechanics predicts $g = 1$, whereas to fit the data Goudsmit and Uhlenbeck postulated $g = 2$. But despite these objections, their model stubbornly continued to agree with experimental results!

Enter Dirac. His system of equations allowed a class of solutions for small velocities, in which only two of the four functions appearing in his equations are appreciable. This was duplexity, but with a difference. Here it fell out automatically as a consequence of implementing general principles, and most definitely did not have to be introduced. Better yet, using his equation Dirac could calculate the magnetism of electrons, also without further assumptions. He got $g = 2$. Dirac's great paper of 1928 wastes no words. Upon demonstrating this result, he says simply, 'The magnetic moment is just that assumed in the spinning electron model.' And a few pages later, after working out the consequences, he concludes: 'The present theory will thus, in the first approximation, lead to the same energy levels as those obtained by [C. G.] Darwin, which are in agreement with experiment.'

But his results spoke loudly for themselves, with no need for amplification. From then on, there was no escaping Dirac's equation. Whatever difficulties arose – and there were some big and obvious ones – they would be occasions for struggle, not desertion. Such gleaming jewels of insight would be defended at all costs.

Although his intellectual starting point, as I mentioned, was quite different and more abstract, Dirac begins his paper by referring to Goudsmit, Uhlenbeck and the experimental success of their model. Only in the second paragraph does he reveal his hand. What he says is quite pertinent to the themes I've been emphasizing.

> The question remains as to why Nature should have chosen this particular model for the electron instead of being satisfied with a point-charge. One would like to find some incompleteness in the previous methods of applying quantum mechanics to the point-charge such that, when removed, the whole of the duplexity phenomena follow without arbitrary assumptions.

Thus Dirac is not offering a new model of electrons, as such. Rather, he is defining a new *irreducible* property of matter, inherent in the nature of things, specifically in the consistent implementation of relativity and quantum theory, that arises even in the simplest possible case of structureless point particles. Electrons happen to be embodiments of this simplest possible form of matter. The valuable properties of Goudsmit and Uhlenbeck's 'spin', specifically its fixed magnitude and its magnetic effects, which aid in the description of observed realities, are retained, now based on a much deeper foundation. The arbitrary and unsatisfactory features of their model are bypassed.

We were looking for an arrow that would be a necessary and inseparable part of elementary bits of matter, like polarization for photons. Well, there it is!

The spin of the electron has many practical consequences. It is responsible for the phenomenon of ferromagnetism, and the enhancement of magnetic fields in the core of electric coils, which forms the heart of modern power technology (motors and dynamos). Active manipulation of electron spins allows us to store and retrieve a great deal of information in a very small volume (magnetic tape, disk drives). Even the much smaller and more inaccessible spin of atomic nuclei plays a big role in modern technology. Manipulating such spins with radio waves and magnetic fields, and sensing their response, is the basis of the magnetic resonance imaging (MRI) so useful in medicine. This application, among many others, would be inconceivable (literally!) without the exquisite control of matter that only fundamental understanding can bring.

Spin in general, and Dirac's prediction for the magnetic moment in particular, has also played a seminal role in the subsequent development of fundamental physics. Small deviations from Dirac's $g = 2$ were discovered by Polykarp Kusch and collaborators in the 1940s. They provided some of the first quantitative evidence for the effects of virtual particles, a deep and characteristic property of quantum field theory. Very large deviations from $g = 2$ were observed for protons and neutrons in the 1930s. This was an early indication that protons and neutrons are not fundamental particles in the same sense that electrons are. But I'm getting ahead of the story . . .

The Dramatic Surprise: Antimatter

Now for the 'bad' news.

Dirac's equation consists of four components. That is, it contains four separate wave functions to describe electrons. Two components have an attractive and immediately successful interpretation, as I just discussed, describing the two possible directions of an electron's spin. The extra doubling, by contrast, appeared at first to be quite problematic.

In fact, the extra equations contain solutions with *negative* energy (and either direction of spin). In classical (non-quantum) physics the existence of extra solutions would be embarrassing, but not necessarily catastrophic. For in classical physics, you can simply choose not to use these solutions. This of course begs the question why *nature* chooses not to use them, but it is a logically consistent procedure. In quantum mechanics, even this option is not available. In quantum physics, generally 'that which is not forbidden is mandatory'. In the case at hand, we can be quite specific and precise about this. All solutions of the electron's wave equation represent possible behaviours of the electron that will arise in the right circumstances. Assuming Dirac's equation, if you start with an electron in one of the positive-energy solutions, you can calculate the rate for it to emit a photon and move into one of the negative-energy solutions. Energy must be conserved overall, but that is not a problem here – it just means that the emitted photon has higher energy than the electron that emitted it! Anyway, the rate turns out to be ridiculously fast, with the transition taking place in a small fraction of a second. So you can't ignore the negative-energy solutions for long. And since an electron has never been observed to do something so peculiar as radiating more energy than it began with, there was, on the face of it, a terrible problem with the quantum mechanics of Dirac's equation.

Dirac was well aware of this problem. In his original paper, he simply acknowledged:

> For this second class of solutions W [the energy] has a negative value. One gets over the difficulty on the classical theory by arbitrarily excluding those solutions that have a negative W. One cannot do this on the quantum theory, since in general a perturbation will cause transitions from states with W positive to states with W negative . . . The resulting theory is therefore still only an approximation, but it appears to be good enough to account for all the duplexity phenomena without arbitrary assumptions.

And he left it at that. This was the situation that provoked Heisenberg's outbursts to Pauli, quoted earlier.

By the end of 1929 – not quite two years later – Dirac had a proposal. It exploited the Pauli exclusion principle, according to which no two electrons obey the same solution of the wave equation. What Dirac proposed was a radically new conception of empty space. He proposed that what we consider 'empty' space is in reality chock-a-block with negative-energy electrons. In fact, according to Dirac, *'empty' space actually contains electrons obeying all the negative-energy solutions.* The great virtue of this proposal is that it explains away the troublesome transitions from positive to negative solutions. A positive-energy electron can't go to a negative-energy solution, because there's always another electron already there, and the Pauli exclusion principle won't allow a second one to join it.

It sounds outrageous, on first hearing, to be told that what we perceive as empty space is actually quite full of stuff. But on reflection, why not? We have been sculpted by evolution to perceive aspects of the world that are somehow useful for our survival and reproductive success. Since unchanging aspects of the world, upon which we can have little influence, are not useful in this way, it should not seem terribly peculiar that they would escape our untutored perception. In any case, we have no warrant to expect that naive intuitions about what is weird or unlikely provide any reliable guidance for constructing models of fundamental structure in the microworld, because these intuitions derive from an entirely different realm of phenomena. We must take it as it comes. The validity of a model must be judged according to the fruitfulness and accuracy of its consequences.

So Dirac was quite fearless about outraging common sense. He focused, properly, on the observable consequences of his proposal.

Since we are considering the idea that the ordinary state of 'empty' space is far from empty, it is helpful to have a different word for it. The one physicists like to use is 'vacuum'.

In Dirac's proposal, the vacuum is full of negative-energy electrons. This makes the vacuum a medium, with dynamical properties of its own. For example, photons can interact with the vacuum. One thing that can happen is that if you shine light on the vacuum, providing photons with enough energy, then a negative-energy electron can absorb one of these photons and go into a positive-energy solution. The positive-energy solution would be observed as an ordinary electron, of course. But in the final state there is also a *hole* in the vacuum, because the solution originally occupied by the negative-energy electron is no longer occupied.

The idea of holes was, in the context of a dynamical vacuum, startlingly original, but it was not quite unprecedented. Dirac drew on an analogy with the theory of heavy atoms, which contain many electrons. Within such atoms, some of the electrons correspond to solutions of the wave equation that reside nearby the highly charged nucleus, and are very tightly bound. It takes a lot of energy to break such electrons free, and so under normal conditions they present an unchanging aspect of the atom. But if one of these electrons absorbs a high-energy photon (an X-ray) and is ejected from the atom, the change in the normal aspect of the atom is marked by its *absence*. The absence of an electron, which would have supplied negative charge, by contrast looks like a positive charge. The positive effective charge follows the orbit of the missing electron, so it has the properties of a positively charged particle.

Based on this analogy and other hand-waving arguments – the paper is quite short, and practically devoid of equations – Dirac proposed that holes in the vacuum are positively charged particles. The process where a photon excites a negative-energy electron in the vacuum to a positive energy is then interpreted as the photon creating an electron and a positively charged particle (the hole). Conversely, if there is a pre-existing hole, then a positive-energy electron can emit a photon and occupy the vacant negative-energy solution. This is interpreted as the annihilation of an electron and a hole into pure energy. I referred to a photon being emitted, but this is only one possibility. Several photons might be emitted, or any other form of radiation that carries away the liberated energy.

Dirac's first hole-theory paper was entitled 'A theory of electrons and protons'. Protons, which form the nuclei of hydrogen atoms and are building blocks of more complicated nuclei, were the only positively charged particles known at the time. It was natural to try to identify the hypothetical holes with them. But severe difficulties with this identification rapidly appeared. For example, the two sorts of process we just discussed – production of electron-proton pairs, and annihilation of electron-proton pairs – have never been observed. The second is especially problematic, because it predicts that hydrogen atoms spontaneously self-destruct in microseconds – which, thankfully, they do not.

There was also a logical difficulty with the identification of holes with protons. Based on the symmetry of the equations, one could demonstrate that the holes must have the same mass as the electrons. But a proton has, of course, a much larger mass than an electron.

In 1931 Dirac withdrew his earlier identification of holes with protons, and accepted the logical outcome of his own equation and the dynamical vacuum it required: 'A hole, if there was one, would be a new kind of elementary particle, unknown to experimental physics, having the same mass and opposite charge of the electron.'

On 2 August 1932 Carl Anderson, an American experimentalist studying photographs of the tracks left by cosmic rays in a cloud chamber, noticed some tracks that lost energy as expected for electrons, but were bent in the opposite direction by the magnetic field. He interpreted this as indicating the existence of a new particle, with the mass of the electron but the opposite charge. Ironically, Anderson was completely unaware of Dirac's prediction.

Thousands of miles away from his rooms at St John's College, Cambridge, Dirac's holes – the product of his theoretical vision and revision – had been found, descending from the skies of Pasadena.

So in the long run the 'bad' news turned out to be 'even better' news. Frog into prince – magic at its most magical.

By now Dirac's holes, now called positrons, are no longer a marvel, but a tool. A notable use is to take pictures of the brain in action – PET scans (positron-electron tomography). How do positrons get into your head? Well, by reading this article, I hope. But how do they get into your head not only conceptually, but physically? They are sneaked in by injecting molecules containing atoms whose nuclei are radioactive and decay with positrons as one of their decay products. These positrons do not go very far before they annihilate some nearby electron, usually producing two photons, which escape your skull and can be detected. Then you can reconstruct where the original molecule went, to map out metabolism, and you can also study the energy loss of the photons on the way out, to get a density profile, and ultimately an image, of the brain tissue.

Another notable application is to fundamental physics. You can accelerate positrons to high energy, as you can of course electrons, and bring the beams together. Then the positrons and electrons will annihilate, producing a highly concentrated form of 'pure energy'. Much of the progress in fundamental physics over the past half-century has been based on studies of this type, at a series of great accelerators all over the world, the latest and greatest being the LEP (large electron-positron) collider at CERN, outside Geneva. I'll be discussing a stunning highlight of this physics a little later.

The physical ideas of Dirac's hole theory, which as I mentioned had some of its roots in the earlier study of heavy atoms, fed back in a big way into solid-state physics. In solids one has a reference or ground configuration of electrons, with the lowest possible energy, in which electrons occupy all the available states up to a certain level. This ground configuration is the analogue of the vacuum in hole theory. There are also configurations of higher energy, wherein some of the low-energy states are not used by any electron. In these configurations there are vacancies, or 'holes' – that's what they're called, technically – where an electron would ordinarily be. Such holes behave in many respects like positively charged particles. Solid-state diodes and transistors are based on clever manipulation of holes and electron densities at junctions between different materials. One also has the beautiful possibility to direct electrons and holes to a place where they can combine (annihilate). This allows you to design a source of photons that you can control quite precisely, and leads to such mainstays of modern technology as LEDs (light-emitting diodes) and solid-state lasers.

In the years since 1932 many additional examples of antiparticles have been observed. In fact, for every particle that has ever been discovered, a corresponding antiparticle has also been found. There are antineutrons, antiprotons, antimuons (the muon itself is a particle very similar to the electron, but heavier), antiquarks of various sorts, even antineutrinos, and anti-π mesons, anti-K mesons . . .[3] Many of these particles do not obey the Dirac equation, and some of them do not even obey the Pauli exclusion principle. So the physical reason for the existence of antimatter must be very general – much more general than the arguments that first led Dirac to predict the existence of positrons.

In fact, there is a very general argument that if you implement both quantum mechanics and special relativity, every particle must have a corresponding antiparticle. A proper presentation of the argument requires either a sophisticated mathematical background or a lot of patience, neither of which I'm assuming you have. So we'll have to be content with a rough version, which shows why antimatter is a plausible consequence of implementing both relativity and quantum mechanics, but doesn't really nail the case.

Consider a particle, let's say a shmoo, to give it a name (while emphasizing that it could be *anything*), moving east at very nearly the speed of light. According to quantum mechanics, there is actually some uncertainty in its position. So there's some probability, if you measure it, that you will find that the shmoo is slightly west of its expected mean position at an initial time, and slightly east of its expected mean position at a later time. So it has travelled further than you might have expected during this interval – which means it was travelling more quickly. But we postulated that its expected velocity was essentially the speed of light, so a faster speed violates special relativity, which requires that particles cannot move faster than the speed of light. It's a paradox.

Here's how you can escape the paradox, with antiparticles: the shmoo you observe is not necessarily the same as the original shmoo! It's also possible that at the later time there are two shmoos, the original one and a new one. There must also be an antishmoo, to balance the charge and to cancel out any other conserved quantities that might be associated with the additional shmoo. Here, as often in quantum theory, to avoid contradictions you must be specific and concrete in thinking about what it means to measure something. One way to measure the shmoo's position would be to shine light on it. But to measure the position of fast-moving shmoos accurately we have to use high-energy photons, and there's also then the possibility that such a photon will create a shmoo-antishmoo pair. And when you report the result of your position measurement, you may be talking about the wrong shmoo.

The Deepest Meanings:
Quantum Field Theory

Dirac's hole theory is brilliantly clever, but nature goes deeper. Although hole theory is internally consistent, and can cover a wide range of applications, there are several important considerations that force us to go beyond it.

First, there are particles that do not have spin, and do not obey the Dirac equation, and yet have antiparticles. The existence of antiparticles is a general consequence of combining quantum mechanics and special relativity, as we've seen. Specifically, for example, positively charged π^+ mesons (discovered in 1947) or W^+ bosons (discovered in 1983) are quite important players in elementary-particle physics, and they do have antiparticles π^- and W^-. But we can't use hole theory to make sense of these antiparticles, because π^+ and W^+ particles don't obey the Pauli exclusion principle. So there is no possibility of interpreting their antiparticles as holes in a filled sea of negative-energy solutions. When there are negative-energy solutions, whatever equation they satisfy,[4] occupying them with one particle will not prevent another particle from entering the same state. Thus catastrophic transitions into negative-energy states, which Dirac's hole theory prevents for electrons, must be banished in a different way.

Second, there are processes in which the number of electrons minus the number of positrons changes. An example is the decay of a neutron into a proton, an electron and an antineutrino. In hole theory the excitation of a negative-energy electron into a positive-energy state is interpreted as creation of a positron-electron pair, and de-excitation of a positive-energy electron into an unoccupied negative-energy state is interpreted as annihilation of an electron-positron pair. In neither case does the *difference* between the number of electrons and the number of positrons change. Hole theory cannot accommodate changes in this difference. So there are definitely important processes in nature, even ones specifically involving electrons, that do not fit easily into Dirac's hole theory.

The third and final reason harks back to our initial discussion. We were looking to break down the great dichotomies light/matter and continuous/discrete. Relativity and quantum mechanics, separately, brought us close to success, and the Dirac equation, with its implication of spin, brought us closer still. But so far we haven't quite got there. Photons are evanescent. Electrons . . . well, they're evanescent too, as a matter of experimental fact, as I just mentioned, but we haven't yet adequately fitted that feature into our theoretical discussion. In hole theory electrons can come and go, but only as positrons go and come.

These are not so much contradictions as indications of missed opportunity. They indicate that there ought to be some alternative to hole theory that covers all forms of matter and treats the creation and destruction of particles as a primary phenomenon.

Ironically, Dirac himself had earlier constructed the prototype of such a theory. In 1927 he applied the principles of the new quantum mechanics to Maxwell's equations of classical electrodynamics. He showed that Einstein's revolutionary postulate that light comes in particles – photons – was a consequence of the logical application of these principles, and that the properties of photons were correctly accounted for. Few observations are so common as that light can be created from non-light, say by a flashlight, or aborted and annihilated, say by a black cat. But translated into the language of photons, this means that the quantum theory of Maxwell's equations is a theory of the creation and destruction of particles (photons). Indeed, the electromagnetic field appears, in Dirac's theory, primarily as an agent of creation and destruction. The particles – photons – we observe result from the action of this field, which is the primary object. Photons come and go, but the field abides. The full force of this development seems to have escaped Dirac, and all his contemporaries, for some time, perhaps precisely because of the apparent specialness (dichotomy!) of light. But it is a general construction, which can be applied to the object that appears in Dirac's equation – the electron field – as well.

The result of a logical application of the principles of quantum mechanics to Dirac's equation is an object similar to what he found for Maxwell's equations. It is an object that destroys electrons and creates positrons.[5] Both are examples of *quantum fields*. When the object that appears in Dirac's equation is interpreted as a quantum field, the negative-energy solutions take on a completely different meaning, with no problematic aspects. The positive-energy solutions multiply electron-destruction operators, while the negative-energy solutions multiply positron-creation operators. In this framework, the difference between the two kinds of solution is that negative energy represents the energy you need to borrow to make a positron, while positive energy is what you gain by destroying an electron. The possibility of negative numbers is no more paradoxical here than in your bank balance.

With the development of quantum field theory, the opportunities that Dirac's equation and hole theory made evident, but did not quite fulfil, were finally met. The description of light and matter was put, at last, on a common footing. Dirac said, with understandable satisfaction, that with the emergence of quantum electrodynamics physicists had attained foundational equations adequate to describe 'all of chemistry, and most of physics'.

In 1932 Enrico Fermi constructed a successful theory of radioactive decays (beta decays), including the neutron decay I mentioned before, by exporting the concepts of quantum field theory far from their origin. Since these processes involve the creation and destruction of protons – the epitome of 'stable' matter – the old dichotomies had clearly been transcended. Both particles and light are epiphenomena, surface manifestations of the deeper and abiding realities, quantum fields. These fields fill all of space, and in this sense they are continuous. But the excitations they create, whether we recognize them as particles of matter or as particles of light, are discrete.

In hole theory we had a picture of the vacuum as filled with a sea of negative-energy electrons. In quantum field theory, the picture is quite different from this. But there is no returning to innocence. The new picture of the vacuum differs even more radically from naive 'empty space'. Quantum uncertainty, combined with the possibility of processes of creation and destruction, implies a vacuum teeming with activity. Pairs of particles and antiparticles fleetingly come to be and pass away. In 1987, I wrote about this in a sonnet that I called 'Virtual Particles':

Beware of thinking nothing's there –
Remove what you can; despite your care
Behind remains a restless seething
Of mindless clones beyond conceiving.
They come in a wink, and dance about;

Whatever they touch is seized by doubt:
What am I doing here? What should I weigh?
Such thoughts often lead to rapid decay.

Fear not! The terminology's misleading;
Decay is virtual particle breeding
And seething, though mindless, can serve noble ends,
The clone-stuff, exchanged, makes a bond between friends.

To be or not? The choice seems clear enough,
But Hamlet vacillated. So does this stuff.

Aftermaths

With the genesis of quantum field theory, we reach a natural intellectual boundary for our discussion of the Dirac equation. By the mid-1930s the immediate paradoxes this equation raised had been resolved, and its initial promise had been amply fulfilled. Dirac received the Nobel Prize in 1933, Anderson in 1935.

In later years the understanding of quantum field theory deepened, and its applications broadened. Using it, physicists have constructed, and established with an astonishing degree of rigour and beyond all reasonable doubt, what will stand for the foreseeable future – perhaps for all time – as the working theory of matter. How this happened, and the nature of the theory, is an epic story involving many other ideas, in which the Dirac equation as such plays a distinguished but not a dominant role. But some later developments are so closely linked to our main themes, and so pretty in themselves, that they deserve mention here.

There is another sense in which the genesis of quantum field theory marks a natural boundary. It is the limit beyond which Dirac himself did not progress. Like Einstein, in his later years Dirac took a separate path. He paid no attention to most of the work of other physicists, and dissented from the rest. In the marvellous developments that his work commenced, Dirac himself played a marginal role.

QED and Magnetic Moments

Interaction with the ever-present dynamical vacuum of quantum field theory modifies the observed properties of particles. We do not see the hypothetical properties of the 'bare' particles, but rather the physical particles, 'dressed' by their interaction with the quantum fluctuations in the dynamical vacuum.

In particular, the physical electron is not the bare electron, and it does not quite satisfy Dirac's $g = 2$. When Polykarp Kusch made very accurate measurements, in 1947, he found that g is larger than 2 by a factor 1.00119. Now this is not a very large correction, quantitatively, but it was a great stimulus to theoretical physics, because it provided a very concrete challenge. At that time there were so many loose ends in fundamental physics – a plethora of unexpected, newly discovered particles including muons, π mesons and others, no satisfactory theory explaining what force holds atomic nuclei together, fragmentary and undigested results about radioactive decays, anomalies in high-energy cosmic rays – that it was hard to know where to focus. In fact, there was a basic philosophical conflict about strategy.

Most of the older generation, the founders of quantum theory, including

Einstein, Schrödinger, Bohr, Heisenberg and Pauli, were prepared for another revolution. They thought it was fruitless to spend time trying to carry out more accurate calculations in quantum electrodynamics (QED), since this theory was surely incomplete and probably just wrong. It did not help that the calculations required to get more accurate results are very difficult, and that they seemed to give senseless (infinite) answers. So the old masters were searching for a different kind of theory, unfortunately with no clear direction.

Ironically, it was a younger generation of theorists – Julian Schwinger, Richard Feynman, Freeman Dyson in the US, and Sin-Itiro Tomonaga in Japan – who played a conservative role.[6] They found a way to perform the more accurate calculations, and get meaningful finite results, without changing the underlying theory. The theory, in fact, was just the one Dirac had constructed in the 1920s and 1930s. The new result of a calculation including the effects of the dynamic vacuum was Schwinger's calculation of the correction to $g = 2$, also reported in 1947. It agreed spectacularly well with Kusch's contemporary measurements. Other triumphs followed. Kusch received the Nobel Prize in 1955, Schwinger, Feynman and Tomonaga received theirs jointly in 1965 (the delay is hard to understand!).

Strangely enough, Dirac did not accept the new procedures. Caution was perhaps justified in the early days, when the mathematical methods being used were unfamiliar and not entirely well defined, involving a certain amount of inspired guesswork. But these technical difficulties got cleaned up in due course. Although QED does have problems of principle, if it is regarded (unrealistically!) as a completely closed theory, they are problems at a different level from what troubled Dirac, and they are very plausibly solved by embedding QED into a larger, asymptotically free theory (see below). This has very little practical effect on most of its predictions.

Feynman called QED 'the jewel of physics – our proudest possession'. But in 1951 Dirac wrote, 'Recent work by Lamb, Schwinger and Feynman and others has been very successful . . . but the resulting theory is an ugly and incomplete one.' And in his last paper, in 1984, 'These rules of renormalization give surprisingly, excessively good agreement with experiments. Most physicists say that these working rules are, therefore, correct. I feel that this is not an adequate reason. Just because the results happen to be in agreement with experiment does not prove that one's theory is correct.'

You might notice a certain contrast in tone between the young Dirac, who clung to his equation like a barnacle because it explained experimental results, and the older inhabitant of the same body.

Today the experimental determination of the magnetic moment of the electron (i.e. the electron's magnetic strength) is

$$(g/2)_{\text{experiment}} = 1.001\ 159\ 652\ 188\ 4\ (43)$$

while the theoretical prediction, firmly based on QED, calculated to high accuracy, is

$$(g/2)_{\text{theory}} = 1.001\ 159\ 652\ 187\ 9\ (88)$$

where the uncertainty in the last two digits is indicated. It is the toughest, most accurate confrontation between an intricate – but precisely defined! – theoretical calculation and a wizardly experiment in all of science. That's what Feynman meant by 'our proudest possession'.

As this article is written, still more accurate determinations of the magnetic moment of the electron, and of its kindred particle the muon, remain an important frontier of experimental physics. With the accuracies now achievable, the results will be sensitive to effects of quantum fluctuations due to hypothetical new heavy particles – in particular, those expected to be associated with supersymmetry.

QCD and the Theory of Matter

The magnetic moment of the proton does not satisfy Dirac's $g = 2$, but instead has $g \approx 5.6$. For neutrons it is worse. Neutrons are electrically neutral, so the simple Dirac equation for neutrons predicts no magnetic moment at all. In fact the neutron has a magnetic moment about ⅔ that of a proton, and with the opposite orientation relative to spin. It corresponds to an infinite value of g. The discrepant values of these magnetic moments were the earliest definite indication that protons and neutrons are more complicated objects than electrons.

With further study, many more complications appeared. The forces among protons and neutrons were found to be very complicated, depending not only on the distance between them, but also on their velocities, and spin orientations, and all combinations of these together, in a bewildering way. In fact, it soon appeared that they are not 'forces' in the traditional sense at all. To have a force between protons, in the traditional sense, would mean that the motion of one proton can be affected by the presence of another, so that when you shoot one proton by another, it swerves. What you actually observe is that when one proton collides with another, typically many particles emerge, most of which are highly unstable. There are π mesons, K mesons, ρ mesons, Λ and Σ baryons, their antiparticles, and many more. All these particles interact very powerfully with each other. And so the problem of nuclear forces, a frontier of physics starting in the 1930s, became the problem of understanding a vast new world of particles and reactions, the most powerful in nature. Even the terminology changed. Physicists no longer refer to nuclear forces, but to the strong interaction.

Now we know that all the complexities of the strong interaction can be described, at a fundamental level, by a theory called quantum chromodynamics, or QCD, a vast generalization of QED. The elementary building blocks of QCD are quarks and gluons. There are six different kinds, or 'flavours', of quarks: *u, d, s, c, b, t* (up, down, strange, charm, bottom, top). Reminiscent of the charged leptons,[7] the quarks are very similar to one another, differing mainly in their mass. Only the lightest ones, *u* and *d,* are found in ordinary matter. Making an analogy to the building blocks of QED, quarks play roughly the role of electrons, and gluons play roughly the role of photons. The big difference is that whereas in QED there is just one type of charge, and one photon, in QCD there are three types of charge, called colours, and eight gluons. Some gluons respond to colour charges, similarly to the way photons respond to electric charge. Others mediate transitions between one colour and another. Thus (say) a *u* quark with blue charge can radiate a gluon and turn into a *u* quark with green charge. Since all the charges overall must be conserved, this particular gluon must have blue charge +1, green charge −1. Since gluons themselves carry unbalanced colour charge, in QCD there are elementary processes where gluons radiate other gluons. There is nothing like this in QED. Photons are electrically neutral, and to a very good approximation they do not interact with other photons. Much of the richness and complexity of QCD arises because of this new feature.

Described thus baldly and verbally, without grounding in concepts or phenomena, QCD might seem both arbitrary and fantastic. In fact QCD is a theory of compelling symmetry and mathematical beauty. Unfortunately, I won't be able to do justice to those aspects here. But I owe you some explanations . . .

How did we arrive at such a theory? And how do we know it's right? In the case of QCD, these are two very different questions. The historical path to its discovery was tortuous, with many false trails and blind alleys. But in retrospect, it didn't have to be that way. If the right kind of ultra-high-energy accelerators had come on line earlier, QCD would have stared us in the face.[8] The *Gedanken*-history that follows brings together most of the ideas I've been discussing, and forms a fitting conclusion to this article's physical part.

When electrons and positrons are accelerated to ultra-high energy and then made to collide, two kinds of events are observed. In one kind of event the particles in the final states are leptons and photons. For this class of events, usually the final state is just a lepton and its antilepton; but in about 1 per cent of the events there is also a photon, and in about 0.01 per cent of the events there are also two photons. The probability for these events, and for the various particles to come out at various angles with different energies, can all be computed using QED, and it all works out very nicely. Conversely, if you hadn't known about QED, you could have figured out the basic rules for the fundamental interaction of QED – that is, the emission of a photon by an electron – just by studying these events. The fundamental interaction of light with matter is laid out right before your eyes.

In the other kind of event, you see something rather different. Instead of just two or at most a handful of particles coming out, there are many. And they are different kinds of particles. The particles you see in this second class of events are things like π mesons, K mesons, protons, neutrons, and their antiparticles – all particles that, unlike photons and leptons, have strong interactions. The angular distribution of these particles is very structured. They do not come out independently every which way. Rather, they emerge in just a few directions, making narrow sprays or (as they're usually called) 'jets'. About 90 per cent of the time there are just two jets, in opposite directions; roughly 10 per cent of the time there are three jets; 1 per cent four jets – you can guess the pattern.

Now if you squint a little and don't resolve the individual particles, but just follow the flow of energy and momentum, then the two kinds of events – the QED 'particle' events, and the 'jetty' events with strongly interacting particles – look just the same!

So (in this imaginary history) it would have been hard to resist the temptation to treat the jets as if they were particles, and propose rules for the likelihood of different radiation patterns, with different numbers, angles, and energies of the jet-particles, in direct analogy to the procedures that work for QED. And this would work out very nicely, because rules quite similar to those for QED actually do describe the observations. Of course, the rules that work are precisely those of QCD, including the new processes where glue radiates glue. All these rules – the foundational elements of the entire theory – could have been derived directly from the data. 'Quarks' and 'gluons' would be words with direct and precise operational definitions, in terms of jets. As I said, it stares you in the face – once you decide what you should be looking at!

Still, there would have been two big conceptual puzzles. Why do the experiments show 'quarks' and 'gluons' instead of just quarks and gluons – that is, jets, instead of just particles? And how do you connect the theoretical concepts that directly and successfully describe the high-energy events to all the other phenomena of the strong interaction? The connection between the supposedly foundational theory and the mundane observations is, to say the least, not obvious. For example, you would like to construct protons out of the 'quarks' and 'gluons' that appear in the fundamental theory. But this looks hopeless, since the jets in terms of which 'quarks' and 'gluons' are operationally defined often contain, among other things, protons.

There is an elegant solution to these problems. It is the phenomenon of *asymptotic freedom* in QCD. According to asymptotic freedom, radiation events that involve large changes in the flow of energy and momentum are rare, whereas radiation events that involve only small changes in energy and momentum are very common. Asymptotic freedom is not a separate assumption, but a deep mathematical consequence of the structure of QCD.

Asymptotic freedom neatly explains why there are jets in electron-positron annihilations at high energies, in the class of events containing strongly interacting particles. Immediately after the electron and positron annihilate, you have a quark and an antiquark emerging. They are moving rapidly, in opposite directions. They quickly radiate gluons, and the gluons themselves radiate, and a complicated cascade develops, with many particles. But despite all this commotion the overall flow of energy and momentum is not significantly disturbed. Radiations that disturb the flow of energy and momentum are rare, according to asymptotic freedom. So there is a large multiplicity of particles all moving in the same direction, the direction originally staked out by the quark or antiquark. In a word, we've produced a jet. When one of those rare radiations that disturbs the flow of energy and momentum takes place, the radiated gluon starts a jet of its own. Then we have a three-jet event. And so forth.

Asymptotic freedom also indicates why the protons (and the other strongly interacting particles) that we actually observe as individual stable, or quasi-stable, entities are complicated objects. For such particles are, more or less by definition, configurations of quarks, antiquarks and gluons that have a reasonable degree of stability. But since the quarks, antiquarks and gluons all have a very high probability for radiating, no simple configuration will have this property. The only possibility for stability involves dynamic equilibrium, in which the emission of radiation in one part of the system is balanced by its absorption somewhere else.

As things actually happened, asymptotic freedom was discovered theoretically (by David Gross and me, and independently by David Politzer) and QCD was proposed as the theory of the strong interaction (by Gross and me) in 1973, based on much less direct evidence. The existence of jets was anticipated, and their properties were *predicted* theoretically, in considerable detail, before their experimental observation. Based on these experiments and many others, today QCD is accepted as the fundamental theory of the strong interaction, on a par with QED as the description of the electromagnetic interaction.

There has also been enormous progress in using QCD to describe the properties of protons, neutrons and the other strongly interacting particles. This involves very demanding numerical work, using the most powerful computers, but the results are worth it. One highlight is that we can calculate from first principles, with no important free parameters, the masses of protons and neutrons. As I explained, from a fundamental point of view these particles are quite complicated dynamical equilibria of quarks, antiquarks and gluons. Most of their mass – and therefore most of the mass of matter, including human brains and bodies – arises from the pure energy of these objects, themselves essentially massless, in motion, according to $m = E/c^2$. At this level, at least, we are ethereal creatures.

Dirac said that QED described 'most of physics, and all of chemistry'. Indeed, it is the fundamental theory of the outer structure of atoms (and much more). In the same sense, QCD is the fundamental theory of atomic nuclei (and much more). Together, they constitute a remarkably complete, well tested, fruitful and economical theory of matter.

The Fertility of Reason

I've now discussed in some detail how 'playing with equations' led Dirac to an equation laden with consequences that he did not anticipate, and that in many ways he resisted, but that proved to be true and enormously fruitful. How could such a thing happen? Can mathematics be truly creative? Is it really possible, by logical processing or calculation, to arrive at essentially new insights – to get out more than you put in?

This question is especially timely today, since it lies at the heart of debates regarding the nature of machine intelligence – whether it may develop into a species of mind on a par with human intelligence, or even its eventual superior.

At first sight, the arguments against appear compelling.

Most powerful, at least psychologically, is the argument from introspection. Reflecting on our own thought processes, we can hardly avoid an unshakeable intuition that they do not consist exclusively, or even primarily, of rule-based symbol manipulation. It just doesn't feel that way. We normally think in images and emotions, not just symbols. And our streams of thought are constantly stimulated and redirected by interactions with the external world, and by internal drives, in ways that don't seem to resemble at all the unfolding of mathematical algorithms.

Another argument derives from our experience with modern digital computers. For these are, in a sense, ideal mathematicians. They follow precise rules (axioms) with a relentlessness, speed and freedom from error that far surpass what is possible for humans. And in many specialized, essentially mathematical tasks, such as arranging airline flight or oil-delivery schedules to maximize profits, they far surpass human performance. Yet by common, reasonable standards even the most powerful modern computers remain fragile, limited and just plain dopey. A trivial programming mistake, a few lines of virus code, or a memory flaw can bring a powerful machine to a halt, or send it into an orgy of self-destruction. Communication can take place only in a rigidly controlled format, supporting none of the richness of natural language. Absurd output can, and often does, emerge uncensored and unremarked.

Upon closer scrutiny, however, these arguments raise questions and doubts. Although the nature of the map from patterns of electrical signals in nerve cells to processes of human thought remains deeply mysterious in many respects, quite a bit is known, especially about the early stages of sensory processing. Nothing that has been discovered so far suggests that anything more exotic than electric and chemical signalling, following well-established physical laws, is involved. The vast majority of scientists accept as a working hypothesis that a map from patterns of electric signals to thought must and does exist. The pattern of photons impinging on our retina is broken up and parsed out into elementary units, fed into a bewildering series of different channels, processed and (somehow) reassembled to give us the deceptively simple 'picture of the world', organized into objects in space, that we easily take for granted. The fact is, we do not have the slightest idea how we accomplish most of what we do, even – perhaps especially – our most basic mental feats. People who've attempted to construct machines that can recognize objects appearing in pictures, or that can walk around and explore the world like a toddler, have had a very frustrating time, even though they can do these things very easily themselves. They can't teach others how they do these things because they don't know themselves. Thus it seems clear that introspection is an unreliable guide to the deep structure of thought, both as regards what is known and what is unknown.

Turning to experience with computers, any negative verdict is surely premature, since they are evolving rapidly. One recent benchmark is the victory of Deep Blue over the great world chess champion Gary Kasparov in a brief match. No one competent to judge would deny that play at this level would be judged a profoundly creative accomplishment if it were performed by a human. Yet such success in a limited domain only sharpens the question: What is missing that prevents the emergence of creativity from pure calculation over a broad front? In thinking about this tremendous question, I believe case studies can be of considerable value.

In modern physics, and perhaps in the whole of intellectual history, no episode better illustrates the profoundly creative nature of mathematical reasoning than the history of the Dirac equation. In hindsight, we know that what Dirac was trying to do is strictly impossible. The rules of quantum mechanics, as they were understood in 1928, cannot be made consistent with special relativity. Yet from inconsistent assumptions Dirac was led to an equation that remains a cornerstone of physics to this day.

So here we are presented with a specific, significant, well-documented example of how mathematical reasoning about the physical world, culminating in a specific equation, led to results that came as a complete surprise to the thinker himself. Seemingly in defiance of some law of conservation, he got out much more than he put in. How was such a leap possible? Why did Dirac, in particular, achieve it? What drove Dirac and his contemporaries to persist in clinging to his equation, when it led them out to sea?[9]

Insights emerge from two of Dirac's own remarks. In his characteristically terse essay 'My Life as a Physicist', he pays extended tribute to the value of his training as an engineer, including: 'The engineering course influenced me very strongly . . . I've learned that, in the description of nature, one has to tolerate approximations, and that even work with approximations can be interesting and can sometimes be beautiful.' Along this line, one source of Dirac's (and others') early faith in his equation, which allowed him to overlook its apparent flaws, was simply that he could find approximate solutions of it that agreed brilliantly with experimental data on the spectrum of hydrogen. In his earliest papers he was content to mention, without claiming to solve, the difficulty that there were other solutions, apparently equally valid mathematically, that had no reasonable physical interpretation.

Along what might superficially seem to be a very different line, Dirac often paid tribute to the heuristic power of mathematical beauty: 'The research worker, in his efforts to express the fundamental laws of Nature in mathematical form, should strive mainly for mathematical beauty.' This was another source of early faith in Dirac's equation. It was (and is) extraordinarily beautiful.

Unfortunately, it is difficult to make precise, and all but impossible to convey to a lay reader, the nature of mathematical beauty. But we can draw some analogies with other sorts of beauty. One feature that can make a piece of music, a novel or a play beautiful is the accumulation of tension between important, well-developed themes, which is then resolved in a surprising and convincing way. One feature that can make a work of architecture or sculpture beautiful is symmetry – balance of proportions, intricacy towards a purpose. The Dirac equation possesses both these features to the highest degree.

Recall that Dirac was working to reconcile the quantum mechanics of electrons with special relativity. It is quite beautiful to see how the tension between conflicting demands of simplicity and relativity can be harmonized, and to find that there is essentially only one way to do it. That is one aspect of the mathematical beauty of the Dirac equation. Another aspect, its symmetry and balance, is almost sensual. Space and time, energy and momentum, appear on an equal footing. The different terms in the system of equations must be choreographed to the music of relativity, and the pattern of 0s and 1s (and *i*s) dances before your eyes.

The lines converge when the needs of physics lead to mathematical beauty, or – in rare and magical moments – when the requirements of mathematics lead to physical truth. Dirac searched for a mathematical equation satisfying physically motivated hypotheses. He found that to do so he actually needed a system of equations, with four components. This was a surprise. Two components were most welcome, as they clearly represented the two possible directions of an electron's spin. But the extra doubling at first had no convincing physical interpretation. Indeed, it undermined the assumed meaning of the equation. Yet the equation had taken on a life of its own, transcending the ideas that gave birth to it, and before very long the two extra components were recognized to portend the spinning positron, as we saw.

With this convergence, I think, we reach the heart of Dirac's method in reaching the Dirac equation, which was likewise James Clerk Maxwell's in reaching the Maxwell equations, and Einstein's in reaching both the special and the general theories of relativity. They proceed by *experimental logic*. That concept is an oxymoron only on the surface. In experimental logic, one formulates hypotheses in equations, and experiments with those equations. That is, one tries to improve the equations from the point of view of beauty and consistency, and then checks whether the 'improved' equations elucidate some feature of nature. Mathematicians recognize the technique of 'proof by contradiction': to prove A, you assume the opposite of A, and reach a contradiction. Experimental logic is 'validation by fruitfulness': to validate A, assume it, and show that it leads to fruitful consequences. Relative to routine deductive logic, experimental logic abides by the Jesuit credo: 'It is more blessed to ask forgiveness than permission.' Indeed, as we have seen, experimental logic does not regard inconsistency as an irremediable catastrophe. If a line of investigation has some success and is fruitful, it should not be abandoned on account of its inconsistency or its approximate nature. Rather, we should look for a way to make it true.

With all this in mind, let us return to the question of the creativity of mathematical reasoning. I said before that modern digital computers are, in a sense, ideal mathematicians. Within any reasonable, precisely axiomatized domain of mathematics, we know how to program a computer so that it will systematically prove all the valid theorems.[10] A modern machine of this sort could churn through its program, and output valid theorems, much faster and more reliably than any human mathematician could. But running such a program to do advanced mathematics would be no better than setting the proverbial horde of monkeys to typing, hoping to reproduce Shakespeare. You'd get a lot of true theorems, but essentially all of them would be trivial, with the gems hopelessly buried amid the rubbish. In practice, if you peruse journals of mathematics or mathematical physics, not to speak of literary magazines, you won't find much work submitted by computers. Attempts to teach computers to do 'real' creative mathematics, like the attempts to teach them to recognize real objects or navigate the real world, have had very limited success. Now we begin to see that these are closely related problems. Creative mathematics and physics rely not on perfect logic, but rather on an experimental logic. Experimental logic involves noticing patterns, playing with them, making assumptions to explain them, and – especially – recognizing beauty. And creative physics requires more: abilities to sense and cherish patterns in the world, and to value not only logical consistency, but also (approximate!) fidelity to the world as observed.

So, let us return to the central question: Can purely mathematical reasoning be creative? Undoubtedly, if it is used *à la* Dirac, in concert with the abilities to tolerate approximations, to recognize beauty, and to learn by interacting with the real world. Each of these factors has played a role in all the great episodes of progress in physics. The question returns as a challenge to ground those abilities in specific mechanisms.

Acknowledgments

This work is supported in part by funds provided by the US Department of Energy (DOE) under cooperative research agreement #DF-FC02-94ER40818. Marty Stock helped with the LaTeX.

Understanding Information, bit by bit

Shannon's Equations

Igor Aleksander

Information is now a commodity like metal or oil, a utility like water or electricity. Politicians, stock-exchange pundits, future watchers and the rest of us wave our arms and say that something is a sign of 'living in the information age'. Although not everyone knows what the information age actually is, there are signs of it everywhere: faxes in waste-disposal tips, e-mails from aircrafts, and mobile phones even on nudist beaches.

It is debatable whether we all have a lot more to say to one another, so it's not the actual information that is the focus of the information age. The hallmark of this revolutionary age is, rather, the amazing opportunity for us to connect to one another or to computers from almost anywhere. Fifty years ago we had just the telephone or the wireless. Now there are globally networked computers (the Internet), digital mobile phones and fibre-optic cables. Even consumer entertainment products have changed out of all recognition. We have gone from the 78-rpm black vinyl record to the digital video disc, from the Brownie box camera to the digital models that grace the shelves of camera shops.

This all points to huge developments in the technology that underpins the transmission of information. But what is information? What inhibits its transmission? Why have these developments required massive industrial investment? Why does the word 'digital' (meaning that which is represented by separate symbols such as numbers) appear so often in the names of these technologies?

I do not intend to provide a detailed description of how this technology works. Rather, my aim is to rediscover a hero of the information age, a man without whose razor-sharp insight none of this technology would work. Claude Shannon was both a mathematician and an engineer, and he brought these disciplines together in a way that changed the world forever.

Shannon's name is attached to two equations that underlie the theory of communication. They have a somewhat forbidding notation:

$$I = -p \log_2 p$$

and

$$C = W \log_2 (1 + S/N)$$

The first of these tells us that the amount of information in any message can be measured as a quantity labelled I, where the unit of measurement is the 'bit'. While bits and the word 'digital' appear often in these descriptions, the two equations are continuous and not digital, which means that the theory applies to old telephone lines as well as the latest digital versions. The statement made by the first equation is that the amount of information I depends on the *surprise* that the message holds. This is because the mathematical way of expressing surprise is as a probability p; the less probable an event is, the more surprising it is and the more information it conveys. Where the $\log_2$ comes from we shall see later. Suffice it to say that without this equation, the world would be without a major unit of measurement, a measure which is as important as the gallon, the litre, the watt or the mile.

The second of Shannon's equations is a 'quality' indicator of a transmission medium such as a telephone line or a cable for a television aerial. It tells us that C (in bits per second), the amount of information that can be transmitted through a line or other medium, depends on two major factors: W, the bandwidth (or range of frequencies that can get through), and S/N, the signal-to-noise ratio. We get a feel for this when, at a noisy cocktail party, we need to shout (or increase S, the signal, to beat N, the noise). If talking to a partially deaf person (someone whose W is restricted), we need to shout even harder. So, using the analogy of miles and gallons, C in bits per second is a quality factor in the same way that 'miles per gallon' is a quality factor for a motor vehicle. These laws are very general: they apply to anything from the simple telephone connection that transmits voice signals translated into electrical quantities, to the latest digital high-definition television where visual scenes are translated from strings of numbers.

Shannon's thought and work transcend the equations themselves: they are merely the symbols that distil an exceptional insight into the nature and harnessing of information.

Shannon's very anonymity is evidence of his success. Even the most hardened Internet aficionado just sits there, switches on the machinery and expects text and pictures to appear on the screen of his personal computer. Let us say that his friend Jill has promised to send him the latest digital portrait of her face. She clicks on this image on her computer, 'attaches' it to an e-mail message and clicks on the 'send' button. But how can the two computers be connected? Is it via a telephone cable? Were this true, our recipient Jack would have to wait over thirty minutes for Jill's face to appear on his screen. I know that delays on the Internet sometime feel terribly long, but the very fact that this transmission would, at worst, take about a minute and a half, is due to Shannon's discoveries. Indeed the very fact that I could predict the thirty minutes, knowing something about telephone cables, is because Shannon taught us *how* to assess the picture and the cable in order to work out how well the job could be done. Enormous design efforts based on such assessments went into the Internet so that we can send each other pictures of our lovers and children.

The Internet is a system of interconnections between hundreds of millions of computers all over the world. It spews information into our computers in much the way that a tap fills our bath. Just as we buy ½-inch or ¾-inch piping for our bathrooms, so we buy appropriate links for our computers to the Internet network. Our Shannon-inspired calculation tells us that bare telephone lines are not appropriate because of their inadequate capacity to handle the large amount of information contained in something like a photograph. Leaving aside the question of what a 'bit' is for later, capacities are measured in terms of bits per second: the higher the value, the faster will my computer fill up with information. Jill's picture contains, say, 20 million of these bits and a telephone line has a capacity of 10,000 bits per second.[1]

$$20,000,000/10,000 \text{ seconds} = 2,000 \text{ seconds} = 33.3 \text{ minutes}.$$

And yet what connects our home PCs to the Internet is nothing but a telephone line. Imagine a world in which we could not measure the gallons of water that we consume or the kilowatts of electricity supplied by the local electricity board. That would have been the world of information without the work of Claude Shannon: the world without the 'bit'. Shannon gave us the measure of information, but he also founded the entire subject of information theory, which anyone designing a communication network needs to know.

Shannon was born in 1916 in Gaylord, Michigan, the son of a businessman and a schoolteacher. He soon demonstrated an aptitude for mathematics and engineering and, like many young people in that era, enjoyed tinkering with radios, the hottest technology of the time. He even earned money from the local department store by repairing radio sets.

When he was sixteen, he entered the University of Michigan to study mathematics and engineering. Four years later he became a research assistant at the Massachusetts Institute of Technology (MIT) in Cambridge, where he worked on early computer projects with the charismatic Vannevar Bush, who later became President Roosevelt's scientific guru and – in some people's eyes – a founder of the Internet.

When Shannon arrived at MIT just before the outbreak of World War II, computers had hardly been invented. The word 'computer' was rarely used. Calculating machines were largely mechanical devices using gearwheels, springs and the like. Some laboratories were looking at electronic or mixed mechanical and electronic calculating machines. Vannevar Bush was one of the very few people who had the confidence to restate Charles Babbage's dream, enunciated a hundred years earlier, that mechanical machinery could take over from human beings some of the drudgery of doing repetitive calculations. Such notions were considered by some at the time to be fanciful daydreams. However, Bush and, later, John von Neumann, grandfather of the current style of computer design and, according to Einstein, one of the most agile minds that had ever graced Princeton, were both strategists highly regarded by US government agencies. So they were able to impress on the government the importance of mechanized computation, thereby unlocking early support for the design of computers. Without Bush and von Neumann, electronic computers would not be nearly so advanced as they are today.

As an ironic contrast, when the British computer pioneer Maurice Wilkes at Cambridge in the UK attempted to obtain funding in the late 1940s to build a computing machine, he received a dusty response from what was then the Department of Industrial and Scientific Research. The bureaucrats effectively suggested that if Wilkes and his colleagues were to sit down with a few mechanical calculators, they could solve all the world's computational problems, avoiding the need to build fancy computing machines.

So Shannon was lucky to be taught by a powerful visionary, a factor that must have contributed to his own lack of fear in undertaking ambitious engineering challenges. But MIT was then, as it is now, an expensive place even for someone endowed with a grant that would cover his fees. Vannevar Bush had invented a calculating engine called a 'differential analyser' (DA). This machine stores numbers on rotating geared cylinders, a bit like mechanical mileage indicators on old cars. To help the young Shannon scrape a few dollars together, Bush offered him part-time work on the DA, which Shannon took up with some gusto. The DA was an experimentalist's dream. It was a large collection of rotating cylinders, gearwheels and electrical control switches. Its main function was to find the solutions to mathematical equations. These were set up by interconnecting the parts to suit the equation. The answer was given by a reading on the mileage-meter-like device. It could take days to set up the machine just to work on one problem. It would then have to be dismantled and reconstructed to work on the next one. So Shannon became one of the world's first programmers, setting up the DA to meet various scientists' needs.

These were formative times for the young Shannon. He became aware of the need to understand two major relationships in information. First, there was an *amount* of information which the DA computations generated, and, second, *a limiting speed* with which the output indicators could accept the information that had been calculated. Amount of information and speed of transmission, these were to become the pillars of Shannon's information theory of the future, the two topics of his two celebrated equations.

Another major factor that influenced Shannon was his fascination with electrical switches and the complex systems for routing electricity that could be designed just with a handful of such switches (think of the way that a light can be switched on from two sides of the same room). He had studied the laws of logic as set out by that British pioneer George Boole who a century earlier at Queen's College, Cork, in the south of Ireland had presented these as 'the laws of thought'. For example, were you to say, 'Alvin and Bob were not both at the party', this is the same as saying, 'Either Alvin was not at the party or Bob was not at the party.' Boole had suggested a notation (now called Boolean algebra) in which the statements above could be turned into a rule which is always true –

$$Not(A \text{ and } B) = (NotA) \text{ or } (NotB)$$

where A and B are statements that are either true or false. Boolean algebra has many such rules.

The reason for mentioning all this is that switches form the basis for both routing and storing information, and Shannon made the following daring intellectual leap. A closed switch is like a 'true' statement and an open switch is like a 'false' statement in logic. So if A and B were switches instead of statements, Boolean algebra could apply to the way that communication networks can be switched to connect communicators and switches organized to store messages. Indeed, the mass of switching that needs to be done inside a computer is now regularly designed or analysed using Boolean algebra. So before the end of his first year at MIT, Shannon had written his master's thesis on the application of Boolean algebra to switching circuits and, in 1938, published a paper called 'Analysis of relay and switching circuits'. This became one of the classic papers in computer literature, and Boolean algebra is now regularly taught to first-year engineering students as the standard way of designing switching circuits in both computers and telecommunication systems. Not a bad discovery for a twenty-year-old! Indeed, Shannon's mind at this early age must have been driven by the passionate desire to tie together the nature of switching with the nature of information and to understand the limits on how fast information can be transmitted from one geographical point to another.

Shannon left MIT in 1940 with both a master's degree and a doctorate in mathematics. MIT now honours this association by having a regular Shannon Day, on which the latest advances in telecommunication are discussed. After a year at the prestigious Institute for Advanced Studies at Princeton, Shannon joined the premier industrial research establishment in the United States: the Bell Telephone Laboratories at Murray Hill in New Jersey. Here, urged on by colleagues, he published in 1948 his internal reports on a statistical theory of communication. This was the celebrated 'Mathematical Theory of Communication'. As an example of the logic that led Shannon to quantify communication, in 1950 he wrote the very first chess-playing program which incorporated a clever way of cutting down on the number of board positions that the machine has to search to find a good move. This 'algorithm' was used in the programming of IBM's Deep Blue machine that beat the grand master Gary Kasparov in 1997 – the first time a reigning chess champion had been beaten by a machine.

By 1957, Shannon was widely recognized in the United States as one of its leading scientists. He was identified as one of the nine leading lights of American science in a special feature in *Time* magazine, published six weeks after the USSR successfully launched the first artificial satellite Sputnik, which had caused a panic in the United States as it appeared they were falling behind their cold-war rivals. In his profile, we learned of his addiction to jazz, his enjoyment of science fiction and his working habits: 'like many scientists, [he] works best at night, with plenty of cigarettes and coffee.'

In World War II movies such as *The Cruel Sea*, the Morse-key operator taps SOS as di-di-di, dat-dat-dat, di-di-di to tell the world through his radio transmitter that the ship is in trouble and may be passing its last minutes afloat. But why was the operator not merely picking up a microphone and using his voice, through his radio transmitter, to tell the world exactly what was happening? The answer is that the 'di' and 'dat' pattern of two simple tones stands a much better chance of getting through the crackle and hiss of the radio transmitter than spoken words, whose multifarious and subtle tones would be lost in what engineers call electronic noise. It is easier to see a dim light far away than to make out the details of a picture at the same great distance. A flashing light with a proper code can tell a story that a poorly seen picture cannot. But this vague notion needs a theory, and this is the theory that Shannon worked out during his early years at Bell Labs.

All communication systems suffer from noise: it sounds like crackle in a telephone, it looks like snow on a television screen. Noise distorts unpredictably the information the sender is trying to transmit to the receiver. This can make the received information unintelligible, hence useless. There is another limitation, something that specialists call bandwidth. Most buyers of hi-fi are aware of this. They first ask what is the 'bass' response – that is, what are the lowest frequencies (rumbling tuba sounds) that can be heard on the equipment? – and then what is the 'treble' or high-frequency response (top notes on a violin) that the equipment can handle? Subtracting the low limit (say 25 cycles per second, called 25 hertz) from the high-frequency limit (say 5,000 hertz) gives the *bandwidth* of the equipment. In other words, poor equipment with low bandwidth will not allow a listener to delight in the full glory of an orchestra's range. In technical terms, like noise, bandwidth makes the received information somewhat less than the transmitted version. Every communication link between a transmitter and a receiver is characterized by some value of its bandwidth and Shannon wanted to predict exactly how these losses in information could be calculated.

Shannon summarized the situation in a way that became the basis of information theory itself. To tidy things up, he imagined that every link between the source of information and the destination of its messages has five major components. First, there is the source. In the case of someone who wishes to transmit a digital picture across the Internet, the source is a computer in which the picture is stored as 20 million 0 or 1 states (or bits) of a memory.

The second element is an encoder. This is the sum total of equipment that prepares the picture to be transmitted in a reasonable time over, say, low-bandwidth telephone lines. As a first step in encoding, a modern computer has a program which 'compresses' the picture. The picture sits in the machine's memory as a sequence of numbers, each number representing the colour and intensity of a dot on the picture. Compression removes redundancies in these sequences of numbers.[2] The next part of encoding is to turn the numbers that represent the picture into 'tones' that can be transmitted down the telephone line. This is called 'modulation' and is needed because the telephone channel is designed to carry audible signals; that is, the human voice. Most telephones now use tones to go with each dialled number. This is an example of modulation.

The third element of the communication system is the telephone cable itself, with its inherent noise and restricted bandwidth. The fourth part is a decoder that restores what is received to a state as close as possible to what was transmitted. In the case of a picture, the decoder must first take the tones and turn them back into numbers, and then it must interpret the numbers so that they reconstruct the picture in the fifth part of the system, which is the receiver: the computer screen in this case. Anyone who has bought a modem for their computer will in effect have bought a box that contains both an encoder (*mo*dulator) and a decoder (*dem*odulator).

If this is a crude picture of a communication system, it is also worth taking a romp through a kind of caricature of the theory that explains the system, and then looking at the effect of these equations on what we now know of communication systems.

First, we need a measure of information.[3] We already know that this is measured in bits, each of which has two values, 0 and 1. The bit, or '*bi*nary uni*t* of information', was one of Shannon's key proposals. Let's work backwards and see why this makes so much sense.

Suppose someone wants to transmit a picture of herself, but also those of her father, mother, two brothers, the family dog, cat and a picture of the family house: eight pictures in all. Having stored them on his computer once, her boyfriend has acquired the information. If she wants him to see any one of these pictures, all she needs to do is to number them from 1 to 8 and transmit the appropriate number. The computer merely places the appropriate picture on the screen without it needing to be transmitted. Now, one bit could specify two numbers, 0 and 1. Two bits can do four (00, 01, 10, 11). Three bits can do eight (000, 001, 010, 011, 100, 101, 110, 111). So here is an illustration of one of Shannon's major insights: information is proportional to how much you don't know. When the man in this woman's life had no information at all about these pictures he needed 20 million bits for each to describe them. Once he has stored them, he needs only three bits to specify one of the pictures. This is how probability creeps into the first of Shannon's equations: in the first instance, had he never seen his girlfriend before, this would have meant an enormously low probability of guessing what she looked like. The less likely an event is, the more information its occurrence carries. Shannon's first equation links the information to something known as the 'logarithm to the base 2' of the probability (written as $\log_2$). This somewhat off-putting jargon is, however, not hard to understand. To take a few simple examples, the logarithm to the base 2 of $2 \times 2 \times 2$ (i.e. 2^3) is 3; the logarithm of $2 \times 2 \times 2 \times 2$ (i.e. 2^4) is 4; the logarithm of $2 \times 2 \times 2 \times 2 \times 2$ (i.e 2^5) is 5. So the logarithm to the base 2 of a number is simply the power to which 2 must be raised to equal that number. It is now possible to return to the complete equation:

$$I = -p \log_2 p \text{ (measured in bits for reasons we shall see).}$$

This reads: 'The amount of information involved in gaining knowledge of an event is dependent on the probability p of that event occurring'. This also leads us neatly to the definition of the bit. Toss a coin. This can result in one of two events: heads or tails. Each event has its own probability of occurrence, and if the coin is unbiased, each of these events has a probability of 1/2. To get the total information involved in the event of tossing this coin we add the information content of both possibilities and get:

$$I = [-(1/2) \log_2 (1/2)] + [-(1/2) \log_2 (1/2)].$$

This turns out to be precisely 1. This is no accident, and to a mathematician it explains why Shannon used $\log_2$ in his equation. That is, the unit of information, the bit, is associated with a switch being on or off, a number being 0 or 1, a coin being heads or tails, and ensures (as we shall see) that *any* other amount of information can be measured in bits. The formula also covers the case of certainty. If the occurrence of an event or its nonoccurrence is certain, the equation tells us that the event yields 0 bits of information. Two bits are like two coins that can result in four messages, so the information content of four equally probable messages is two bits. This means that the equation can be applied to any number of messages. For example, if the upper-case letters of the alphabet are being transmitted from a word processor this implies 26 messages, which requires five bits (which as 2^5 gives us 32 messages, that is, a few spares over the 26). The power of the first little equation thus allows us to measure unequivocally how much information is contained in something we are trying to communicate to someone else.

The words 'crackle and hiss' have already crept into this essay. It is one of the laws of nature that this 'noise' cannot be avoided whenever electricity or wireless media are used to convey information. Were I to connect two computers with the telephone cable, there would be a level of electrical energy arriving at the receiver that is not transmitted by the transmitter. The information we want is sent down the cable encoded as a sequence of voltage levels each representing, say, a picture point in Jill's picture.[4] But electrons in the cable have a habit of jumping about. This random activity modifies the voltage being transmitted so that the receiver may get a randomly modified number. Such random activities exist not only in transmission down cables; the free space that is used for radio transmission also has sufficient randomly moving charged particles to produce significant variations in transmitted signals. These variations will appear in the received picture as dots or 'snow', or a hiss in the messages transmitted by the Morse-code operator on the doomed ship at sea.

This is where Shannon's other equation comes in. The imperfections of a channel in terms of frequency limitation (bandwidth W) and noise (N) may all be incorporated into one statement for the capacity C of a communication medium with respect to the strength of a signal S.

$$C = W \log_2 (1 + S/N) \text{ in bits per second.}$$

We need to unpack this a little, using, again, a picture transmitted over the Internet. First, we make a rough assumption that the bandwidth W is a 'maximum' frequency of transmission by assuming that the lower limit of the bandwidth is zero. Even more roughly we take this to mean 'the maximum number of packets of bits we can transmit in any second'. A packet of bits represents a range of numbers (3 bits give 8 numbers, 4 bits give 16 numbers, and so on). Now, this range of numbers depends on how much noise there is in the system. The term $(1 + S/N)$ tells us how likely noise is to change the number in the packet. So if there is no noise, N is 0 and $(1 + S/N)$ turns out to be infinity (plus 1), which tells us that each packet could be as big as we like. So the whole of our picture could be enveloped into just a single packet which could then be transmitted W times in a second! With values of W of 10,000 or so even for very poor lines, communication would be prodigiously fast in a noise-free line. Sadly, noise is always there, and if, say, the noise were to be $\frac{1}{7}$ of the signal strength, $(1 + 7)$ tells us that anything more than 8 numbers would mean that the transmitted number was changed by noise. So in this case $\log_2(8) = 3$, which means that only three bits per packet can now be transmitted W times a second. So for W being 10,000, even the compressed version of the picture of, say, 1 million bits would take:

$$1,000,000/(3 \times 10,000) \text{ seconds} = 33.3 \text{ seconds}$$

(The uncompressed version would now take about 10 minutes.)

Were the noise to equal the signal strength, then $(1+S/N)$ becomes 2 and only one bit per packet could be sent, making the compressed transmission time about a minute and 40 seconds. As the noise gets even greater, $(1 + S/N)$ tends towards 1, which means that no bits per packet can be transmitted, because $\log_2(1)$ is 0.

For the Morse operator on the sinking ship, it is the high level of noise that does not permit the transmission of speech (which needs about 8,000 bits per second). It does, however, allow the three or four bits per second of the dits and dats which, with Morse's clever encoding, get the essence of the message across.

The simple telephone cable is not the only medium for the transmission of information, with its 10,000 bits per second of information. There are all sorts of cables and other transmission media that have much larger bandwidths. 'Coaxial' has a solid wire in the middle, surrounded by a plastic insulator and sheathed by a flexible outer metal sleeve. The bandwidth of this cable can be 200 million hertz (or in other words 200 megahertz – this is the same as 400 million bits per second). Clearly this allows much faster communication between computers, but it is also a little more expensive. Even larger bandwidths may be obtained with fibre-optical cables, which instead of transmitting electrical pulses transmit optical ones (laser-generated light, actually). FM (frequency-modulated) radios pick up stations around 100 megahertz. This means that the free space, in which information travels as electromagnetic waves, has a vast bandwidth.

That's all very well, but even the most exquisite broadcast of classical music needs only a bandwidth of 30,000 hertz, so how are much larger bandwidths exploited? Shannon's concept of the encoder or the modulator explains how. To reduce the problem to that of simple numbers, our digital pictures come in handy again. We saw earlier (see note 2) that each picture point requires 256 numbers, that is 8 bits (simply because $\log_2 (256) = 8$). Say we have a bandwidth/noise situation that is plentiful and we wish to transmit, say, eight times that number in the same time. This suggests that we might try to transmit eight pictures simultaneously if we could only use the bandwidth as eight separate channels rather than just one. It can be easily done. Anything that goes into channel 1 will be given the number 1 as a prefix. Channel 2 will get 2 and so on. So for every period of time we transmit a group of eight numbers, each prefixed by its channel number. At the receiving end the decoder must be built so that these channel numbers are detected and the picture points separated. The channel numbers are 'carriers' of the information related to each channel.

Something very similar happens when we tune into the station of a radio. We tune into the carrier of the particular channel and the radio set decodes the content of that channel. So the free space bandwidth of, say, 300 million hertz can cope with 10,000 different radio stations and more (as not all demand a bandwidth of as much as 30,000 hertz). This encoding also takes a particularly interesting form on the Internet. The coding numbers are things like jack@toc.ac.uk, which may be Jack's e-mail address to which Jill sends her pictures so that Jack and only Jack gets them. This curious form of coding means that, in the case of e-mail, Jill's message with its channel code rushes about this vast network in order to find Jack's computer whose address is the destination of the message. The destination computer then decodes Jill's pictures and delivers them to Jack's screen. So this establishes a unique channel between Jack and Jill despite the megajungle of cables, satellite links and radio transmission that the Internet implies.

It should be stressed that 'compression' is very much part of the encoding and decoding process as envisaged in Shannon's five-stage scheme of 'source–encoder–channel–decoder–destination'. Those who use the Internet and download pictures or movies will be aware of standards that have names such as JPEG (for stills) or MPEG (for movies). These are encoding and decoding protocols that save the Internet user hours of waiting for data to be downloaded. So wherever we look in the vast world of modern

telecommunication we find that Shannon's model of the nature of information is enormously helpful in the design of the systems that enable
high-speed communication.

An unexpected spin-off from Shannon's definition of the bit is that not
only is it the unit of transmitted information, it has also become the unit of
information storage or 'memory'. A single switch can be on or off, in accordance with the definition of a bit as being the carrier of only two messages.
A switch therefore records, memorizes or stores one bit of information. Two
switches can be in four combinations of *on*ness and *off*ness and the $\log_2$ element of Shannon's equations again comes in useful because, in order to
store, say, one million messages, the number of switches required is given
by $\log_2 (1,000,000)$, and turns out to be only a small number of 6 bits –
about 20. It is this relationship that gives computers their prodigious powers
of memory. Anyone who uses a reasonably up-to-date computer will be
aware of at least two kinds of memory: hard disk and random access.
Typically a hard disk may store 5 gigabytes. For an unimportant reason
eight bits are called a byte, so 5 gigabytes turns out to be 40,000,000,000
bits. The hard disk is a rotating piece of metal on which a bit is stored by
magnetizing a local patch of the metal through the use of a 'head' which
becomes a magnet (or not, depending whether it is fed an electric current).
The patch is either magnetized or not and hence much like a switch: it
stores one bit. The rotating disk leaves a trace of set and unset switches.
These can be 'read' by the same head because magnetized patches induce a
current in this head that then can be transmitted as an informational bit or
used in the computer in some other way. The reason that a computer also
has a random-access memory is that the hard disk is relatively slow by
virtue of its rotational inertia. It may make it necessary to wait (a hun-
dredth of a second, roughly) before a particular patch of metal can be
accessed. The random-access memory is much faster (it needs a small fraction of a millionth of a second for access). It allows the state of a tiny
silicon switch to be accessed in the way we access a file in a filing cabinet.
We need a tag like 'tax' or 'mortgage' which identifies the desired folder. A
glance is enough to pick out the file. In a similar way, every silicon switch
has a tag called an 'address' which, when applied to the whole bank of
switches, will pick out the right one with the matching address. So the
random-access memory is fast, but not as large as the hard disk.

   With these ideas we can now build a much more complete picture of
what happens when we download a picture from one computer to another.

If it was taken with a digital camera, the picture is first sensed by special light-sensitive electronics and then translated into stored bits in the camera's own random-access memory. This is then transmitted (through the use of software and appropriate cables) to the hard disk of the sender's machine, where it occupies 20 million out of the 40,000 million magnetic switches. If she wants to view the picture on the screen of her machine, she moves it into the random-access memory of her computer. This in turn is made by certain programs to move the bits onto the computer screen, where the electrical energy of the bits is turned back into light patterns. Then, when the recipient asks for the image to be downloaded, it is transmitted over cables and is transferred into his random-access memory and viewed on the screen. To keep it permanently, he transfers it to his hard disk.

The storage capacity of computers has grown prodigiously. Taking just the area of storage devoted to our single picture, with Shannon's $\log_2$ relationship it is possible to ask how many different pictures could be represented in this space. The answer is x in

$$20,000,000 = \log_2 (x).$$

x turns out to be roughly 10 followed by 7 million zeros, an astronomical number.

Shannon's brainchild, the bit, has not only given us a means for measuring information as a utility, it has become the very currency of computation. Technology has thundered on, but Shannon's insight and his formulations of half a century ago remain unshaken. The vast versatility of the computer and the even more awesome power of millions of interconnected computers on the Internet have created organisms where the complexity is beginning to be way beyond our grasp. It's $\log_2$ that does it.

Why is it that everything is going digital? There are now digital telephones where there were ordinary (analogue) ones, there is digital radio and television and we pay more for digitally recorded music on compact discs than reproductions of the older analogue albums. The whole world of consumer communication products is going digital. Governments are behind this move (although they sometime fail to give cogent reasons for their support). Shannon explained this vast wave of change when he defined the maximum capacity of a channel in the second of his equations, and when he defined a standard encoder–channel–decoder structure for any electronic communication system.

'Digital' simply means that data is transmitted as discrete symbols. But to drive this point home it may be easier to say what is not digital. Human communication at its most direct is not digital. When I speak, I create pressure waves in the air by the movement of my vocal cords, the shape of the cavity behind my mouth and the configuration of my tongue and lips. These waves reach my hearer's eardrum, causing his cochleae (coiled-up trumpet-like organs in the ear) to translate these pressure waves into internal neural signals and transmit sensory information to the brain, which the recipient describes as 'hearing'. But as soon as the communication becomes electronically aided, the possibility of translating these waves into sequences of numbers encoded as bits becomes an option. This system is then digital: that is, no more waves, just information as bits.

While Shannon's theory applies to both analogue and digital systems, the theory itself shows that digital is most efficient, and if the digits are binary this is the best of all. The argument is based on cost and goes a little like this. Let us say that a picture point requires 256 numbers. The channel here can be thought of as not being digital, just a box that needs to be a certain size to accommodate all the numbers between 0 and 255 as, say, a number of little cubes. Now we introduce the cost of that channel. This is the cost not of the cubes but of the box that must carry them. It is natural to think that the larger the number it carries, the more it would cost. This channel costs 256 units of some currency.

How much would it cost were we to use two smaller channels to carry the same information? The boxes would have to carry a mere 16 numbers because the combination of the numbers in the two boxes would give us 16×16 possible numbers, which brings us back to the necessary 256. But the total cost of the two channels is now $16 + 16 = 32$ units, rather a worthwhile saving. So why not keep going in the same direction? We note that the process stops when each channel holds only two numbers, when we have eight channels to give us $2 \times 2 \times 2 \times 2 \times 2 \times 2 \times 2 \times 2 = 256$. The cost of this is $2 + 2 + 2 + 2 + 2 + 2 + 2 + 2 = 16$ units. According to the late Norbert Wiener of MIT (the 'grandfather of cybernetics'), this was proof of Shannon's brilliance in defining the bit. Each of the channels in the lowest-cost system is a binary channel that carries only one bit, and this is by far the most economical way of transmitting information.

It is the economic efficiency of binary encoding that is causing the world to go digital. Yet again Shannon's $\log_2$ – which features in both his equations – is at work, because if someone is using nondigital waves, say, that have amplitudes that need to represent some integer number A precisely,

Shannon tells us that it costs a great deal less to use bits to transmit the same information. Probably the most vivid illustration of this is the progression in recorded music from long-playing records that produced waves through wavy grooves and needles which needed a relatively vast area to store thirty minutes of music per side, to the modern digital video disc (via the compact disc). The DVD can store up to four hours of music in $\frac{1}{25}$ of the space of the LP by using pure digital techniques. But the same goes for other things. Mobile phones and cordless phones work noticeably better through being digital. And all this because of the $\log_2$.

So these two equations have changed our world of communication:

$$I = -p \log_2 p$$

$$C = W \log_2 (1 + S/N).$$

I have argued that despite their formidable appearance, the real power of these equations lies in the rough relationship that underpins both:

Information in bits = $\log_2$ (what needs to be communicated).

Shannon's framework (source–encoder–channel–decoder–destination) holds true whether we send digital pictures using the latest Internet technology, whether we are chatting over our mobile phone or we are talking to one another in a noisy pub. It is within this framework that the first equation is born: a very general definition of information based on surprise and probability. But the important message from this equation is that if the probability of an event is 50 per cent, it contains exactly one bit of information. This can be extended to the more general idea that anything that constitutes a real transaction may be broken down into an appropriately sized string of bits. The second equation then concentrates on the nature of the channel: the telephone lines, the free space or the noisy pub. Shannon showed that there is a limit to the number of bits per second that can be transmitted in the given medium, a limit set by the bandwidth and the noise in the channel. The way to exploit this limit with the greatest economy is by digital encoding. The trick in this exploitation is to design ever-improved encoders that take raw information and turn it into optimally encoded strings of bits. Entire industries have been built around this problem of encoding, as can be seen with mobile telephones and the encoding of music and video for entertainment.

Shannon's effect is not restricted to the world of communications. The form of the equation may be found in other fields of science under the heading of 'entropy', that is, in the degree of disorder of a physical system. This, in information theory, is expressed as the degree of surprise. However, Shannon's formulation has shown that information behaves according to the laws that govern physics, thermodynamics, physical chemistry, and are well known to mathematicians. Information theory was a specialism for which only designers of electronic equipment had a vague feel before the 1950s. Shannon showed that it is a material equivalent to the fundamental particles of the universe and that it has a system of laws equivalent to those that rule these particles. In my own work too, which has to do with modelling the intricate architecture of the brain, the language of information theory reigns supreme. The storage capacity of brain cells may be measured in bits, and the anatomy of interconnections between the many modular areas of the brain can be analysed using the ideas of channel capacity.

The retiring person who unleashed this innovative ferment, Claude Shannon, is one of the technological giants of the twentieth century. The word 'technology' here is perhaps inappropriate, because Shannon made a major *intellectual* contribution to our contemporary world. Shannon was curious about complex things. Shannon's equations are not about nature, they are about systems that engineers have designed and developed. They are equations that elegantly capture the complexity of information and the means whereby it can be stored and transmitted. Shannon's contribution lies in making engineering sense of a medium through which we communicate. He shares the same niche as other great innovators such as his boyhood hero Thomas Edison (who turned out to be a distant relative, much to Shannon's delight) and Johann Gutenberg. Like the printing press, the Internet is a celebration of human language, the characteristic feature of conscious human beings. In much the same way Gutenberg's imagination was triggered by the turn of a wine-press screw, Shannon's was stimulated by the click of a differential analyser switch.

Having completed a brilliant academic career, Shannon retired from MIT in 1978, becoming professor emeritus and a deeply respected elder statesman of US science. In 1985 he was awarded the Kyoto Prize, the computer world's equivalent of the Nobel Prize. After he retired, he pursued a wide range of interests, including a mathematical theory of juggling, the design of a motorized pogo stick, and the development of a system for playing the stock market using probability theory. The end of his life was tragically blighted by Alzheimer's disease and he was too ill to attend the unveiling of his statue in his native town of Gaylord, Michigan, in the autumn of 2000. He died on 24 February 2001 in a Massachusetts nursing home. His passing was politely recorded but it was plain that the media – busy participating in the information revolution – had, for the most part, little appreciation of the indubitably great figure the world had lost.

It is true that we nowadays use the word 'intellectual' to refer to those who make contributions to the humanities, philosophy and politics. It was not always so. The ideal Platonic and Aristotelian intellect included practical cunning and mathematical abstraction. Shannon changed the world by being a master of both.

Hidden Symmetry

The Yang–Mills Equation

Christine Sutton

Summertime in New York – hot and steamy, the stuff of movies. The year is 1953: Stalin is dead, Elizabeth II is the newly crowned Queen of England, and a young senator named John Fitzgerald Kennedy is about to marry Jacqueline Lee Bouvier. The paths of two young men cross as they share an office at the Brookhaven Laboratory on Long Island. Like a rare alignment of planets, they pass briefly through the same region of space and time. The juxtaposition gives birth to an equation that could underlie the Holy Grail of physics – a 'theory of everything'.

Robert Lawrence Mills and Chen Ning Yang were born a world apart, but shared a passion for theoretical physics. Yang, who turned thirty-one in September 1953, had come to the United States from China and gained his doctorate at the University of Chicago before joining the Institute of Advanced Study in Princeton, New Jersey. Mills, at twenty-six, was a new research associate at the Brookhaven Laboratory, who had studied at Columbia and Cambridge universities. In 1953 Yang was visiting Brookhaven for the summer and was allocated a space in the same office as Mills. Their paths soon diverged, but the Yang–Mills equation has ensured that after only a brief encounter their names have become inseparable.

Back in the 1950s, the Yang–Mills equation seemed the result of an interesting idea that had little bearing on reality, but by the end of the twentieth century it had come of age. It underlies the work behind the Nobel Prizes in physics in 1979 and 1999, and is important enough in mathematical terms to have been named by the Clay Mathematics Institute as one of seven 'Millennium Prize Problems', the rigorous solution of which will win a prize of $1 million.

So why all the interest? What is it about the Yang–Mills equation that makes it so important? What, indeed, *is* the Yang–Mills equation? To begin to answer these questions we must look first at some of the underlying concepts that physicists use every day to interpret the phenomena of the world about us.

The Forces of Nature

The story of the Yang–Mills equation has roots in the seventeenth century, when Isaac Newton, so the story goes, took inspiration from the fall of an apple and was led to an equation of gravity. Today we launch satellites into orbit about the Earth, and send spacecraft to probe distant planets, all on trajectories calculated courtesy of Newton. His wide-ranging achievements included the publication in 1687 of *Philosophiae Naturalis Principia Mathematica*, known simply as the *Principia*. In this vast work, Newton set out to explain in mathematical terms as much as he could of the physical world, from the paths of the planets to the regularity of the tides. His main tools were equations that relate motions to forces – the same equations that continue to form the basis of the mechanics and dynamics taught in schools and universities throughout the world. However, Newton knew well that his work dealt only with certain aspects of the physical world. In his preface to the *Principia*, he comments:

> I wish we could derive the rest of the phenomena of nature by the same kind of reasoning from mechanical principles; for I am induced by many reasons to suspect that they all depend upon certain forces by which the particles of bodies, by causes hitherto unknown, are either mutually impelled towards each other and cohere in regular figures, or are repelled and recede from each other; which forces being unknown, philosophers have hitherto attempted the search of nature in vain.

Three hundred years later, Newton's wish is on the verge of being fulfilled, as the searches of modern natural philosophers – the physicists – have revealed the structure of previously unknown forces. The Yang–Mills equation seems to articulate mathematically the basic principle that underlies these forces, just as Newton desired. In a sense it is the modern analogue of Newton's equations of motion, a formula to unlock the beauty of natural relationships, and this potential is as highly regarded today as it was by Newton.

Nowadays, once physics students have mastered Newton's mechanics – applying his equations to colliding billiard balls, soaring rockets and the like – they learn that all the matter we see in the world about us, and in the universe at large, is formed from particles that are controlled by forces. The forces are now largely known and, as Newton's tantalizingly prophetic words imply, they give shape and structure to the universe of particles; they provide the invisible skeleton of the universe.

But what do we mean by a force? The various forces drive the interactions between the particles, clustering them together in structures on all scales, from tiny atoms to colossal galaxies. The forces act invisibly, sometimes pulling particles together like people drawn to the music of a street busker, sometime thrusting them apart, like the bell that sounds at the end of the school day. Without the forces there would be nothing but a gas of particles, with no means of interacting with each other, and no means of revealing their existence.

The most familiar of the forces at work in constructing the universe from the fundamental building bricks is gravity, the force mastered by Newton's mathematics three centuries ago. Less familiar is electromagnetism, the single force that underlies the many facets of electricity and magnetism, from the natural phenomena of lightning strikes and lodestones to the modern wizardry of television and radio. Two further forces, called weak and strong, are much less well known, although their effects are just as influential in making the matter of the universe the way it is.

The weak and strong forces act within the nucleus that lies at the heart of every atom of every kind of matter. There these two forces compete with the electromagnetic force for ultimate control over the atom. At times, the strong force wins, binding together the constituents of the nucleus (particles we call protons and neutrons) to form a stable entity. More often, the weak force or the electromagnetic force is victorious, leading to a huge variety of unstable, radioactive nuclei, some of which, such as the nuclei of uranium, occur naturally on the present-day Earth.

The discovery of the weak and strong forces, previously invisible in their submicroscopic nuclear domain, has brought physicists at the start of the twenty-first century to the point where they can claim to be able to derive natural phenomena from the fundamental interactions of forces. No,

physicists cannot 'derive' a cow, or the grass it eats, from these first principles, but they can derive the properties of materials. They can calculate the
collective electromagnetic interactions of the electrons that swirl around in
the atoms in a solid, and use these calculations to invent new materials.
They can use their knowledge of the weak and strong forces within atomic
nuclei to calculate how vital elements, such as carbon, oxygen and iron, are
built in the hearts of stars. Probably most exciting of all for the physicists
themselves, however, is the discovery that they seem to be coming closer to
a single theory, a single set of related equations, to describe all the forces –
and the Yang–Mills equation is fundamental to this 'unification'.

Back in the 1930s, the nature of matter seemed to have been reduced to
a few basic building blocks. By then it was clear that the chemical elements,
from hydrogen and helium, through carbon, oxygen, iron and so on, to uranium, all consist of their own unique atoms. However, this huge variety of
atoms in turn is built from three basic components: negatively charged electrons, positive protons and neutral neutrons. The protons and neutrons reside
together in the nucleus at the heart of the atom, while the electrons circulate
at relatively huge distances and give atoms their size, and ultimately give
matter its shape.

The very existence of the atomic nucleus is at first sight paradoxical, for
the protons, each with the same electric charge, should repel each other.
'Like charges repel,' we learn at school; the electrical forces within a
nucleus should simply blow it apart. As this manifestly does not happen in
the stable matter about us, there must be a force that is stronger than the
electric force. But it must act only across distances comparable to the size
of a nucleus – otherwise, adjacent atoms would be squashed much closer
together and matter would be much denser than it is. This force, which binds
protons and neutrons together in nuclei, became known simply as the strong
force. But what is the origin of this force? Can it be understood in the same
way that physicists have come to understand electromagnetism, the best-
known force of all? This was the challenge that led to the Yang–Mills
equation.

Understanding Electromagnetism

The essence of electromagnetism is electric charge, which can be either positive or negative. Electric currents are electric charges in motion. The radio
waves from a radio station pulse out from electric charges jiggling in

synchronization in devices aptly called oscillators. The carriers of the electric charges are (by and large) electrons, the negatively charged components of atoms. However, the fundamental principles of electromagnetism were understood long before the nature of the atom and its contents became clear. The reason for this is twofold. First, it is the concept of electric charge, rather than the concept of the atom or the electron, that is fundamental to the actions of the electromagnetic force; second, the electromagnetic force has a long range, extending far beyond the boundaries of a single atom, so that its effects could readily be measured more than two centuries ago.

But what *is* electric charge? In a definition that sounds dangerously circular, electric charge is the source of the electromagnetic field. This field is the region of influence around a charge that determines the force that another charge feels at some distance away. The force is the embodiment of the field – in a sense, the real and measurable effect of unseen tentacles extending outwards from the charge. Two charges of the same type – both positive, say – repel each other: the force between them pushes the charges apart. This force decreases rapidly as the charges move apart, until their influence on each other becomes insignificant. The electromagnetic field surrounding each of the charges determines the force that the other charge experiences.

Stationary electric charges give rise to electric fields, but moving charges also create *magnetic* fields. Electrons gyrating in the atoms of a simple bar magnet create the magnetic fields that grip pins or shape characteristic patterns in iron filings in experiments at school. However, 'magnetic charges' do not seem to exist. These would be single magnetic poles, but magnets always have an even number of poles, most commonly two poles, called north and south.

To calculate the electromagnetic force due to an electric charge, or a collection of charges, requires equations that describe the basic electric and magnetic fields. In the 1860s the Scottish physicist James Clerk Maxwell succeeded in synthesizing all existing knowledge of electricity and magnetism in a self-consistent and beautifully succinct set of equations. Like Newton's equations of motion, Maxwell's equations are used to this day. They give the recipes for calculating electric fields due to electric charges or magnetic fields, and for calculating magnetic fields due to electric currents. The equations also embody an important feature of electromagnetism: the conservation of electric charge.

Conservation of charge means simply that charge can neither be created nor destroyed. If you 'charge up' something, as when combing your hair in dry air or revitalizing a car battery, you are simply redistributing existing charges – basically, atomic electrons. There are processes in nature that *can* create a charged particle, such as an electron, but they always create another particle with opposite charge. The electromagnetic force can create a negative electron together with a closely similar positive particle known as the positron or anti-electron. The electron and the positron are always created at the same place and this has an important implication. Charge conservation is more than a 'global' statement about a large system, with positive charge created 'here' balanced by negative charge popping up 'there'. It is a 'local' statement referring to each point in space and time, place to place, moment to moment. One of the beauties of Maxwell's equations is that they ensure this *local conservation* of electric charge, and they do this through an inherent symmetry in the behaviour of the electromagnetic force.

Nearly a century after Maxwell, Chen Ning Yang began to wonder whether he could start from the opposite direction, in order to come to an understanding of the strong force between particles. Could he begin with an appropriate conserved quantity and use symmetry to discover equations for the strong force?

The Importance of Symmetry

In mathematics, by symmetry we mean that something looks the same after a particular action, like a square rotated by 90 degrees or a circle rotated by any amount. In 1918, the young German mathematician Emmy Noether discovered a deeply fundamental relationship between symmetry and the conservation of physical quantities, such as electric charge. She found that for every conserved quantity there is a related symmetry – and vice versa.

In a dynamical system of objects moving under the effects of forces, both energy and momentum are conserved; in other words, the net amounts of these quantities remain unchanged. When a rocket shoots off to the Moon, it has momentum that it did not have when standing on the launch pad. To compensate, the momentum of the Earth changes, albeit imperceptibly since the Earth is so massive. The changes in momentum of each object – the rocket and the Earth – are equal in amount but opposite in direction, so they add up to zero, exactly equal to the situation before the rocket launched; momentum is conserved. But what symmetry is involved in conservation of momentum? It is the symmetry of the equations of motion at different points in space. The movement from the launch pad on Earth to a point en route to the Moon does not change the basic equations – this is the symmetry. Momentum conservation ensures this symmetry, and vice versa.

Noether's theorem tells us that as electric charge is always conserved, there should be a related symmetry in the electromagnetic force. Indeed there is, and it involves something known as the 'potential', which is a way of characterizing the field due to a force, be it an electric field, a gravitational field or some other force field.

The potential provides a more compact, 'shorthand' means of describing a field, rather as a two-dimensional contour map is a more compact representation of a three-dimensional landscape. Contour lines join points of equal elevation, and the more tightly the lines are packed, the steeper is the terrain. The two-dimensional contour map contains all the information that an experienced person needs to make a journey in the mountains. Similarly, the electric potential of a collection of charges, say, contains all the information that a physicist needs to calculate the electric field, and hence the electric forces at work in the system.

Electric potential is familiar to most of us as 'voltage'. A bird on a wire carrying a high voltage can sing just as happily as it can in the apparently safer haven of a branch in a tree. This is because electric fields, which give rise to electric forces, depend on *differences* in voltage, or electric potential. Our whole planet could be raised in potential by 1,000 volts and our power plants and electrical equipment would function just the same. What matters is the difference between 'live' and 'ground' ('earth'), not their absolute values. This 'invariance' is an example of a *global* symmetry: the electric field is unchanged when the same potential is added (or subtracted) everywhere in space and time. Similarly, Maxwell's equations are unchanged by a global shift in electric potential, as they deal principally with fields, not potentials.

However, Maxwell's equations also contain a more stringent *local* invariance or symmetry. The electric potential can be changed by different amounts at different points in space and time, but Maxwell's equations still remain the same. This is a local invariance, and it arises because electric charges underlie magnetic fields as well as electric fields. It turns out that local changes to the electric potential give rise to local changes in another potential, known as the magnetic potential. The net effect of the changes in the two potentials ensures that the electric and magnetic fields described by Maxwell's equations remain the same even when the change in the potentials is local. There is a local symmetry in Maxwell's equations, and it is this symmetry that seems to be related to the conservation of electric charge.

Particles and Waves

This may seem a long way from the strong force between particles, but there is a beauty that commands attention in having a force – in this case the electromagnetic force – described by equations that display an underlying symmetry principle. Indeed, it raises the possibility that the physical processes we see – in other words the very interconnections we observe between electricity and magnetism – arise from the local symmetry. And this brings us back to Yang and Mills, who were hoping to see whether they could start from the principle of local invariance and derive the equations for the strong interactions between particles.

In the century between Maxwell and Mills and Yang, there had been a great revolution in physics with the development of quantum mechanics, which we must use in place of Newton's mechanics when dealing with systems that are very small. On the scale of atoms, it is not possible to know exactly where a particle is and how fast it is moving. This is because the very act of observation disturbs the particle. We can measure the speed of a car by using a radar system to bounce radio waves off the car as it passes a certain point. The energy of the radio waves is so small that it makes no difference to the motion of the car. But change the object from a car to a molecule, and the radio waves will have enough energy to push the molecule around. Quantum mechanics deals with this basic problem of not knowing both position and velocity (or, strictly speaking, momentum) by treating particles as *waves*, and describing a particle mathematically by something called the wave function, which is related to the probability of finding a particle in a particular state.

Rather as voltages can be shifted up or down without changing the electric field between them, a wave can be modified in a way that does not change its overall effect. The property of the wave that we can change is called its phase, and this can be thought of as a measure of whereabouts the wave is in its undulating pattern. The value of the phase at a fixed position changes as the wave goes up and down. A change, or shift, in phase applied to the wave as a whole simply moves the wave pattern along; it does not change important properties such as intensity or wavelength.

In the same way, the wave function describing a particle can be modified by a constant phase shift that does not change the observable behaviour of the particle. Here again is an example of a global symmetry at work. However, is there also a local symmetry, as there is in Maxwell's equations? Suppose the phase shift is a local change, varying at different points in space and time. Do the quantum-mechanical equations that describe the particle remain unchanged – invariant – as a result of this local phase shift?

The straight answer is no, so it might seem that we should forget this line of thought, and abandon this obsession with local symmetry. However, if we try to modify the equations for the particle in such a way that they do not change with a local phase shift, we make a remarkable discovery. The equations are invariant provided the particle is moving under the influence of some force field. The situation echoes the connection between local changes in the electric and magnetic potentials in electromagnetism, only now it is local changes in a particle's phase that are related to local changes in the field through which it moves. The discovery is even more remarkable when we realize that the electromagnetic field supplies precisely the required modifications to the quantum-mechanical equations – provided we let the phase shift for the particle depend on the particle's electric charge. It seems that the principle of local invariance reveals directly the nature of the electromagnetic interactions of charged particles.

Hermann Weyl, a German mathematician, was the first to realize the deep connection between local invariance in a particle's wave function and electromagnetic theory. He called the invariance 'gauge invariance', for originally he had thought about changes in scale, or 'gauge', rather than changes in phase. In his classic paper, which appeared in 1929, he says, 'It seems to me that this new principle of gauge invariance, which follows not from speculation but from experiment, tells us that the electromagnetic field is a necessary accompanying phenomenon . . . of the material wavefield . . .'

Weyl had thus taken the dramatic step of proposing that gauge invariance – a basic symmetry – could be used as a principle from which the theory of electromagnetism could be derived. In the case of electromagnetism, this was an elegant idea, but it led to nothing new, as the electromagnetic force was well known and already understood through Maxwell's equations. Weyl's proposal would be of much greater consequence for a force, such as the strong force, where the equivalents of Maxwell's equations were still unknown. Would it be possible to find these equations by beginning with the appropriate symmetry principle? At the time of Weyl's paper, the composition of the atomic nucleus was still not properly known, and the notion of the strong force had yet to be developed. A new application of Weyl's principle would have to bide its time.

A New Kind of Symmetry

Twenty years later, these profound ideas that link symmetry and electromagnetism entered the mind of the young Chinese physicist who had come to study at the University of Chicago. Chen Ning ('Frank') Yang was the son of a professor of mathematics, and had come to the United States in 1945. He had adopted the name Franklin – hence 'Frank' – in honour of Benjamin Franklin, whose autobiography he had read in China. First at university in Kunming in Yunnan province, and later at Chicago, Yang had thoroughly studied review articles on field theory by Wolfgang Pauli, one of the sharpest theoretical physicists of the time. Yang writes that he had been 'very much impressed with the idea that charge conservation was related to the invariance of the theory under phase changes . . . [and] even more impressed by the fact that gauge-invariance *determined* all the electromagnetic interactions'.

Yang was initially unaware that these ideas were due to Weyl, and still did not realize it when both men were at the Institute of Advanced Study in Princeton and even met occasionally. Weyl had left Germany in 1933 and taken up a position at Princeton, becoming a US citizen in 1939, whereas Yang joined the Institute in 1949. It seems that Weyl, who died in 1955, probably never knew of the remarkable paper that Yang wrote with Mills – the paper that demonstrates for the first time how the symmetry of gauge invariance could indeed specify the behaviour of a fundamental force.

While at the University of Chicago, Yang had begun to apply these ideas to another property of particles that, like electric charge, is conserved in their interactions. His aim was to find the equations that describe the field associated with the gauge invariance of this property, which goes by the rather confusing name of 'isotopic spin' or 'isospin'. Isospin is like a name tag that labels particles that have different electric charges but otherwise appear the same. Imagine a pair of identical twins, Peter and Paul, dressed in the same way except that one wears a coat. Remove the coat and the twins become indistinguishable, although they still have different names. So it is with particles, such as protons and neutrons, where the proton carries a 'coat' of positive charge, while the neutron is 'unclothed'; that is, uncharged. Studies of atomic nuclei in the 1930s revealed that once the differences due to the different electric charges are allowed for – once the imaginary 'coat' of charge is removed from the proton – neutrons and protons, neutrons and neutrons, and protons and protons, all interact in the same way. In other words, the other force between the particles – the strong force – sees no difference between them. The proton and neutron, which have closely similar masses, appear to the strong force as two states of the same particle, the 'nucleon'; like the names of the twins, the values of isospin are now all that differentiates the particles. The situation echoes the way that a particle can appear in different states of the quantum property called 'spin', and the mathematics that describe the spin states of a particle can be used to describe the isospin states.

Mathematically speaking, you can 'rotate' a proton's isospin to turn it into a neutron, and the effects of the strong force on the particle will not change. There is a symmetry in the force, and as Emmy Noether's theorem tells us, something must be conserved; the something is isospin. Now we have all the pieces that came together in the Yang–Mills equation.

A New Kind of Field

From 1949 onwards, Yang tried several times to take the procedures of gauge invariance in electromagnetism and apply them to isospin. These attempts, according to Yang, always led to 'a mess' and he would founder at the same point in the calculations, when he tried to define the strength of the related field. But he was never completely deterred. As he explains in his *Selected Papers*, 'This kind of repeated failure at some seemingly good idea is, of course, a common experience for all research workers. Most such ideas are eventually discarded or shelved. But some persist and may become obsessions. Occasionally an obsession does finally turn out to be something good.' In particular, many kinds of short-lived particle were being discovered in experiments, and there seemed to be almost as many ideas about the forces underlying their interactions. For Yang, 'the necessity to have a *principle* for writing down [these] interactions became more obvious'.

Yang returned to these ideas once again at the Brookhaven Laboratory, in the summer of 1953, and this time Robert Mills, the young physicist who shared his office, caught the obsession. Together they broke through Yang's earlier barrier and discovered the equations for the field associated with the gauge symmetry of isospin.

If we ignore electromagnetic effects, the choice of what we call a proton and what we call a neutron becomes arbitrary – change all neutrons to protons and vice versa and nuclear reactions will remain the same. This corresponds to a global change in the isospin states – we have 'rotated' isospin at all points in space and time by the same amount so that all protons become neutrons, and all neutrons become protons. But what happens, asked Yang and Mills, if we make different changes at different points in space and time? Suppose that the 'rotation' between the two isospin states is completely arbitrary, or 'physically meaningless', as Yang and Mills say in their paper. This is just like the arbitrary phase shift in the wave function of a charged particle, which is compensated by changes in the electromagnetic field. So is there a field that similarly compensates local changes in isospin, and ensures that nuclear reactions always look the same?

The theory with isospin turns out to be more complex than electromagnetic theory in a fundamental way. The compensating field must be able to adjust the local changes or 'rotations' in isospin, so as to keep the identity of a proton or neutron the same everywhere. To do this, the field itself must possess the property of isospin. By contrast, in electromagnetism, local modifications to the wave function of a particle do not change the particle's electric charge. This is reflected in the fact that the electromagnetic field does not change electric charges. Electric charge may be definable as the source of the electromagnetic field, but the electromagnetic field is not itself a source of electric charge. In the Yang–Mills theory, however, in some incestuous-sounding way, the field is a source for itself.

The Yang–Mills equation is the equation of motion of this field. It is the equivalent of Maxwell's equations – or Newton's equations of motion – and can be written down in a similar way. In the notation that Yang and Mills used, the equation is[1]:

$$\partial \mathbf{f}_{\mu\nu} / \partial x_\nu + 2\varepsilon \, (\mathbf{b}_\nu \times \mathbf{f}_{\mu\nu}) + \mathbf{J}_\mu = 0$$

Here $\mathbf{f}_{\mu\nu}$ describes the strength of the Yang–Mills field, with the $\partial/\partial x_\nu$ indicating that the equation depends on how the field strength varies with space and time; ε has the role of 'charge' and $\mathbf{J}_\mu$ gives the related current; $\mathbf{b}_\nu$ is the potential of the field. The term $(\mathbf{b}_\nu \times \mathbf{f}_{\mu\nu})$ supplies the all-important difference from electromagnetism, for it brings about the dependence of the Yang–Mills field on itself. In Maxwell's equations of electromagnetism, the equivalent term is zero because the fundamental fields do not affect each other.

The Mass Problem

There remained one major stumbling block with the new field that Yang and Mills had found, and this concerned the 'field particles'. In the quantum theory of fields – the theoretical framework with which Yang and Mills worked – fields are represented by particles. These 'field particles' are more than a convenient mathematical means of describing the field; in certain circumstances they emerge from the field as detectable entities, as real as electrons or protons. In electromagnetic theory, the field particles are photons, which can emerge from the electromagnetic field as the light we see.

The field particles act like balls in a game of 'quantum catch' between interacting 'matter particles', such as electrons and protons. In the electromagnetic case, charged particles interact by playing 'catch' with photons. The photons have no mass so the interactions can occur over long distances, in principle infinitely long distances. (You can imagine throwing the photon 'ball' infinitely far.) By contrast, the range of the strong force between protons and neutrons appears to be limited to the dimensions of the nucleus. This implies that the strong 'ball' must have some mass to ensure that the exchange – the interaction – always takes place over a limited time; that is, a short distance.

The new field Yang and Mills had found would adjust isospin as necessary at each point in space and time, changing protons to neutrons, neutrons to protons, or leaving them the same. To do this required three carrier particles, with three states of isospin. The field could also change electric charge, from positively charged proton to uncharged neutron, say. So two of the carrier particles had to be charged, positive and negative, while the third remained neutral and would participate in interactions between protons and protons, or neutrons and neutrons. So Yang and Mills knew the charge and the isospin of the new field particles, but they had no idea of their masses, and they recognized this as a weakness in their theory. When Yang presented the theory in a seminar at Princeton in February 1954, he found himself under attack from no less a person than Pauli. As soon as Yang had written on the blackboard an expression involving the new field, Pauli asked, 'What is the mass of this field?' When Yang explained that it was a complicated problem and that he and Mills had come to no definite conclusions, the acerbic Pauli charged, 'That is not sufficient excuse.'

Although they had finished the major part of their work by February 1954, Yang and Mills held back from publishing a paper. As Yang writes, 'The idea was *beautiful* and should be published. But what is the mass of the gauge particle? We did not have firm conclusions, only frustrating experiences to show that [this] case is much more involved than electromagnetism. We tended to believe, on physical grounds, that the charged gauge particles cannot be massless.' Yang himself italicizes the word 'beautiful', and it would seem that beauty overcame doubt. He and Mills submitted their paper to the prestigious journal the *Physical Review* towards the end of June 1954, and it was published three months later on 1 October. Their next-to-last paragraph ends by stating regretfully that they 'have not been able to conclude anything about the mass of the *b* quantum'; in other words, the carrier of their new field.

Electroweak Unification

Progress in understanding the elementary particles and forces comes, as in any science, through the interplay of ideas and discoveries – theories and experiments. Like a duet for musical instruments, the two complement each other; sometimes one leads, sometimes the other. Sometimes one instrument tries out fragmentary new melodies, while the other continues an existing theme. Then later, one of the fragments may turn into the dominant theme. In a similar way, physicists explore different avenues with theoretical ideas and experimental investigations. Some will prove fruitless and be forgotten, while others may return to lead our understanding at a later stage. The Yang–Mills approach may have failed originally to provide insight into the mysterious workings of the strong force, but it now underpins our present understanding of particles and forces. However, it was only through further theoretical advances and experimental discoveries that it became clear exactly how the Yang–Mills approach relates to the nature of the forces between particles.

In October 1979, a quarter of a century after the publication of the paper by Yang and Mills, three theoretical physicists received word from Stockholm that they had been awarded the Nobel Prize for physics. Sheldon Glashow, Abdus Salam and Steven Weinberg had independently put together a new theoretical framework based on the principle of local invariance. The ideas of Yang and Mills, and of Weyl before them, had come of age, but in a rather unexpected way.

The new theory dealt with two forces together – not the electromagnetic and strong forces, as might have emerged from the line Yang and Mills were following, but the electromagnetic and weak forces. The 'electroweak theory' also successfully resolved the problem of mass, and incorporated heavy field particles. More than that, the theory even predicted the masses of these particles (with a little help from some quantities that could be measured).

The weak force underlies certain types of radioactivity, where atomic nuclei 'decay' when a neutron they contain changes into a proton, or vice versa. These processes cause a real alchemy, as they alter the number of protons in the nucleus, and that in turn changes the precise chemical nature of the atom to which the nucleus belongs. Carbon can turn into nitrogen, lead into bismuth, and so on. In a similar way, protons turn into neutrons in the chain of nuclear reactions that release energy at the heart of the Sun and other stars. So although the weak force is some 100,000 times weaker than the strong force within an atomic nucleus, it has a very direct and profound influence on the nature of our universe and, through our Sun, on life itself.

In terms of the phenomena of the everyday world, it seems amazing that the electromagnetic and weak forces should be intimately linked in a fundamental way. The long-range behaviour of electricity and magnetism underlies phenomena on a grand scale, such as thunderstorms and the aurora borealis, while the weak force acts insidiously on a microscopic, subatomic scale. The life-giving energy we receive from the Sun comes to us as photons of light – the electromagnetic field particles – even though the energy is liberated in reactions initiated by the weak interactions of nuclei deep in the Sun's core. It is the link between these seemingly disparate phenomena that Glashow, Salam and Weinberg found, although it is a discovery that none of them set out to make.

In Britain, Abdus Salam was interested in understanding weak interactions between particles in terms of local invariance. The weak force can change the electric charge of particles – when turning neutrons into protons, for example. So Salam proposed that the weak force could result from a field like that described by Yang and Mills, with three 'field particles' of positive, negative and zero electric charge. The positive and negative field particles could readily be associated with known weak interactions that change charge, but the neutral field particle was more problematic. A natural choice was to identify it with a familiar neutral field particle – the photon of electromagnetism – and so the idea of 'electroweak unification' began to grow in Salam's mind.

In the United States, Sheldon Glashow pursued a similar scheme, although for different reasons. The problem he hoped to resolve was that existing theories of the weak force all led to unphysical infinite quantities appearing in the calculations. He believed that by incorporating electromagnetism and the weak force into the same theory, the parts of the calculations that gave embarrassing infinities would cancel out. He chose to base his attempts on the Yang–Mills approach and, like Salam, assumed the neutral particle to be the photon of electromagnetism. However, both Glashow and Salam soon realized independently that a better theory incorporated the symmetries of the electromagnetic and weak interactions in separate ways. The result was a theory with two neutral field particles – the photon of electromagnetism and a different neutral particle for the weak field.

There were several problems with this theory to begin with, one of them being the problem of mass, which had caused Yang and Mills such difficulties. As with the strong force, the range of the weak force appears to be very small, implying that the weak 'balls' in the game of 'quantum catch' must be very heavy. In the electroweak theory, while the photon remained massless, the positive, negative and neutral particles of the weak field all had to have a large mass. But endowing the field particles with mass would destroy the local invariance and with it the *raison d'être* of this approach. Moreover, to Glashow's dismay, the problem of the infinities remained, and crucially there was no experimental case for the existence of the heavy, neutral field particle that the theory demanded.

The solution to the difficulty with mass came from an unexpected quarter – a completely different area of physics dealing with the way that atoms behave collectively in solids. The key was the concept that a physical system can exist in a state that lacks symmetry, even though the underlying equations are symmetric. Atoms in iron, for example, behave like tiny magnets. In an ordinary lump of iron these atomic magnets point in random directions, so there is symmetry, as no direction is preferred to any other. However, the iron can be magnetized, in which case the atomic magnets line up in the direction of the magnetic field. The symmetry seems to disappear, although the equations describing the motions of the atoms still retain their original symmetry. Several theorists, including Peter Higgs from Edinburgh University, realized that they could apply these ideas to allow particles to acquire mass, by introducing an additional field into their equations – a field that has now acquired Higgs's name.

The Higgs field is unusual in that although its related potential is symmetric, the solutions to equations of motion in the field are asymmetric. In effect, the Higgs potential is like the dimpled bottom of a wine bottle – the overall shape is symmetric, but a pea momentarily balanced on top of the dimple will roll off in one direction, and so break the symmetry. The implication for equations describing the interactions of particles is that the particles are like the pea on the dimple – in the initial theory they have no mass, but as they interact with the Higgs field they break the symmetry and acquire mass.

Steven Weinberg in the United States saw promise in using the ideas of symmetry-breaking in a Yang–Mills theory to describe the strong interactions. Initially success eluded him, as he tried to match the massive and massless field particles that appeared in his theory with known strongly interacting particles. Then 'at some point in the fall of 1967', he recalled in his Nobel Prize acceptance speech, 'I think while driving to my office at the Massachusetts Institute of Technology, it occurred to me that I had been applying the right ideas to the wrong problem.' He realized that the massless particle he needed was the photon, and the massive particles were the weak field particles. 'The weak and electromagnetic interactions could then be described in a unified way in terms of an exact but spontaneously broken gauge symmetry.'

Four years later, in 1971, the final theoretical touch came to turn 'the Weinberg–Salam–[Glashow] frog into an enchanted prince', in the evocative phrase used by Sidney Coleman. Gerard 't Hooft, working with Martin Veltman in Utrecht, proved that the infinities in the theory cancelled – in a process known as 'renormalization'. Now Glashow could see how his earlier difficulties had been solved. 'In pursuit of renormalizability,' he has written, 'I had worked diligently but completely missed the boat. The gauge symmetry is an exact symmetry, but it is hidden. One must not put mass terms in by hand' (as he had tried). The work of 't Hooft and Veltman had elevated the electroweak approach to a reputable theory, and in 1999 they were rewarded with the Nobel Prize in recognition of their work in turning the electroweak 'frog' into a prince among theories.

In the course of a decade, from 1973 to 1983, many of the key elements fell into place. In 1973, experiments revealed the first glimpses of 'neutral weak currents'. These reactions, which had not been seen before, reveal the existence of the heavy neutral field particle for the weak force. In 1983, both charged and neutral weak field particles were produced and detected in high-energy collisions, their masses just as calculated from the electroweak theory. This was dramatic confirmation for the basic ideas of Yang and Mills.

The Colour Force

Where did all this progress with the weak and electromagnetic forces leave the strong force, the force that Yang and Mills had been attempting to describe? The 1960s had brought great changes – perhaps appropriately for a decade synonymous with the breakdown of old ideas – not least in our perception of what the fundamental particles really are. The proton, neutron and a host of short-lived particles had been found to consist of more fundamental particles, which became known as quarks. The proton and the neutron, for example, each consist of three quarks, bound together by the strong force. It was clear by now that the strong force lay with some property of quarks.

Theorists began to realize that if three identical quarks are to form a particle similar to the proton, then the quarks must carry a new distinguishing property. To satisfy the rules of quantum theory, this property must differentiate the otherwise identical quarks. In analogy to the three primary colours of light, this three-valued property became called 'colour', and its possible values dubbed red, green and blue. Importantly, it became clear that colour rather than isospin is the 'strong charge' – the source of the strong interactions between quarks.

Remarkably, the theory that physicists have constructed for coloured quarks is of precisely the kind that Yang and Mills had explored, but as colour has three values, rather than the two values Yang and Mills had considered for isospin, the resulting fields are more complicated. Instead of three field particles, there need to be eight. These field particles are known as gluons, and like the quarks, they must have colour, so that the new field that guarantees the local invariance is a Yang–Mills field – it is a source for itself. The theory that describes strong fields arising from 'colour charges' is known as quantum chromodynamics, or QCD, in analogy to quantum electrodynamics, or QED, the quantum theory of the electromagnetic force. QCD has turned out to be a very successful theory, so how does it solve the problem of the massive field particles expected for the short-range strong force?

The answer lies with the complexity of the interactions that can occur between the gluons themselves – a feature that simply does not arise in QED with its uncharged photons. The gluon interactions in QCD make the effective strength of the force around a 'strong charge' – a red quark, say – decrease at small distances. This was a surprising discovery, as physicists had known for two centuries that the force due to an electric charge increases as you approach the charge. However, the new effect seemed to explain paradoxical observations made in the early 1970s, in experiments that probed protons and neutrons with high-energy electrons. These experiments showed that as the electrons probed smaller distances, they began to interact with the quarks inside the nucleons as if they were almost free, or not bound at all within the larger entity. This meshed well with the idea of a strong force that becomes weaker with decreasing distances.

So what happens at increasing distances? The strong force appears to become stronger. The result seems to be that a single quark cannot be plucked from a proton or neutron in the way that an electron can be knocked from an atom. Certainly, there is no evidence that single quarks have ever been observed. So the strong force appears to have a short range, limited to about the size of the particles within which the quarks are incarcerated. There is no need, it turns out, to require that gluons are massive in order to explain the short range of the strong force. The gluons of QCD remain massless, and so pose no problems to the theory's local symmetry.

An Idea of its Time

Like many advances in science, the road from the work of Yang and Mills to the emergence of electroweak theory and QCD in the 1970s was long and complicated. Glashow, in accepting his share in the Nobel Prize in 1979, spoke of how 'the patchwork quilt' of the 1950s had 'become a tapestry' in the 1970s. 'Tapestries,' he continued, 'are made by many artisans working together. The contributions of separate workers cannot be discerned in the completed work, and the loose and false threads have been covered over. So it is in our picture of particle physics.' On the same occasion, when Salam had reached the culmination of the electroweak synthesis nearly halfway through his talk, he noted that he had already mentioned around fifty theorists by name.

No scientist, in recent history more than any time before, works in complete isolation. Moreover, there is often a sense in which scientific progress and discoveries are of their time. Yang and Mills may have been ahead of their time in that it took nearly two decades for their belief in a basic principle to bear fruit, but they were also of their time. In 1953, in different parts of the world, others were also beginning to build the same kind of theory. Pauli, whose articles on field theory had inspired Yang, started to investigate the possibility of extending the local phase transformations of electromagnetism to isospin, but this work was never published and appeared only in letters to Abraham Pais. It appears that Pauli realized the theory would yield massless particles for the fields, in apparent contradiction to the short range of the strong interactions.

In Cambridge, Salam's student Ronald Shaw investigated local invariance for isospin in a similar way to Yang and Mills, and derived the same new field carried by three particles. Shaw wrote later, in 1982, 'My gauge-field work arose out of my fascination with invariance ideas in general prodded by a (rather rough) preprint of Schwinger's which I found lying around in 1953.' He had completed this work by January 1954, but it was to form only one chapter in part II of his PhD thesis. Part II, Shaw explained in 1982, 'consisted of several disjoint bits, including Ch. III on SU(2) gauge fields. I remember feeling inadequate . . . and so I searched around and in late 1954 and in 1955 produced Part I of my thesis.' What price inadequacy! Shaw presented his thesis in September 1955; the paper by Yang and Mills had appeared in *Physical Review* in October 1954. Of course Shaw, like Yang and Mills, believed that his theory described massless particles that did not exist, so we should sympathize with his reticence.

The third derivation of the same theory occurred in Japan, where Ryoyo Utiyama was seeking a mathematical structure that would link gravity and electromagnetism. In March 1954 he completed some work on 'the idea of a general gauge theory'. As the Irish theoretician Lochlainn O'Raifeartaigh explained in his book *The Dawning of Gauge Theory*, 'As he also included gravity, it is fair to say that Utiyama's approach was the broadest and most comprehensive. But from the point of view of priority, his contribution appeared later than that of Yang and Mills . . .' Utiyama had been invited to visit the Institute of Advanced Study in Princeton, and arrived there in September 1954. He soon heard that Yang had announced a theory similar to his, and was sent a copy of the preprint. In a book published in 1983, Utiyama recalled, 'I immediately realized he had found the same theory as I had developed. I was too deeply shocked to examine Yang's paper closely and to compare it carefully with my own work.' Only in March 1955 did Utiyama return to his general gauge theory and look more carefully at what Yang and Mills had done. He realized that his approach was more general and so wrote a paper for *Physical Review*, which was published early in 1956. The originality of Utiyama's work has in general been overlooked, probably, as O'Raifeartaigh writes, 'Because Utiyama considered general groups and quoted Yang and Mills, this 1956 paper is often thought to be a straightforward generalization of the Yang–Mills theory.' Certainly Utiyama had cause to write, 'I very much regret not having submitted my paper to a Japanese journal in March 1954 when I had completed the work.'

Yang and Mills wrote only two papers together – the famous paper of

1954 in which Yang–Mills equation first appeared, and a much less well known one on the photon, which was written in 1966. Moreover, while Yang is nowadays widely regarded among physicists as one of the most brilliant theorists of the second half of the twentieth century, Mills never figured again on the world stage. Only three years after his work with Mills, Yang shared the 1957 Nobel prize for physics with Tsung Dao ('T.D.') Lee, a fellow Chinese American. They had found that the only way to explain the puzzling properties of some unusual subatomic particles was to assume a difference between left and right when particles interact through the weak force. They proposed how such a seemingly outlandish idea might be tested experimentally – and to the great amazement of all physicists, including the redoubtable Pauli, an experiment by Chien Sung Wu and her colleagues showed that the weak force does indeed differentiate between left and right. The collaboration between Yang and Lee produced many important papers over a number of years, although they fell out, sadly, in the late 1970s.

Mills, in contrast to Yang, continued his research in physics in relative anonymity. He joined Ohio State University in 1956, where he remained until his retirement in 1995. However, Yang retained a deep respect for Mills. 'Bob had a brilliant mind. He was very quick at grasping new ideas', Yang wrote on hearing of his colleague's death in 1999, 'I shall treasure the memory of our intensive collaboration and of our many discussions.'

Writing at the same time, Yang also commented that while he and Mills 'were pleased by the beauty of their work', neither of them 'had anticipated its great impact on physics'. Now, at the beginning of the twenty-first century, their work underpins the electroweak theory and the theory of quantum chromodynamics that stand as the key components of the Standard Model of fundamental particles and forces. It seems that beauty, through symmetry, and the workings of the physical world are inextricably linked, as Werner Heisenberg, one of the fathers of quantum theory, had begun to realize. In an essay on 'Beauty and theoretical physics', Yang quotes Heisenberg as saying in 1973, 'We will have to abandon the philosophy of Democritus and the concept of elementary particles. We should accept instead the concept of fundamental symmetries.'

When Yang and Mills set out to understand the strong force in terms of isospin in the summer of 1953, they found instead a principle based on symmetry that yields equations to link elementary particles and forces. With this discovery, they took a major step towards fulfilling Newton's wish, made nearly three hundred years previously. It remains for theorists in the twenty-first century to discover whether a full unification – including gravity – can be derived from the same principle, so that Newton's wish may be realized.

A Mirror in the Sky

The Drake Equation

Oliver Morton

Since Galileo, the astronomer's gaze has always been one of the most symbolically potent forms of perception. The astronomer sees farther, and sees more, than anyone else, and his sort of seeing is more profoundly divorced from touching than any other form of vision. It is undeniable that the power imputed to the astronomer's gaze derives in part from its association with the supposed insights of astrological seers. At the same time, though, astronomy's power as an emblem of the purity of scientific observation depends strongly on the seemingly obvious proposition that there is no connection between the people doing the seeing and the universe being seen. The astronomer sees all, and all he does is see – and the fact that he does it from solitary mountaintops makes it all the more romantic.

In the decades since World War II, astronomy has moved off the mountaintop. Technology has offered more and more ways of seeing through remote instruments and in different wavelengths. One of the first of these new ways of seeing was the use of radio waves. And this use was to lead to the creation of a fascinating new form of astronomy, one which polarized science and appealed to the public in almost equal measure. Radio astronomy provided a new way of looking at the question of intelligent life beyond the Earth, a question simultaneously on the fringes of the natural sciences and at their heart. On the fringes not just because there is almost no certainty to be had on the subject, but also because it goes beyond the normal remit of natural science; by its very nature, it is an inquiry into the artificial. And that, in turn, puts it at the very heart of science. The use of radio telescopes to look for alien civilizations addresses the question of humanity's place in the universe with the tools of a natural science. It seeks to answer the question 'What is man?' not by subjective introspection but by using the astronomer's objective gaze: by looking out, identifying some distant phenomenon and saying, 'Man is a thing like that.'

Radio astronomy was still in its infancy when, in the late 1950s, a young practitioner called Frank Drake realized that there was something unique about it. He and his fellow radio astronomers were the first people who could, in principle, detect the presence of people like themselves going about their business around other stars. The Earth, thanks to military radars and commercial television broadcasts, was already giving off more energy at the radio end of the electromagnetic spectrum than the Sun around which it orbited. A signal transmitted in a tight beam from a radio telescope such as the one Drake worked on, the 85-foot dish of the National Radio Astronomy Observatory at Green Bank, nestled in the hills of West Virginia, could be picked up by a similar telescope in another nearby solar system. The observatory he worked at could be set to the task of detecting not just stars and galaxies but civilizations.

Drake was not the only person to be struck by this idea; in the same year, 1959, two Cornell physicists, Giuseppe Cocconi and Philip Morrison, published a paper on the subject in *Nature*.[1] But Drake has greater claims to fame than priority in publication. He was the first person to put the idea into practice. Before dawn on 8 April 1960 he crawled into the canister at the dish's focus, five storeys above the ground, and installed an amplifier particularly well suited to boosting signals in the narrow range of frequencies he deemed of interest. (The amplifier was on loan from MIT, driven down specially in the passenger seat of a vintage Morgan sports car by its creator Sam Harris, an engineer and revered radio ham.) For most of an hour Drake fiddled with the tuning knobs; then, once he was safely back in the control room, the telescope was aimed at Tau Ceti, 12 light years away. Nothing was heard. When Tau Ceti set, the dish was realigned to scan Epsilon Eridani, 10.5 light years away, and picked up a strong but fleeting signal. For days the observers didn't know what, if anything, the signal had been: five days later it returned and was clearly identified as a passing aircraft, thus becoming the first of the false alarms that have dogged the science ever since.

Drake did not just pioneer the practice of attempting communication with extraterrestrial intelligence (soon to be known as CETI, and later as the search for extraterrestrial intelligence, SETI). The year after his first search he put together a magnificently durable rhetorical device with which to structure future discussions on the topic. Although Drake's search – referred to as 'Project Ozma', after the Wizard of Oz's daughter – had found nothing, along with the *Nature* paper by Cocconi and Morrison it had generated enough interest to make a small meeting on the subject an attractive idea. This meeting took place at Green Bank in late 1961, and Drake had the job of putting together the scientific programme. He decided to structure it as an inquiry into the likely number of radio-emitting civilizations in the galaxy. Counting sources of different sorts was a typical radio-astronomical approach to the sky. And an estimate of the number of sources – N – would not only set limits on the chances of radio astronomers picking up signals from alien intelligences; it would also determine the best method of looking. If N was large, then looking at individual nearby stars was a worthwhile strategy. If N was small, it would be best to sweep the whole sky.

If the meeting was to be about a number, the number would have to be calculated. Drake's starting point for the calculation was the rate at which stars reasonably similar to the Sun are created in the galaxy, R^*. Once that was settled, one would need to move on to the fraction of those stars possessing planets – a research question on which Otto Struve, Drake's boss and supporter, had done ground-breaking work. Then would come the number of the planets around a given star that would be habitable. Then there was the fraction of those habitable planets on which life originated. Then there was the fraction of those Earth-like planets on which life has originated where intelligence evolves. Then there was the fraction of those intelligences that would produce technological civilizations. Civilization was defined, in the gloriously pragmatic piece of disciplinary self-regard that made the whole idea possible, as the infrastructure that allows radio astronomy.

Each of these considerations can be expressed as a simple number. Drake decided that the fraction of stars with planets would be f_p; the average number of habitable planets around such stars would be n_e; the probabilities of life, intelligence and civilization would be f_l, f_i and f_c. Last of all there was the average lifetime of technological civilizations: L. With very little real effort, Drake found he had produced a formula with which to calculate N. Broken down simply, it is the rate at which habitable planets become available multiplied by the probability of a technological civilization on any given habitable planet multiplied by the length of time those civilizations typically keep on communicating. Written out in full – written out the way that Drake wrote it on the blackboard at the Green Bank meeting – it looks like this:

$$N = R^* \times f_p \times n_e \times f_l \times f_i \times f_c \times L$$

This simple expression became known as the Drake equation, though Drake formula would probably be more accurate. It doesn't express a law of nature; it just tells you how to calculate a number by multiplying seven numerical factors such as R^*, f_p, n_e etc. Most equations sit at the end of a creative process, encapsulating hard-won insights, generalizing them and extending their reach; the Drake equation, on the other hand, is just a starting point. It is not an analytical tool but a pedagogical one; not an equation to be used but one to be talked about. 'It didn't take any deep intellectual effort or insight on my part,' Drake later recalled. 'But . . . it expressed a big idea in a form that a scientist, even a beginner, could assimilate.'[2] Drake's distillation of the issues into a set of respectable factors had a rhetorical elegance that endured. It put the subject into a truly scientific-looking format, and in so doing seemed to make the huge unknowns involved more bearable. And it provided the illusion of a bridge between the very different types of problem posed by thinking about alien civilizations, a bridge between questions about where radio waves come from and questions about where radio astronomers come from.

Factor by factor, the Drake equation slides from questions of astronomy to questions of biology to questions of sociology. As it does so, it moves from familiar types of question to unfamiliar ones. Astronomers often set out to discover things such as the rate of star formation in a galaxy similar to our own; sociologists never consider the likelihood of a given intelligent species developing a civilization. In ignoring this difference the Drake equation exhibits the classic technocratic lapse of mistaking the ability to be able to state a question in the language of science with the ability to solve it using the practices of science. The first takes only a mind; the second takes a body of knowledge and a way of mobilizing it.

The eleven scientists at the Green Bank meeting warmed to the equation as soon as Drake wrote it up, and spent most of three days assigning values to its coefficients. It was agreed that there were about 10 billion Sun-like stars in the galaxy, and since the galaxy was about 10 billion years old, it would be fair enough to put the rate of formation down at one a year. Struve reckoned half of them might have planets. Phil Morrison thought the fraction lower, about 20 per cent. Opinions on the number of planets per system that might be habitable varied from one (on the basis that in this system, the Earth alone is habitable) to five (on the basis that Mars, Jupiter and some of the moons of the outer planets might be habitable, too). The youngest of the scientists there, Carl Sagan, argued that life's arising on a suitable planet was a certainty; life had arisen on Earth just through the workings of predictable physics and chemistry on cosmically common-place materials, and it would arise elsewhere in just the same way. The respected biologist Melvin Calvin backed Sagan up. Intelligence was also argued to be more or less certain; on the basis of his studies of dolphins, John Lilly argued that intelligence had arisen on Earth not once but twice, though, as Morrison pointed out, it was hard to imagine dolphins doing much by way of radio astronomy.

The second day's session began with a bang – or at least a pop – as news reached Green Bank that Calvin had been awarded the Nobel Prize. In a champagne toast over breakfast Struve nominated Calvin an honorary dol-phin. The honour was instantly spread around and the Order of the Dolphin was founded; the entire assembled company was inducted by acclamation, as was Elvar, John Lilly's top research dolphin and the creature that put the cetacean into CETI. When they got back to work, Calvin claimed that all intelligent creatures would eventually make use of the electromagnetic spectrum. The question of whether they would develop civilizations that could deploy this ability in a technological manner was not really looked into. The idea that some might not want to communicate was raised, as was the idea that even Trappist aliens would give themselves away because sig-nals they created for their domestic use would leak out willy-nilly. Eventually f_c was tentatively assigned a value of one tenth. Then came L, the lifetime of a technological species.

It would have been obvious to the Dolphins that the value of L had a pressing relevance in the cold-war world even if one of their number, Phil Morrison, had not been in charge of arming the Nagasaki bomb on the remote Pacific island of Tinian. All of them had doubts about the nuclear future – not to mention worries about overpopulation and pollution. A life-time for technological civilizations of less than a hundred years seemed depressingly plausible, though not unavoidable. In the end they settled on a broad range, from a thousand years to a hundred million. Taking all their estimates together ($R^* = 1$ to $10, f_p = 0.5, n_e = 1$ to $5, f_l = 1, f_i = 1, f_c = 0.1$, $L = 10^3$ to 10^8), the Dolphins put N somewhere between a thousand and a billion, a number that they felt comfortable with. Such an estimate sat easily with the astronomical assumption – ascribed to Copernicus – that the Earth should not be seen as a particularly special place. If humanity was one civilization in a million in this galaxy alone, that principle of mediocrity was clearly being met. Redefining humanity's home as a middling sort of place might be off-putting to some; but to astronomers and biologists tracing their roots to the great displacements of the Copernican and Darwinian revolutions, anything else would have been improper.

For some – notably Sagan – the clearest insight offered by the Drake equation was that SETI could be seen as a way to measure the likelihood of human survival. If SETI succeeded because N was high, it would mean that L was high too, and thus that technological civilizations were not doomed to self-destruct. Barney Oliver, a research head at Hewlett-Packard and another veteran of the Green Bank meeting, made the point unmissable in the report on 'Project Cyclops' he edited in 1971. Cyclops was envisaged as a multi-billion-dollar array of hundreds of radio telescopes devoted to SETI research. In the introduction to the project's sales pitch Oliver chose his estimates of the number of planets, the rate of star formation, the origin of life and so on in such a way that they all cancelled out. In the Cyclops report, the Drake equation is simplified to $N = L$. And L 'turns out to be the most uncertain factor of all!'[3]

This argument was not strong enough to win funding for Cyclops. (The only thing that could have got such a huge undertaking funded, as the astute commentator on the politics of space John Pike has remarked, would have been a convincing rationale for distributing the dishes so that each congressional district got one.) But L was still capable of having a political impact. In 1978 Senator William Proxmire gave NASA's proposal for a (much smaller) SETI radio-astronomy programme one of his 'Golden Fleece' awards for wasting taxpayers' money, and resolved to have all its funding withdrawn. Sagan, by then famous, went to see the senator, a man he knew to be worried about nuclear arms, and took him through the Drake equation term by term, laying particular emphasis on L. According to Sagan's wife Ann Druyan, Proxmire's attitude shifted from 'I'm going to hear this egghead out and then get him out of my office' to 'being slack-jawed with fascination and amazement . . . And he admitted he was wrong, very gracefully.'[4]

Thought-provoking as they may have been, these arguments based on L had their flaws. For one thing, SETI programmes can't really provide a value for L because they don't really try to measure N. A single contact was all that was needed: no one was planning on a census. Leaving that aside, the argument that L has some relevance to humanity implicitly assumes that the lifespans of civilizations are evenly distributed. But what if the distribution is very uneven? This was in fact what the Green Bank attendees tended to suspect, and why their estimate for L covered such a wide range. Sebastian von Hoerner, one of the Green Bank radio astronomers, pointed out that if even a few civilizations lasted for billions of years, then L, the average lifespan, would have a high value even if most civilizations burned out like febrile nuclear mayflies. If 1 per cent of technological civilizations last a billion years, and the other 99 per cent last only a century, then L comes out at an impressive sounding 10 million years even though the overwhelming probability for any given civilization at the nuclear threshold is a speedy demise. It seems a fair bet that Sagan did not go into these finer points with Senator Proxmire.

SETI's relevance to the end of the world was not limited to questionable calculations about its imminent likelihood. The first popular book to deal with the Green Bank meeting and its ideas, *We Are Not Alone* by Walter Sullivan of the *New York Times*,[5] is dedicated to 'those everywhere who seek to make "*L*" a large number'. And that was part of what the Dolphins hoped to do, in various ways. The most obvious of these was by establishing contact. If L was more than a few decades, then most technological civilizations would be older than the Earth's. So if contact were ever made, it would almost certainly be with an older and wiser civilization, wiser both in technology and in morality. Wiser in technology simply because it had had more time to develop things; wiser in morals because it would have survived the dangers of the nuclear age by somehow eradicating the causes of conflict. So while the cosmology of the expanding universe made telescopes into time machines capable of looking at the past, SETI made them instruments for observing a hopeful future. As A.G.W. Cameron, an early supporter of the idea, put it, 'If we now take the next step and communicate with some of these societies, then we can expect to obtain an enormous enrichment of all phases of our sciences and arts. Perhaps we shall also receive valuable lessons in the technique of world government.'[6] Even if the aliens did not say anything of practical use, their very presence, it was argued, would make the peoples of the Earth come together. The aliens would not be a threat to unite against – early SETI discussions were unanimous on this, on the basis that no bellicose civilization could survive long after developing nuclear technologies – but in their majestic and profound otherness they would make us give up our petty squabbling.

The writer Ed Regis was later to point out that SETI proponents –

notably Sagan – deployed arguments like this last one in haphazard and seemingly contradictory ways. On the one hand they said that finding life out there would make us forget our differences and thus less likely to exterminate ourselves. On the other hand they said that not finding life out there would make us value ourselves more deeply and for that reason renounce warfare. As Regis was quick to see, neither argument has the ring of practical truth – are we to imagine LBJ, his finger pulling back from the button, explaining to his aides, 'If there were other intelligent species out there, I'd press this thing, but as we're alone in the universe, I won't'? The fact that Sagan saw the conclusion that humanity should eschew weapons of mass destruction as following from either a successful SETI search or an unsuccessful one certainly suggests that it was the conclusion that he started out with (as indeed it was).

The link between SETI and salvation was deeper than dodgy rhetoric, though. For Sagan SETI was not just a way of measuring the likelihood of survival. It was a way of promoting the sort of values that might increase that likelihood: values of scientific rationality and international cooperation. The sort of civilization that weathered the nuclear crisis would be the sort of civilization that pulled together and looked at the big picture. It would be one that understood its place in the cosmos, one that looked far into space and far into the future. SETI, conceived as a global undertaking in which Americans and Russians could unite, was a way not just of detecting such civilizations but of trying to create one.

In this, the SETI of the 1960s bears a striking resemblance to the world of *Star Trek*. Unlike most written science fiction, *Star Trek* had a strong streak of utopianism. By conflating Kennedy's 'New Frontiers' rhetoric with the 'Endless Frontier' of Vannevar Bush's iconic 1946 report on science and government, *Star Trek*'s 'final frontier' provided an uplifting scientific future. As Constance Penley writes in her study of the show's influence, *NASA/TREK*, the programme was instrumental in the process by which '"Going into space" [became] the prime metaphor through which we try to make sense of the world of science and technology and imagine a place for ourselves in it.'[7] In their worries about nuclear warfare, overpopulation, depletion of resources, toxic pollution and so on, the SETI scientists were very much trying to make sense of the world of science and technology, and 'going into space' – with the astronomers' gaze, if not with the body – was the way they did it.

Like the Order of the Dolphin, the crew of the *Enterprise* was at the same time both 'seeking new civilizations' and embodying one. The starship's bridge held a Russian and an African as well as Americans, not to mention a Vulcan. 'This approach,' said the creator of the series, Gene Roddenberry, 'expresses the "message" of the series: we must learn to live together or most certainly we will soon all die together.'[8]

The role of Russians in SETI was rather more substantial than that of Ensign Chekov on the *Enterprise* (which was basically to attract teenage girls who had soft spots for mop-tops like the Monkees' David Jones). During the 1960s, Soviet scientists too became interested in listening for voices from beyond. While in America SETI was basically just a discussion – the only actual search undertaken was Drake's – the Russians repeatedly turned their antennas to the sky (though the antennas were pretty poor things). Even more explicitly than in America, space-faring civilizations were seen in Russia as part of a necessary historical progression – a rather more theorized one in which the Soviet Union was leading the peoples of the Earth towards the future with its Sputniks. According to the influential theories of Nicolai Kardashev, the Earth was in the process of becoming a 'type I' civilization, in which one species has come to control the energy resources of a whole planet. Far in the future lay the possibility of becoming a 'type II' civilization with control over the entire energy output of a star, or a 'type III' civilization with control over a whole galaxy. Though type I civilizations might be the most common, it was type II and type III civilizations that would be most noticeable; they could be thousands of times scarcer yet millions of times brighter. Freeman Dyson, a physicist and mathematician at Princeton, had also speculated about civilizations using most of the energy from a star, and had suggested that their waste heat might be detectable by satellites sensitive in the infrared. Star-encompassing type II technologies became known as 'Dyson spheres', though Dyson later wrote that the idea actually came from Olaf Stapledon's novel *Star Maker*, where in the far future 'every solar system [was] surrounded by a gauze of light traps, which focused the escaping solar energy for intelligent use, so that the whole galaxy was dimmed'.[9] Stapledon foreshadowed SETI's spiritual hopes as well as its technological speculations. The motivation behind seeking out the eponymous intelligence in *Star Maker* 'was not merely scientific observation, but also the need to effect some kind of mental and spiritual traffic with other worlds, for mutual enrichment and community'.

Seized with the need to make SETI truly global and the Earth's emerging type I civilization stable, Sagan devoted some of his boundless energies to arranging a follow-up to the Green Bank conference in the East. Together with Kardashev he put together a meeting at the Byurakan observatory in Armenia in 1971. The American contingent consisted both of Green Bank veterans and new talent, including Francis Crick, Thomas Gold, William McNeil and Marvin Minsky (who brought some Frisbees, never before seen in Armenia and a great hit). They were joined by about thirty Soviet scientists, including Kardashev's colleague Iosif Shklovskii, who on the occasion of the fifth anniversary of the first Sputnik launch had written a book called *Vselennaia, Zhizn, Razum* (Universe, Life, Mind). Sagan had arranged for its translation into English and added a large amount of his own material to it at the same time. The resulting book, *Intelligent Life in the Universe*, which gave pride of place to the Drake equation, was at the time the only full-length treatment of the topic by scientists available, and, in the historian Steven J. Dick's phrase, 'the bible of the SETI movement'.[10]

In Byurakan the Drake equation again formed the basis of the discussions. Crick was not as easily convinced as Green Bank's astronomers had been of Sagan's claim that life was inevitable. The later parts of the equation were, as usual, hard to discuss. David Hubel, a neuroscientist, said that even before he had been plied with Armenian cognac over breakfast he had had no idea why some creatures evolved intelligence while others did perfectly well without it. The anthropologist and ethnographer Richard Lee said that language was crucial for technological civilization but not sufficient, as the culturally rich but gadgetry-poor lives of Bushmen showed (he also gave a toast at the conference dinner in the language of the !Kung). Despite these drawbacks, a figure for N of a million was reached, just as it had been in Sagan and Shklovskii's *Intelligent Life in the Universe*.

While the Drake equation served as a starting point for the week-long meeting, the participants moved beyond it in discussions to which the term 'wide-ranging' scarcely does justice. An extraordinary selection of the fancies and fears of the free-ranging scientific mind were on offer: vast machines burrowing 60 kilometres under the Earth's surface; life on neutron stars; artificial intelligences; universes trapped inside elementary particles; new laws of physics; the effect of sunspots on creativity; tachyons; the 'genetic self-elimination of reason' (that is, the unlimited reproduction of the stupid); the coming golden age; nanotechnologies; black holes; white holes; antimatter rockets. The valiant stenotypist Floy Swanson took it all down for posterity.[11]

Throughout the first decade of SETI studies, radio astronomers managed to keep a fairly good grip on the terms of the debate they had started. It's not that there weren't further-out ideas; Byurakan was full of them. But as Phil Morrison argued there, adding untestable possibilities to the debate was not particularly helpful. SETI had started as a radio-astronomy project because Morrison, Cocconi and Drake had seen that radio astronomy provided the only way that a civilization such as ours could be detected by another civilization such as ours. Aliens that were not bright radio sources – because they were intelligent clouds of gas, or sybarites in a post-technological golden age, or creatures of nuclear matter crawling around on the surfaces of neutron stars – just didn't cut it. The technicalities of SETI were thus technicalities of radio astronomy, such as the choice of appropriate frequencies (frequencies that would seem 'natural' for radio astronomers of all species at all times) and signalling strategies. The difference between type I and type II civilizations, for example, was basically a difference in the power of their radio transmitters, and a difference in the strategies that would be suitable for detecting them. (From around the time of the Byurakan conference Drake himself became increasingly interested in type IIs – supercivilizations – and in 1974 he and Sagan mounted a search for them. Instead of the 85-foot telescope at Green Bank that Drake's first search had used, they used the newly refurbished Arecibo radio telescope, a vast bowl in the hills of Costa Rica. Instead of looking at individual stars, they looked at whole galaxies, simultaneously sampling the emissions of billions of far-off stars on the basis that what the supercivilizations lacked in numbers they would make up for with the sheer power of their beacons.)

The fact that the radio astronomers were the only people who could actually do anything more than talk about civilizations beyond the solar system did not mean that everyone paying attention deferred to them. Many biologists felt both that the origin of life was less of a sure-fire thing than Sagan claimed, and that, given life, the likelihood of people like us being created by evolution was vanishingly small. The most influential version of this argument was 'The nonprevalence of humanoids' by the eminent evolutionary theorist George Gaylord Simpson.[12] Evolution, said Simpson, was contingent, not determinist. It had no interest in producing people, and had done so only by chance. It would not produce people again, because it never produced anything twice. And if it would not produce people again on Earth, why would it do so elsewhere? Sagan's response was that though the specific evolutionary history of humans might be unlikely, there were many possible evolutionary histories leading to creatures as smart as humans. Even though each of those histories might be highly unlikely in itself, if they were numerous enough, the chances of one of them being followed would be quite high.

Answering the question of how high those chances were – of how many evolutionary histories could be expected to converge on intelligence – more or less came down to a matter of taste. Some evolutionists, like Stephen Jay Gould, felt that while evolution certainly does not repeat itself in its details, intelligence might be one of the things that evolution could be expected to invent again and again, like the eye or the wing.[13] If so, it would be reasonable to think it might evolve on other planets as well, and thus that SETI might be worthwhile. Other evolutionary scholars, such as Jared Diamond, thought that while human-style intelligence of the sort that makes radio transmitters clearly has some use, that did not mean that it would evolve repeatedly. Woodpecking, Diamond pointed out, is a very successful ecological strategy, but also one that is hard to evolve. Only one clade of creatures in the Earth's history – the woodpeckers – has ever made use of it. If woodpeckers had not come along, there is no reason to believe that their niche would ever have been filled; in places the birds never reached, it never was. The sort of intelligence SETI was looking for, Diamond argued, was much more like woodpecking than it was like seeing or flying.[14]

Everyone from Simpson onwards realized that though one might have an opinion on this subject, there was no real way of estimating the frequency of intelligence. But that uncertainty in itself makes the Drake equation problematic. If you are being rigorous, multiplying together factors like f_i and f_c means multiplying together the uncertainties in your estimates of those factors. The uncertainties in the Green Bank estimates for R^* (between 1 and 10) and for L (between 1,000 and 10,000,000) meant there was a great deal of uncertainty in the final answer, which was put as somewhere between a thousand and a billion. And as Francis Crick pointed out in Byurakan, the uncertainties in the estimate of things like the probability of life originating were huge. It might be a certainty; it might be a trillion-to-one long shot. There was no reliable way of knowing. And that uncertainty should properly be carried through to the final estimate of N. The mathematician Alfred Adler fell on these huge uncertainties with glee in an acerbic *ad hominem* review of the conference proceedings for the *Atlantic*.[15] Adler quotes Crick ('It is not possible . . . to make any reasonable estimate whatsoever of the factor f_l'), L. M. Mukhin ('I do not quite understand how we can estimate f_i') and Sagan ('We are faced . . . with very difficult problems of extrapolating . . . in the case of L, from no examples at all') and concludes that the 'purpose of the conference, a determination of estimates for the number N, is quite clearly a total fraud'.

'It is almost incredible,' Adler went on, 'that the truly distinguished scientists among the conferees (and there were indeed several of these) could be willing, almost eager, participants in a travesty of all that is taken seriously by men and women who love and value science and intellect.' Adler blames this willingness on the uncouth influence of the intellectual fashions of the young, personified by Sagan.

> The modern technologist is a gifted, highly trained, opportunistic, humourless and unimaginative ass . . . He charges through subtleties and profundities where wise men hesitate to walk on tiptoe; he usurps domains about which he knows nothing . . . Rational, civilized man appears to have become very tired, no longer able to withstand the onslaughts of the manic young masters who promote large grants of money and influence, leave him breathless at conferences, lavish the currency of vague new ideas upon all those around them and exhaust their weakened elders with pure, assured, unself-conscious power.

While Adler was beyond compare in his spleen, resentment against the arrogance and the funding of technocratic space scientists was a theme in other critiques of SETI and related projects in the nascent field of exobiology, such as Simpson's.

Adler was right that the presence of massive uncertainties means that, in any strict sense, the Drake equation is more or less meaningless. But to criticize it on those grounds was to mistake its underlying purpose. The Drake equation was a way of saying that the universe produces civilizations just as it produces planets and stars, and that the tools that detect these other aspects of the universe might, as a matter of fact, also detect civilizations. Everyone concerned was aware of the equation's limitations, and some were impatient with the sociological pretensions it made room for: 'To hell with philosophy – I came here to learn about observations and instruments,' said Dyson at the Byurakan meeting. But they were also aware that it stood not for an answer but for a type of approach. To break its hold it was not enough to show that it could never provide answers. You would have to show that the approach it stood for would never find answers. An argument that the universe had not produced civilizations for us to contact would count as a far more powerful argument against SETI than any critique of methodology could be.

That apparently powerful argument appeared in the mid-1970s, when the astronomers' view of the universe was challenged by the ideas of space flight. One of the basic assumptions of SETI was that civilizations stayed in the stellar systems where they had evolved, and thus would behave like point sources spread across the sky, twinkling on and off over the millions of years as the L tolled for each of them. But what if civilizations moved? If aliens could travel between the stars, even doing so quite slowly, the time it would take them to visit all the solar systems in the galaxy would be only 100 million years or so. If that was the case, then, given that the galaxy is a good 10 billion years old, one might expect them to have already reached us. This insight is normally traced back to a question asked by Enrico Fermi, the great experimental physicist and builder of the first nuclear reactor, over lunch at Los Alamos one day in 1950. According to Edward Teller, Fermi suddenly 'came out with the quite unexpected question "where is everybody". The result of his question was general laughter because of the strange fact that in spite of Fermi's question coming from the clear blue, everybody round the table seemed to understand at once that he was talking about extraterrestrial life.' (In fact there had been a preprandial discussion of extraterrestrials occasioned by a *New Yorker* cartoon suggesting that the city's dearth of rubbish bins was due to the fact that flying saucers were taking them away; Fermi affected to find it a good theory because it explained both the presence of the flying saucers and the absence of the rubbish bins.)[16]

Among the founders of SETI this 'Fermi paradox' was taken as evidence that interstellar space travel was a practical impossibility. In a study of 'the limits of space travel' published the year after the Green Bank meeting, Sebastian von Hoerner wrote, 'I personally draw this conclusion: space travel, even in the most distant future, will be confined to our own planetary system, and a similar conclusion will hold for any other civilization, no matter how advanced it may be. The only means of communication between different civilizations thus seems to be electromagnetic signals.'[17] Of the original Dolphins, only the Stanford professor Ron Bracewell saw direct contact as a useful way for old civilizations to contact young ones, imagining probes sent out by advanced civilizations to nearby star systems to look for intelligent life and, if it was found, to converse with it. Countering the arguments that launching such probes would be prohibitively expensive, Bracewell stressed that the probes would be small, 'something about the size of a football . . . with intelligence able to emulate a human packed into something the size of a human head',[18] and that a single probe would be able to visit a number of different stars if it lasted long enough. Although as reasonable as many of the other speculations on offer, Bracewell's idea was too far from the basic 'civilizations are like stars' approach to become part of SETI's shared world view.

Bracewell's idea had another problem, too: it brought the aliens close to home. Even the most imaginative seem to have shied away from the thought of alien civilizations encroaching on the solar system. When, during the Byurakan meeting, Dyson expounded his view that the largest and most appealing habitat available in the galaxy was not on the surfaces of planets but on the surfaces of comets, and that life might slowly spread from cometary cloud to cometary cloud without ever coming near any hot disturbing stars, Thomas Gold immediately asked him if he thought life from elsewhere could have already spread to the cloud of comets around our own sun. Even a thinker as supple as Dyson had to admit that he had not considered the possibility.

Bracewell continued to wander from SETI's straight and narrow, speculating in a Fermi-like way that the first space-faring civilization in the galaxy would occupy the whole thing. But the idea was 'hidden on one or two pages [of a book he wrote in 1974] because the editor felt that it was better to plug the idea of widespread intelligence than to plug uniqueness and colonization'.[19] The next year, though, a similar version of Fermi's argument was made by the physicist Michael Hart.[20] Hart started from what he called 'Fact A': 'There are no intelligent beings from outer space on Earth now.'

Hart imagined four classes of explanation for this fact. The first was physical – interstellar travel was too hard. This he disputed on the grounds that long-lived aliens, or hibernating aliens, or robots, or self-contained city-ships would not find trip times of centuries all that off-putting, and so could make the journey at realistically achievable speeds well below the speed of light. The second class of explanation was 'sociological': that the aliens would not care about exploring, that their civilizations would necessarily stagnate, that they would all self-destruct, that they would choose not to disrupt the development of primitive societies such as ours. Hart argued that to explain a lack of extraterrestrials on Earth, such explanations would have to apply to all alien civilizations; if even one was dynamically expansionist, in the way we might imagine ourselves to be, it would be here by now. The temporal explanations – that there had not been time for them to get here – he dismissed on the basis that the time taken to colonize a galaxy is trivial compared to the age of the galaxy or the time-spans of evolution. The fourth class of explanations – that they *had* visited – fell to similar arguments. If they had only just visited, why had they stayed away so long? If they had visited long ago, why were they not still here? Hart concluded that 'the idea that thousands of advanced civilizations are scattered throughout the Galaxy is quite implausible in the light of Fact A . . . Our descendants might eventually encounter a few advanced civilizations . . . but their number should be small, and could well be zero.'

A few years later, the physicist Frank Tipler put forth similar arguments in a less nuanced but peculiarly striking form.[21] Tipler argued that advanced civilizations would be likely to build 'von Neumann machines' – self-reproducing robots, in this case also equipped for interstellar flight – and send them out into the galaxy. As soon as this happened, evidence of the machines would be unmistakable and everywhere, since they would spread to all solar systems in just a few tens of millions of years. No von Neumann machines are visible; no other civilizations exist. Tipler took this argument seriously enough to campaign actively against any more money being 'wasted' on SETI, since its basic assumptions had been demonstrated to be false. Hart, on the other hand, integrated his criticism into the language of the SETI community, taking 'Fact A' as proof that N is very small and re-examining the Drake equation on that basis. He pursued research on the likelihood of habitable planets that seemed to back this up by suggesting they were rare. At a symposium devoted to SETI held during the 1979 meeting of the International Astronomical Union in Montreal,[22] Hart presented a range of values for factors in the Drake equation and argued that making merely moderate or slightly pessimistic assumptions quickly brought N down to one or less.[23]

N had never really been argued about before; it was always somewhere between 10,000 and 10 million, and no one ever took a strong line favouring one particular figure. As long as it was a number similar in size to the number of entries in a typical astronomical catalogue counting galaxies or nebulae or radio sources or what have you, it was fulfilling its role of making SETI look like normal astronomy. The power of the challenge laid down to SETI by Tipler and Hart is reflected in the fact that the 1979 International Astronomical Union symposium on the topic featured six papers advocating widely different ranges of value for N, and they are given pride of place. The 'Where is everybody?' arguments struck at the core of SETI in a way evolutionary arguments never had, and they gained converts. Sagan's co-author Shklovskii came to agree with them. So did von Hoerner, boiling the issue down to the pithy observation that humans would doubtless wish to explore and colonize, and so, 'if we were typical, we should not exist'. SETI had been predicated on the idea that there were other radio users more or less like us all over the sky – that you could point to the stars and say, 'Man is a thing like that.' If your human archetype was an explorer rather than a radio astronomer, the Fermi paradox hit home.

In the 1970s, space flight seemed real in a way it had not before, and thus so did the idea of humanity as a species of galactic explorers. As Hart wrote, 'After the success of *Apollo 11* it seems strange to hear people claim that space travel is impossible.' NASA had already launched four interstellar spacecraft by the time Tipler made his contributions, in that *Pioneer 10*, *Pioneer 11*, *Voyager 1* and *Voyager 2* were all on course to leave the solar system. Another explanation for why the Fermi paradox became a factor in revising estimates of N downward at this time is that exponential population growth had become a major concern. The nuclear bomb had been joined by the 'population bomb'; the Club of Rome had estimated that the world was fast running out of nonrenewable resources. To the sort of people who both thought about these things and took SETI seriously, expansion into space looked like a necessary next step – and thus a necessary step for any other advanced civilization, a step that would in time lead to scarcity-induced expansion across the galaxy. Finally, a low N brought SETI in line with the mood of the rest of the emergent field known as exobiology. SETI's first decade, the 1960s, had coincided with the first plans to search for life on other planets; Drake was at one of the first meetings of this new exobiological community, and Sagan was to be its most prominent champion. By the late 1970s, though, exobiology was in the doldrums. The *Viking* missions to Mars, their landers geared primarily towards the search for life, had found nothing of clear exobiological interest. Life in the solar system seemed to be confined to Earth, an impression strengthened by the images of the home planet taken by Apollo astronauts, icons of fragile life in the midst of lifelessness. In a universe that seemed locally sterile, cosmic loneliness was credible, if not palatable.

The high-N camp had various counter-arguments, but none was utterly compelling. Hart and his colleagues might be underestimating the sheer difficulty of interstellar travel. Sagan, borrowing the mathematics that was used to explain the spread of muskrats, pointed out that the spread of civilizations through the galaxy might be far slower than the speed of their starships, because the time it would take to fill up the new worlds acquired would be far greater than the time it would take to get to the next solar system. Advanced civilizations might find their interactions with each other considerably more interesting – or even destructive – than any other activities, in a way that would check expansion. No one would build von Neumann machines because they might get out of control and eat up the whole galaxy.[24]

Potentially the most powerful counter-argument, though, was mostly eschewed. This was that Hart's 'fact A' was not a fact; that the absence of extraterrestrials did not need explaining because they were not absent. Hart equated believing in the presence of extraterrestrials with believing that UFO sightings were evidence for that presence, and dismissed the idea with the somewhat uncompelling observation that 'Since very few astronomers believe the UFO Hypothesis it seems unnecessary to discuss my own reasons for rejecting it.' But believing that there are unrecognized aliens in the solar system, or even on Earth, seems at first glance no more far-fetched than many of the other speculations accepted as part of the SETI debate. Some people took the notion seriously enough to suggest searching for industrial waste in the asteroid belt. But the idea of a presence closer than the astronomical realm was one that few could countenance.

Part of the reason for this was undoubtedly that it was very hard in practice to distinguish such a belief from UFO theories, and these were deemed unscientific in a way that other SETI ideas were not. Another reason for not accepting extraterrestrials nearby may have been the mythic power of the astronomer's gaze. If there were alien spacecraft nearby, we would surely see them through our telescopes. In fact this is not necessarily the case. There is an enormous amount of our solar system that we have never looked at. There are millions of unclassified asteroids that are large enough to be interstellar spacecraft. But one assumes they are not, or they would have been seen. Accustomed to the astronomer's gaze, one assumes that the visible universe is seen, even though most of it has never been inspected. Working by analogy with everyday life, one assumes that the astronomical foreground is more easily seen than the background, even though this is not the case. Just as distant type II civilizations would outshine nearby type Is, so almost everything we see in the sky, though much farther away, is brighter than a nearby chunk of rock – or a nearby spaceship.

There is also a more profound objection to the idea that 'fact A' is false. In the presence of undetectable intelligences, science itself breaks down: one can no longer reliably establish what is natural. SETI's subject was always a curious blend of the natural and the artificial. Alien civilizations were treated in many ways as natural objects (their likely distribution definable through a scientific calculation such as the Drake equation) but at the same time seen as artificial. The skill of SETI observers lay in distinguishing the alien signal from both the truly natural and the terrestrially artificial. This was a conceptually difficult task since, as Marvin Minsky pointed out in Byurakan, Shannon's law states that efficiently coded communications are indistinguishable from random noise if the coding scheme is not known.

The artificial could thus look natural; it would look unnatural only if the aliens wished to make it look so – if they designed their emissions to be a beacon. But nature could look like a beacon, too. When pulsars were discovered, it took time for it to become clear that nature could be responsible for these staccato emissions; the Cambridge team that discovered them jokingly referred to the signals' sources as LGMs, for little green men. In 1965 Kardashev claimed that radio emissions from quasar CTA-102 were artificial, a claim quickly retracted when optical evidence linked them to fluctuations in the quasar's brightness. And every SETI search from Ozma on has picked up signals of human origin, from aircraft location beacons to spy satellites.

The distinctions between the natural, the artificial and the oddly in-between world of intelligent aliens were troublesome but manageable in the highly specialized field of SETI radio astronomy, which dealt only with objects too far away to have any causal effects on the world. If they were to come nearer to home, though, they would be worrying. Scientists rely on the fact that they can identify what is natural. Aliens among us throw that reliance into doubt. A world where hidden aliens lurked would become unnatural; it would develop an agenda; it would deceive. In such a world the most basic scientific assumption – the idea that one can arrange for 'all other things to be equal' – could no longer be made. Shortly after the first Green Bank meeting, and shortly before his untimely death, Otto Struve wrote, 'There can be little doubt today that the free will of human beings is not something that exists only on the earth. We must adjust our thinking to this recognition.' But such an adjustment is not, in practice, possible in the natural sciences; not unless the aliens are willing to enter our social world through communication.

Alien free will made its bearers unacceptable on Earth and inscrutable beyond. Post-Hart arguments about the value of N were beset by a new need to explain the aliens' strategies; to make assumptions about their tactics, their intentions, their fears and aversions and requirements. Such behavioural assumptions proved far harder to reconcile with the logic of the natural sciences than the relatively minor assumptions about wavelength choice and beacon strategies that SETI had always had to make (though in truth the difference was one of magnitude rather than of kind). From being a natural science without data, SETI became something yet harder to carry off with conviction: a social science without communication. For many this was hard to take; if it was not natural science, it was uninteresting, as

Dyson explained when replying to a plea for support from Sagan. 'I wonder why you take Tipler so seriously. I think his arguments get more attention than they deserve. I cannot take seriously any of his numbers, nor yours either. Any specific model of the future has to be absurdly narrow and unimaginative.'[25]

Despite the problems with N, SETI searches continued into the 1980s and 1990s. As Stephen Jay Gould, one of many eminent speakers on science that Sagan asked to sign his 'SETI petition' in 1982, put it, 'I am selfish enough to want to see some exobiological results . . . in my lifetime [and] SETI is all we have for now.' More recently, though, new approaches and discoveries have made non-SETI exobiology a much more promising topic again. As part of this process, the field has been reconceptualized by its primary sponsor, NASA, as 'astrobiology': the primary difference is that astrobiology seeks to include aspects of the study of life on Earth in the same framework as the study of life elsewhere. This is a logical step that also lays to rest the old criticism that exobiology is a science with no subject matter. Within the solar system, possible new subject matter includes fossilized – as opposed to extant – life on Mars, interest in which has been growing since the mid-1980s, long before the much-debated evidence for ancient bacteria in the Martian meteorite ALH 84001. There is great interest in the possibility of life in the ice-covered ocean of Europa, one of Jupiter's moons. There has been much excitement over the discovery of planetary systems around other stars, and a telescope system for detecting Earth-like planets and looking for chemical evidence of life in their atmospheres is now the centrepiece of NASA's long-range astronomical programme. (Interestingly, such systems are derived from yet another suggestion of Ron Bracewell's; the technique that they might use actually to detect life on those far-off planets dates back to a suggestion made by James Lovelock in the 1960s, when he was sharing an office with Sagan at NASA's Jet Propulsion Laboratory.)

As far as NASA is concerned, astrobiology is a leading candidate to be the revolutionary new science of the twenty-first century. But astrobiology, unlike exobiology, has no place for SETI. This is in part because of a political battle that the agency lost in 1993, when the US Senate removed all funding from the ambitious SETI programs which NASA had finally got round to undertaking. But if SETI is no longer politically feasible, it may also no longer be intellectually necessary. If life on Earth-like planets can be detected directly, through Martian palaeontology or Europan oceanography or extra-solar infrared spectroscopy, there is no longer quite such a profound need for it to talk to us. Astrobiology can carry on happily in the space defined by the first four terms of the Drake equation, concerning itself with the frequency of habitable planets and the likelihood of life. The later terms can, for the moment, be left in abeyance, along with all of SETI's worrying dependence on the unnatural. If this new astrobiological research also provides strong arguments for a small N – a case made in Peter Ward and Donald Brownlee's *Rare Earth* (Copernicus, 2000) – then all the better, at least from the point of view of disciplinary coherence. An astrobiologically certified low N would make SETI just a historically interesting blind alley.

SETI, however, continues; indeed it grows. Though there is no centralized funding, philanthropy and discretionary spending by universities have allowed a range of searches, including a large part of the axed NASA programme, funded privately through the SETI Institute in Mountain View, California. On Project Ozma's fortieth anniversary, Drake – now chairman of the SETI Institute's board of trustees – was happy to be able to point out that thanks to improvements in radio technology and in signals processing, which allow vast numbers of frequencies to be scanned simultaneously, today's search technology is 100 trillion times more powerful than the original equipment he had at Green Bank. SETI capabilities have grown even faster than computer power, doubling every ten months or so. The astronomer's gaze grows ever more powerful as the long watch drags on.

But perhaps it is not really a watch at all. Radio astronomy, simply through the connotations of the word 'radio', has always been torn between the opposed metaphorical modalities of vision and hearing. When radio astronomy produces images – jets from quasars, discs around black holes – it is a form of the astronomer's gaze. But SETI produces no images. And so SETI has always had a particular affinity with listening, not seeing, especially in its popular presentations – and since SETI exists now as a popular pursuit, funded by subscription, facilitated by thousands of volunteers sharing out data-processing tasks on their home computers, it is worth taking that popular presentation into account. One of the first realistic SETI novels was James Gunn's *The Listeners*. In the most successful mass-media representation of SETI research – the 1997 film of Carl Sagan's novel *Contact* – the audience is immediately struck by the fact that one of the researchers is blind (as, indeed, is one of the members of staff at the SETI Institute, Kent Cullers). The film's most powerful epiphany is not visual but aural, a visceral throb that saturates the senses. The protagonist is presented as being too dazzled by her intellect to hear her heart; her final vindication, though, is noise. Unlike the novel on which it is based, the film looks favourably on that famously blind phenomenon, faith; this has upset some of Sagan's more scientistic supporters, but it is hard to see it as wholly unfair comment on SETI as it is understood by some of its supporters.

The astronomer's gaze is powerful because vision is the sensory metaphor that defines objective knowledge: 'Seeing is believing'. Hearing, on the other hand, is a primary metaphor for comprehension. The experience of seeing is necessarily embedded in a spatial model of the world, with distance between the beheld and the beholder always already there. Hearing is direct, immediate; we feel it in our ears. Vision objectifies the world; hearing opens the door to language and feeling. To listen to someone and to look for them are very different things, for no one can be invisible, and yet everyone, to be heard, must first speak. To listen is to be passive, to look, active. In all these ways, SETI seems more auditory, less visual. It needs no numbers or catalogues; it needs no equations or calculations. In the end it is not a study of the universe, but a communication with something we cannot know until it speaks.

When we pray, we close our eyes.

Equations of Life

The Mathematics of Evolution

John Maynard Smith

It is often supposed that biologists can do without mathematics. After all, there is not a line of algebra in Darwin's *Origin of Species*. But this impression is false, as soon becomes apparent if we follow the subsequent history of evolutionary biology. The theory of evolution by natural selection works only if offspring resemble their parents, but Darwin himself did not understand the processes of heredity that bring this about. When, following the rediscovery of Gregor Mendel's genetic laws in 1900, a science of genetics began to be developed, students of evolution soon split into two camps. On one side the 'biometricians', a group of statisticians led by the combative Karl Pearson, argued that what was important in evolution was the natural selection of continuously varying characters such as size or bodily proportions, and that genes were an irrelevant figment of the Mendelians' imagination. On the other side, many practical animal and plant breeders accepted Mendel's theory of discrete hereditary factors, or genes, and went further, arguing that new species arose by genetic mutations of large effect, and that natural selection was irrelevant.

With hindsight the argument seems absurd. But it needed mathematical studies – primarily by the founding fathers of population genetics, the Britons R. A. Fisher and J. B. S. Haldane, and the American Sewall Wright – to show how the two views can be reconciled. The observations of the biometricians concerning continuously varying characters, the correlations between relatives, and the effects of selection, can be explained as the effects of many genes, each of small effect.

In this essay, I start by describing a very simple equation that predicts the rate of evolution when selection is not operating, and explains how it has been used to date past evolutionary events. But my main theme concerns those many cases in which it is hard to predict the effects of natural selection because there is no one 'best' thing for an animal or plant to do: instead, the best thing to do depends on what other members of the population are doing. I shall describe how a new branch of mathematics, 'evolutionary game theory', has been used to solve such problems.

I first discuss a quantitative argument relating to the process of genetic mutation and evolution. The founding fathers had shown that it is selection and not mutation that determines the direction of evolution. In the 1950s, it was discovered that a gene consists of a string of four kinds of molecule, the bases A, C, G and T. Fortunately, you need not know the chemical formulae of these bases – I don't know them myself. What matters is that the precise sequence of the bases in a gene determines the sequence of amino acids in the protein that is made, and the proteins in turn determine the kind of organism that develops.

Soon, data became available on evolutionary changes in the amino acid sequence of proteins (and later in the sequence of the bases A, C, G and T that code for the amino acid sequence). Looking at these data, the Japanese geneticist Motoo Kimura came up with a novel, and to some people shocking, idea. Although he accepted that adaptive evolution is the consequence of natural selection acting on effectively random mutation, he thought that many – indeed most – of the evolutionary changes in amino acids were non-adaptive, or 'neutral'. That is, the evolutionary substitution throughout a population of one amino acid by another – say leucine by threonine – occurs not because selection favours threonine, but by sheer chance. Thus suppose that in a population, some proteins have leucine at a particular site, and others have threonine at the same site. Then, because the genes present in the population in one generation are a random sample of those present in the previous generation, the proportions of threonine and leucine will gradually change, and ultimately one of them will be lost, and the other fixed as the only one in the population.

Kimura pointed out that his assumption of neutrality has an interesting consequence for the rate of evolution. To follow his argument, we must first understand what is meant by the 'mutation rate'. Suppose that you inherit a particular gene from one of your parents – your mother, say. The mutation rate is simply the probability that when you pass that gene on to a child, it will have undergone a genetic change, or mutation. Usually, much the commonest kind of change is the alteration of a single base (A, C, G or T) to another: usually but not always this alters an amino acid. Kimura argued as follows. Suppose that the 'neutral mutation rate' of the gene (the probability, in one generation, of a new mutation arising that has no effect on the chances of survival) is m. Most such neutral mutations will be lost by chance in a few generations. Very occasionally, after many generations, the new mutation will be 'fixed' in the population: that is, every gene in the population will be a direct descendant of the original mutant gene, and an evolutionary change will have occurred.

What is the chance of this happening? Obviously, it depends on the size of the population: the chance will be greater in a small population. Thus imagine a population of 1,000 rats. There are two sets of chromosomes in each rat, so there are 2,000 copies of each gene in the population. Now suppose that the chance of a new neutral mutation is 1/100 (of course this is far too high). Then there are 2,000/100, or 20, new neutral mutations each generation. Each new neutral mutation has exactly the same chance of being ultimately fixed as all the others – after all, that is what 'neutral' means. Because there are 2,000 genes, each new mutant has a 1/2,000 chance of being fixed. Hence the number of neutral mutations that arise, and are fixed, in each generation is 20 × 1/2,000 = 1/100. Notice that this answer is equal to the mutation rate, and independent of the population size, which cancels out: if we doubled the population size, that would double the number of mutants, and halve the chance that each would be fixed. If the population size is N, and the neutral mutation rate per generation is m, then the number of neutral mutations that are fixed in each generation is $2Nm \times \frac{1}{2}N = m$. Expressed as an equation:

$$\text{number of neutral mutations fixed in each generation} = \text{neutral mutation rate per generation}$$

If only all mathematical biology were so easy!

So Kimura's theory says that the rate of neutral molecular evolution depends only on the mutation rate, and is independent of the population size. This is important because typically we have no idea of past population sizes, but we can plausibly assume the mutation rate to be roughly constant. Kimura's idea was the starting point of a large body of mathematical theory, which has been applied in two ways. First, it provides a 'nul hypothesis' against which selection can be measured: departures from the predictions of the neutral theory indicate that there has been selection. Second, there are changes which probably were nearly neutral (the best candidates are so-called 'synonymous' base changes; that is, changes that do not alter an amino acid, and so do not alter the functioning of a protein). We can use these changes to date past events. If we can estimate the mutation rate, then, by comparing the DNA sequence of a gene from two present-day animals, and counting the number of synonymous base changes between them, we can date the latest common ancestor of the two animals – for example, of humans and chimps, or of birds and mammals.

The neutral theory of molecular evolution resembles a theory in physics – for example, Newtonian mechanics – in being a body of mathematical predictions derived from a small number of simple assumptions. It differs in that the assumptions are at best only very approximate: for example, mutation rates are not constant, and even synonymous mutations are not absolutely neutral. For this reason, the way we use mathematical theories in biology is different. If a physical theory – for example, Newton's law of universal gravitation – makes predictions that differ even slightly from observation, physicists get worried, and seek for a reason for the discrepancy. For example, Newton's laws predict that the planets should travel in elliptic orbits, with the Sun at one focus. Twice in history, careful measurement has revealed that the motion of a planet departs slightly from that predicted. On the first occasion, irregularities in the motion of Neptune suggested that the planet was being disturbed by attraction by a previously unknown celestial body; following this clue, Pluto was discovered. The second discrepancy, in the motion of Mercury, was later found to be in exact accord with Einstein's general theory of relativity.

In biology, we use equations in a rather different way. We do not have the luxury of studying the interactions between only two bodies, each of which can be treated as if its whole mass is concentrated at a single point. Instead, we study the interactions between a vast number of organisms, each of them in itself of enormous complexity. How can simple equations help us in the face of all these complications? First, we isolate some phenomenon for study. The rhythmic beating of the heart, the daily rhythm of waking and sleeping, and the ten-year cycle in the numbers of hares, lynxes, game birds and other animals in the Canadian Arctic, are all periodic oscillations, but they can hardly have the same underlying mechanism. So we study one of them at a time. Then, by a mixture of experiment and intuition, we guess, or in more pompous terms we hypothesize, a mechanism. To find out whether our guess is right, it is often helpful to write down equations describing our proposed mechanism, and, by solving (or simulating) these equations, find out whether they generate the *kind* of behaviour that we observe. In other words, we hope that our equations will predict qualitatively the right behaviour. Precise numerical fit is usually too much to hope for. For example, we would hope that a mathematical model of the Canadian ecological cycle would predict periodic (regular) oscillations in numbers, of roughly the right period, but would not expect to get the period or amplitude exactly right (in nature, they are pretty variable anyway).

One reason why we can expect only qualitative predictions is that in any particular model, we leave out so much. For example, in modelling the Canadian cycle, we would probably include only a few of the most abundant species (and our guess about which to leave out could easily be wrong), and ignore fluctuations in the climate from year to year (although some theories have regarded such fluctuations as crucial) and spatial differences in the environment. I have often been asked by students starting research in evolutionary biology what justification there can be in leaving out of a model something that must surely affect the outcome. The answer is, first, that if I leave out something that is truly important, the model will not give the right predictions, even qualitatively, and second, if you try to put everything into a model, it will be useless. You will have to use computer simulations to study its behaviour, and you will have little idea which of the features you have incorporated are important. As the anthropologists Robert Boyd and Peter Richerson once remarked, 'To replace a world you do not understand by a model of the world you do not understand is no advance.'

So, in biology, only rather simple models are useful. The price we pay for simplicity is a lack of quantitative accuracy in our predictions. But if we are lucky, we may find that our model will explain features of the world we had not thought of when we were constructing it. I will try to illustrate this by describing a model of animal behaviour which, with George Price, I thought up almost thirty years ago. Its original version was so simple as to be almost trivial. However, even in that simple form it explained some aspects of behaviour which, although familiar to biologists, had not been in our minds when we invented the model. More important, the method we adopted, of constructing an 'evolutionary game', has proved to be of much wider application than we had ever dreamed of. It has been used to analyse topics as various as animal signalling, the growth of plants, the evolution of viruses, and the ratio of the numbers of males and females in a population: I shall discuss the last of these problems, and the original problem in behaviour that started the whole thing off.

First, the original problem. When I was a zoology student back in the 1940s, my fellow students and I were fascinated by the recent discoveries of the animal behaviouralists Konrad Lorenz and Niko Timbergen – partly, I suspect, because our teachers seemed never to have heard of them, and it was nice to feel one up. Lorenz, in particular, emphasized how, when two animals compete for a resource, often they do not use all their weapons – teeth, horns, tusks – in an all-out fight, but indulge in ritualized displays, and often settle the contest without an escalated fight. At that time, such ritualized behaviour was explained by saying that it was good for the species; as one distinguished zoologist put it, if escalated fights were common, many animals would be injured, and that 'would militate against the survival of the species'.

Even as a student, I knew that had to be wrong. Darwin's theory of evolution by natural selection – the overarching theory that does for biology what Newton's laws did for physics for three hundred years – is primarily a theory of individual selection. In any given environment, some kinds of individuals are more likely to survive and reproduce than others. When they reproduce, they pass their characteristics on to their offspring. The result is a population consisting of individuals with characteristics that make for survival. If this is bad for species survival, so be it. How, then, can one explain ritualized contest behaviour?

Although I was aware of the problem in 1950, I did not think seriously about it for twenty years. Then, rather by accident, I decided to learn some 'game theory', to see whether it would help with the animal fighting problem. Game theory had been developed, first by John von Neumann and Oskar Morgenstern in the early 1940s, to study human 'games'; that is, human interactions in which the best thing to do depends on what your 'opponent' does. For example, in a game of poker, when does it pay to bluff? The snag with this classical game theory, for a biologist, is that it assumes that each contestant behaves rationally, and calculates that his opponent will also be rational. Obviously, one cannot make such assumptions about animals. However, game theorists had come up with one simple idea that I found very helpful: this is the 'payoff matrix'.

Thus imagine the following 'game'. Two animals are competing for some resource (a territory, an item of food, a mate), of value V. An individual can choose one of two 'strategies', Hawk or Dove. (A strategy here means just a kind of behaviour, but in later applications it means any heritable characteristic. The terms Hawk and Dove seemed natural because the game was invented in Chicago during the Vietnam War – they are not particularly appropriate for the birds in question.) In a contest, a Hawk fights with all its weapons, until it wins and gains the resource (value V), or is seriously injured (cost C; the meaning of value and cost is discussed further below). A Dove displays; if its opponent escalates, it runs away before it is hurt, but if it meets another Dove they share the resource, each getting value $V/2$. Given these strategies, we can write down the 'payoffs' to an individual, depending on its own strategy and that of its opponent, as follows:

i) A Hawk, playing against a Dove, get the resource, without injury: its payoff is V. The Dove gets nothing, but is not injured: its payoff is 0.

ii) If a Hawk meets another Hawk, it has an evens chance of winning (payoff V), or of losing (payoff $-C$): on average, its payoff is $\frac{1}{2}(V-C)$.

iii) If two Doves meet, they share the resource and each gets a payoff of $V/2$.

These payoffs can be summarized in a 'payoff matrix', giving the payoff to an individual adopting the strategy on the left, if its opponent adopts the strategy above:

	opponent's strategy	
	Hawk	Dove
payoff to Hawk	$\frac{1}{2}(V-C)$	V
payoff to Dove	0	$V/2$

Imagine a population playing this game: how will it evolve? When a Hawk meets a Dove, it wins, but it does not follow that Hawks will replace Doves in the population, because when a Hawk meets another Hawk, it may do very badly. We want to know not the outcome of a single encounter, but how the population will evolve over time. I make the following assumptions:

i) Each individual engages in one contest against a random opponent; we would get the same answers if individuals engaged in a series of contests against random opponents.

ii) An individual then produces a number of offspring that depends on the payoff it receives in the contest.[1] In other words, we interpret V and C as the change in the expected number of offspring resulting from the fight.

iii) When individuals reproduce, Hawks produce Hawks, and Doves produce Doves. In effect, this assumes asexual reproduction, and ignores the details of Mendelian genetics. At the time, the justification for this was that, usually, little is known of the genetics of specific behavioural traits, but that there is some resemblance between parent and offspring for almost every trait ever studied. Since then, some rather advanced mathematics has been deployed to show that the results of our original assumptions agree rather well with what happens in a sexual population, although there can be differences.

So, we have our model. What does it predict? First, suppose that the value V is larger than the cost C: for example, $V = 10$ and $C = 4$. The payoff matrix is then:

	opponent's strategy	
	Hawk	Dove
payoff to Hawk	3	10
payoff to Dove	0	5

What will happen? For this matrix, the answer is easy. We need to know the relative numbers of offspring produced, respectively, by Hawks and Doves. Hawk does better than Dove, no matter what its opponent does (3 is greater than 0, and 10 is greater than 5). In other words, whatever the proportions of Hawks and Doves in the population, Hawkish individuals have, on average, more offspring. The population will come to consist entirely of Hawks.

More interesting is the case where the cost C is greater than the value V: it no longer pays to risk injury to get the resource. For example, let $V = 4$ and $C = 10$. The payoff matrix is now:

	opponent's strategy	
	Hawk	Dove
payoff to Hawk	−3	4
payoff to Dove	0	2

Now, if the population consists mainly of Hawks, it pays to play Dove, but if the population is mainly Doves, it pays to play Hawk. In other words, a population of Hawks could be invaded by a Dove mutant, and a population of Doves could be invaded by a Hawk mutant. What will happen? It seems that we might end up with a population of individuals that sometimes play Hawk and sometimes Dove – a so-called 'mixed strategy' – just as a good poker player will sometimes bluff and sometimes not. But is this so? If it is, in what proportions will Hawk and Dove be played?

The trick is to look for an 'evolutionarily stable strategy', or ESS. An ESS is an 'uninvadable' strategy, in the following sense. Suppose almost all members of a population adopt some strategy, say S. Then, typically, an S individual will meet another S, and get the 'payoff to S against S'. Now imagine a rare mutant, Y. It too will typically meet an S, and get the 'payoff to Y against S'. Suppose that the 'payoff to Y against S' is less than the 'payoff to S against S', for all possible mutants. Then no mutant can invade the population, and we say that S is an ESS. Informally, an ESS is a strategy that does better against itself than any strategy can do against it. More formally, an ESS can be defined as a strategy S such that, if almost all the population adopts S, no other mutant strategy Y can invade: this will be so if strategy S does better against itself than any mutant, Y, does against S. Thus, returning to the first matrix, for $V = 10$ and $C = 4$, Hawkishness is an ESS because Hawk does better against Hawk than Dove does against Hawk. For our present matrix, we are looking for a strategy 'play Hawk with probability p and Dove with probability $1-p$'. To find p, we use the fact that, at an ESS, the payoffs when playing Hawk and Dove must be equal, on average: otherwise the population would not be in equilibrium. This gives the value $p = 0.4$. That is, the ESS is for an individual to play Hawk on 40 per cent of occasions and Dove on 60 per cent of occasions. The conclusion is that if the only tactics possible in a contest are 'fight to the death' and 'display, but run away if your opponent attacks', the only ESS is a mixed one: sometimes do one thing and sometimes the other. There is one slight complication. There are two possible stable states for a population playing this game. One is for all individuals to adopt the mixed strategy; the other is for 40 per cent of the population always to play Hawk, and 60 per cent always to play Dove.

Is this conclusion qualitatively right for animals in the real world? There are a number of situations in which organisms do adopt a mixed ESS of the kind predicted. For example, in bluegill sunfish there are two kinds of males. One kind grows without breeding for over five years, and then establishes a breeding territory, fertilizing the eggs of females that enter the territory. A second kind, called 'sneakers', hide in the territory of a breeding male; when females enter the territory and deposit eggs, the sneaker males emerge from hiding, shed sperm on the eggs, and flee. But I do not think the solution applies to the kind of contest that the Hawk–Dove game was intended to model. What is wrong? Price and I suggested other possible strategies an animal could adopt; I shall discuss two of them.

The first is 'Retaliator', 'Play Dove, but if your opponent escalates, fight back'. For our numerical example, the payoff matrix is:

| | *Opponent's strategy* | | |
	Hawk	Dove	Retaliator
Payoff to Hawk	−3	4	−3
Payoff to Dove	0	2	2
Payoff to Retaliator	−3	2	2

Retaliator is almost an ESS. A population of Retaliators cannot be invaded by Hawks (2 is greater than –3) or by any mixed strategy, but the evolution of a population containing only Retaliators and Doves is ambiguous, because they behave identically. To get round this, Price and I introduced a more complicated strategy, 'Prober-Retaliator', which, against Doves, staged an occasional 'trial escalation', returning to display if its opponent escalated in response. This strategy did prove to be an ESS, and may be rather a good description of some animal contests. Our Prober-Retaliator strategy has since become popular in discussions of evolutionary games under the name 'Tit-for-Tat'.

There is, however, another solution which is both more elegant and a better description of what usually happens. Suppose two humans were to play this game. Surely they would agree to share the resource? But what if the resource cannot be divided? You may already have thought that the assumption that the resource can be shared is often unrealistic. I think that two humans would choose to toss a coin, and perhaps arrange for a friend to be present to enforce the coin's verdict. One cannot imagine two animals tossing a coin. But what does the coin actually do? All it does is introduce an asymmetry into an otherwise symmetrical situation. This suggested to us that animals might rely on some asymmetry to settle contests. The obvious asymmetry is that between the 'owner' of a resource and an intruder. Of course we did not suppose that animals have a concept of ownership; it is sufficient that an animal should fight harder for a resource, such as territory, that it has occupied for some time than for one it has only just found.

Consider the strategy that, for obvious reasons, we called 'Bourgeois': fight hard for a resource if you already possess it, but not otherwise. That is, 'play Hawk if owner, play Dove if intruder'. If, reasonably, we assume that a Bourgeois strategist will find itself an owner and an intruder with equal probability, the payoff matrix is:

	opponent's strategy		
	Hawk	Dove	Bourgeois
Payoff to Hawk	–3	4	0.5
Payoff to Dove	0	2	1
Payoff to Bourgeois	–1.5	3	2

Thus Bourgeois is an ESS: in a population consisting mainly of Bourgeois, the average payoff to Bourgeois is 2 and to Hawk or Dove respectively only

0.5 or 1, respectively. So Bourgeois cannot be invaded by either Hawk or Dove. Note that the conclusion does not require any assumption that the owner of a resource is more likely to win an escalated contest.

Many animals follow this simply strategy. The strongest evidence for this conclusion comes if, when two animals both perceive themselves as owners of the same resource, an escalated fight ensues: it is sometimes possible to create this situation experimentally. Stimulated by the model just described, the zoologist Nick Davies investigated territorial behaviour in male speckled wood butterflies. Males hold patches of sunlight in a wood as breeding territories. If an intruder enters a patch of sunlight already held by a male, there is a brief contest in which the two males fly up in a spiral, and the intruder retreats. Davies removed a territory holder, and allowed a second male to occupy the patch of sunlight. He then released the original male. Both males then behaved as owners. The result was a prolonged spiral flight, much more costly than a typical encounter. Complications arise if other asymmetries, such as differences in size or age, are also present. One can also ask, what of the alternative strategy, 'if owner, play Dove; if intruder, play Hawk'? If animals played just one game in their lives, this would seem to be a valid alternative to Bourgeois, although I have never been sure what to call it. But if there are repeated games, there is an obvious snag. Once an animal has become the owner of a resource, it must surrender it to the next intruder. I know of one report of an animal, a semi-social spider, that appears to adopt this strategy.[2] These spiders make a communal sheetlike web, with separate holes, each occupied by a single spider. If a spider is driven from its hole and the hole is destroyed, it runs over the surface and enters another hole. The occupant of the hole emerges, and seeks yet another hole; the process is repeated until a spider finds an unoccupied hole. I doubt if such paradoxical behaviour can be common.

Before leaving these simple contests, I'd like to look at one last game, called the 'War of Attrition' – mathematically rather more difficult than the games I've so far discussed. Suppose two individuals are competing for an indivisible resource of value V, and that they lack the physical ability to escalate. All they can do is to continue to display until they or their opponent gives up. How long should an individual go on for? Clearly, if displaying is truly cost-free, and if V is greater than zero, the game collapses because they would go on displaying for ever, which is absurd. We must suppose that displaying is costly, and that cost increases with time, equally for both contestants: let us assume a contest lasts a time t, at a cost kt to both contestants. Suppose that two contestants choose times t_1 and t_2 respectively. If t_1 is greater than t_2, the contest continues for a time t_2, at a cost kt_2 to both contestants, and the first contestant gets the resource: the payoffs are $V\text{-}kt_2$ to the first contestant (he only has to pay kt_2, although he was willing to pay kt_1), and $-kt_2$ to the second.

How should an individual behave? More precisely, is there any behavioural rule that is evolutionarily stable, in the sense that, if everyone follows the rule, no 'mutant' rule would be better? It turns out that the ESS is for each individual to have the same constant probability of giving up per second, regardless of how long the contest has already lasted: for example, 'give up during the next second with probability one in a hundred'. The value of the probability depends on the value V: the larger it is, the smaller the probability of giving up per second, and the longer the average length of a contest. The result of following such a rule is shown in Figure 1, known as an exponential decay.

This kind of decay describes the level of radioactivity, as time passes, of a piece of radioactive material such as nuclear waste. It is the distribution expected if every radioactive atom in the sample has a fixed probability per second of 'decaying' and splitting into smaller atoms: similarly, in the 'War of Attrition', individuals have a fixed probability per second of giving up.

Why the similarity? Why should competing individuals have a fixed probability of giving up? I was led to the solution by the following reasoning. Consider an individual in the course of such a contest. He has already displayed for a time t_1: how much longer should he go on for? The answer is that he should continue for exactly as long as he was willing to go on for at the start of the contest. After all, he still faces the same rewards and

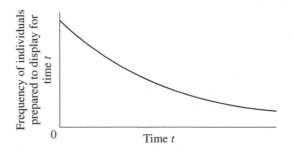

Figure 1 The 'War of Attrition' game. Distribution of frequencies in a population of individuals displaying for different lengths of time. The shape of the graph is predicted by a simple equation.

punishments that he did then: he still has resource of value V to win, and it will still cost him kt for every extra element of time t he continues. It is true that he has already spent kt_1, but that is water under the bridge – there is nothing he can do about it now. If his future potential gains and losses are exactly what they were at the beginning of the contest, his future behaviour should be exactly what it was at the start. In other words, his probability of giving up in any time interval should remain constant, just as a radioactive atom has a fixed chance of decaying per unit time, no matter how long it has been in a radioactive state.

Do animals play this game? I think not: it will always pay them to find an asymmetry to settle the contest. But there is a similar but slightly more difficult game, in which a number of animals compete simultaneously for a resource. My favourite example is a study by Geoff Parker, when he was a graduate student at Bristol, of male dung flies, waiting at a cow pat for the arrival of virgin females. How long should they stay? Geoff found that the flies adopted a rather accurate quantitative solution to the evolutionarily stable waiting time: of course, to discover this, Geoff had to wait there too.

A final suggestion before leaving animal contests. If you enjoy programming a computer, you can have fun analysing the dynamics of a population playing the 'Hawk–Dove–Retaliator–Bully' game. The first three strategies have already been described. Bully is the opposite of Retaliator; that is, 'play Hawk against Dove; Play Dove against Hawk'. The dynamics of this game are weird.

In a game such as the simple Hawk–Dove game, in which the possible strategies consist of the two pure strategies Hawk and Dove, and a set of mixed strategies with varying frequencies of Hawk and Dove, there is always an ESS. Depending on the payoff values, this will be Hawk, or Dove, or some mixed strategy, or both Hawk and Dove; in the last-named case, the population will evolve to all Hawk, or all Dove, depending on the initial frequencies. But not all games have an ESS. At first sight this seems odd; surely the population must go somewhere? It certainly does go somewhere, but not necessarily to a stable point; it may continue to evolve cyclically for ever (or, in practice, until circumstances change). But for a game to have no ESS there must be more than two pure strategies.

A very simple game that may have no ESS is the child's game Rock–Scissors–Paper. In this game, Rock beats Scissors (Rock blunts Scissors), Scissors beats Paper (Scissors cut Paper), and Paper beats Rock (Paper wraps Rock). The payoff matrix in this case is:

	opponent's strategy		
	Rock	Scissors	Paper
Payoff to Rock	1+e	2	0
Payoff to Scissors	0	1+e	2
Payoff to Paper	2	0	1+e

I have assumed here that winning is worth 2 units of payoff. If two opponents adopt the same strategy, they each get a payoff of $1 + e$ (assuming that they can share the reward) where the additional payoff is denoted by e, perhaps for not quarrelling. But e may not be positive: one could equally well assume that it is negative, implying a small cost to individuals adopting the same strategy. As we will see, the outcome crucially depends on whether e is positive or negative.

It is clear that no pure strategy – R, S or P – can be an ESS: a population playing R could be invaded by P, and so on. The only candidate for an ESS is the mixed strategy 'play R, S and P with equal probability'. Checking the stability conditions shows that if e is positive, this mixed strategy is *not* an ESS; it can be invaded by any of the three pure strategies. But the pure strategies are not stable either. So what will happen? The behaviour of a population can be plotted as a trajectory in 'state space', as shown in Figure 2(b). The system cycles for ever.

However, if the reward e is negative, then the mixed strategy is an ESS, and the dynamics are as shown in Figure 2(c). What if the reward is zero, $e = 0$? The equilibrium is now 'neutrally stable', and has the dynamics shown in Figure 2(e). Such dynamics, with a set of closed cycles, are said to be 'conservative'. We do not expect to find such systems in the real world, because the least change in circumstances will convert the dynamics into those shown in Figure 2(b) or 2(c). It is possible to find permanently cycling dynamic systems with constant amplitude – the living world is full of them. But their dynamic behaviour is that shown in Figure 2(d): the system settles down to a cycle of fixed amplitude, whatever its starting point.

I used the Rock–Scissors–Paper game in 1982 to illustrate the theoretical possibility of a game with no ESS; I never expected to find that animals play so simple a game. I was therefore astonished, fourteen years later, to come across an article in the scientific journal *Nature* entitled 'Lizards play Rock–Scissors–Paper'.[3] The article described a species of lizard with three types of male. Males with orange throats each hold a territory containing several females. A population of such males can be invaded by 'sneaker' males with green throats, which wait until an orange-throated male's back

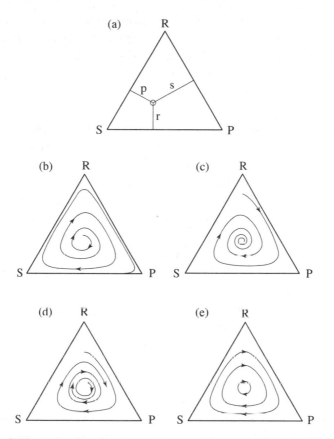

Figure 2. The dynamics of the Rock–Scissors–Paper game. The state of the population at any moment can be expressed as the values of r, s and p, the frequencies of R, S and P in the population. Because $r + s + p = 1$, this state can be represented, as in (a), by a point in an equilateral triangle, and by the dynamics of this point (i.e. how it changes with time); (b) shows what happens when the additional payoff e is positive; (c) shows what happens when e is negative; (d) shows a limit cycle – a behaviour not exhibited by the simplest form of R–S–P game, but one that is likely if more realism is incorporated; (e) shows what happens when e is zero.

is turned, and mate with one of his females. But once the population consists mainly of green-throated males, it can be invaded by blue-throated males, each of which holds a territory large enough to contain one female. And, of course, once blue-throated males are common, the original orange-throated males can invade, and the cycle is complete.

For a theorist, there is a special satisfaction when an animal is found doing something that has been predicted by theory but seemed too strange actually to occur. However, I think that continually cycling games, with no ESS, may be more common than we have thought. A model of a situation that may be widespread is called the 'Red Deer Game'. Imagine a species – the red deer is an example – in which males grow without reproducing until they have reached a certain size. Then, in the breeding season, they enter a competitive arena, in which the strongest males accumulate a harem of females with which they mate. During this competition, a male uses up so much energy that from then on he grows little or not at all. At what age, or size, should a male enter the arena? If he entered too soon, he would fail to hold a harem; if too late, he may not survive to breed at all – after all, there are continuing risks of death from predation, disease and starvation. A model of such a game, assuming that a male's breeding success is an increasing function of the proportion of the breeding population smaller than himself, suggests that there may be no ESS. The males will increase in size, and in age at first breeding, until most males are ageing before they breed. Then one of two things will happen. The population may be invaded by 'sneakers', who steal matings without attempting to hold a harem. There are surprisingly many species with a mixture of harem-holding males and sneakers – it is not clear whether this is a stable mixed ESS or a transitory stage in a cycle. Alternatively, the species may go extinct in competition with an ecologically similar species of smaller size and greater ecological efficiency.

There is a curious fact about mammals which suggests that the Red Deer Game may be telling us something. Fossil evidence shows that most mammalian lineages increase in size: for example, the earliest horses were no larger than a medium-sized dog. Yet mammals as a whole are not bigger today than they were 50 million years ago (and this is true even if we ignore the recent extinction of many large species, probably resulting from human hunting). A possible explanation is that many species increased in size because of male–male competition, as suggested by the Red Deer Game, and then went extinct in competition with smaller species. But this is speculation.

I now turn to a topic in which quantitative predictions are sometimes possible. This is the evolution of the sex ratio. Why, in most species, are there equal numbers of males and females, and why, sometimes, are there not? For example, some wasps produce ten times as many daughters as sons.

The basic answer to this question was given by R. A. Fisher in 1930. Although he did not explicitly use an analogy with human games (at least not in this context), his argument is essentially an ESS argument. It can be paraphrased as follows. Suppose a female could choose the sex of her off-spring, which sex should she choose? On Darwinian grounds, she should choose that sex ratio that will maximize the number of grandchildren to whom she transmits her genes. How can she do that? Her choice is determined by a very simple argument. If every child has one father and one mother, then, *on average*, members of the rarer sex will have more children. So she should choose to have children of whichever is the rarer sex.

Obviously, the only stable state, or ESS, is that with equal numbers of males and females. Hence the 1:1 sex ratio. In fact, Fisher went a little further, and allowed for the possibility that it might be more costly to produce girls than boys, or vice versa. Then, he argued, parents should expend equal resources producing males and females.

This is a prediction that is often realized with great accuracy. But, you may say, this is because the mechanism of sex determination – usually by producing equal numbers of gametes with X and Y chromosomes – is such that it is only able to generate a 1:1 ratio. This argument has some validity. Sex ratio is one of only two traits of the fruitfly *Drosophila*, out of many hundreds, that have been exposed to prolonged artificial selection with no effect whatever. However, I refuse to believe that natural selection could not have altered this mechanism if it paid to do so. There are, in fact, some species in which females can and do alter the sex ratio among their offspring in response to changed circumstances. In mammals, a female can alter the sex ratio among her offspring at relatively little cost by selecting which of the fertilized blastocysts should implant and grow in her uterus. However, the clearest case of maternal control of the sex ratio occurs in the Hymenoptera (ants, bees, wasps). In this order, as in many insects, the female stores sperm after mating. If she fertilizes an egg, it develops as a female; if she does not, it develops as a male, with only a single set of chromosomes. Experiments show that females can and do choose the sex of individual offspring. This remarkable arrangement provoked the ecologist Eric Charnov to dedicate a scientific paper 'To the Almighty, for creating the Hymenoptera, and thus making sex-ratio theory testable'.

What use do female Hymenoptera make of this ability? One example, first elucidated by the mathematical biologist Bill Hamilton, concerns the parasitic wasps that lay eggs in moth caterpillars. The larvae develop within the caterpillar, killing the host, and often mate with one another immediately after emergence; the males then die, and the females disperse to seek another caterpillar. If, typically, only one female wasp lays eggs in a single caterpillar, what sex should they be? As in Fisher's argument, females should act so as to maximize the number of their grandchildren. Since one male can produce enough sperm for many females, she should produce one male, and all the rest females. Highly female-biased sex ratios are in fact found in such parasites.

But things are not so simple. Suppose a second female lays eggs in the same caterpillar. She could gain by producing several males. Consider the following oversimplified model. Each female wasp lays all her eggs in only one caterpillar, and each caterpillar is parasitized by two wasps. It is then reasonable to assume that the number of eggs laid by each female is constant and equal to n (the number cancels out, but I find it easier to write down the equations with it included). Mating within a caterpillar is random, and each female mates only once. What is the stable sex ratio? This is quite a tricky problem: it requires a knowledge of the mathematics of differential calculus, but it is not obvious how the calculus should be applied. Following the ESS procedure, one looks for a sex ratio such that if all the females in a population adopt it, no mutant female with a different sex ratio can do better – that is, pass on her genes to more grandchildren. The answer is that females should produce a ratio of 1 male to 3 females. In other words, the ESS for this game is for females to produce a male-biased sex ratio, but not so extreme as it would be if only one female lays eggs in a caterpillar. If you can work this out from first principles, you should consider becoming a biologist – we need you.

The application of game theory to sex-ratio evolution does yield quantitative predictions. But testing is still difficult because, as will be obvious from the model I have just described, one has to make a number of assumptions. In practice, it would not be the case that each caterpillar was parasitized by exactly two wasps, that parasite mortality was independent of the number of eggs laid, that mating was random, and so on. So, even in sex-ratio theory, tests are usually qualitative.

Evolutionary game theory can be applied whenever the best thing for an individual to do – its 'fitness' – depends on what others are doing. Its range of application is therefore wide; for example, it has been applied not only to animals but also to plants, and even to 'selfish' genetic elements that replicate out of time with the rest of the genome. A topic that has attracted a lot of attention recently is that of animal communication. The problem is in principle very simple: why don't animals lie? Thus suppose, in the Hawk–Dove game, an animal could signal 'I am going to escalate'. If the signal was honest, the sensible thing for its opponent to do would be to retreat. So making the signal would be a cheap way of getting the resource without fighting. Soon everyone would signal whether they intended to escalate or not. Soon after that, no one would believe the signal, and communication would have collapsed. This difficulty has been tackled by treating communication as a two-person asymmetric game, with considerable success.

It should be clear that what is being tested by these models is not the theory of evolution by natural selection itself. That theory must be tested by other means. The theory that evolution has in fact happened can best be tested by the fossil record: as J. B. S. Haldane once remarked, a single fossil rabbit in Cambrian rocks would show that evolution had not happened. The theory that the mechanism of evolution is natural selection could be disproved if it could be shown that offspring do not resemble their parents (a theoretical possibility, but hardly likely to happen), or that acquired characteristics are often inherited (since this might offer an alternative mechanism). Game-theory models assume the correctness of natural selection. What they are testing is a particular explanation of the evolution of a particular characteristic, whether it be fighting behaviour, the sex ratio or warning colouration.

The usefulness of a mathematical as opposed to a verbal model is twofold. First, to write down a model you have to be absolutely clear about what you are assuming. Or rather, even if the author of the model makes an assumption of which he is not aware, it is always possible for others, looking at the model, to see that it only holds if this unconscious assumption has been made. For example, when George Price and I first wrote down the payoff matrix for the Hawk–Dove game, we did not state explicitly that we were assuming that the resource could be shared, but the matrix implies it. This role of model-making is, I think, very important. Often, when thinking about a problem in biology, I find that I only begin to understand it when I have written down a mathematical model. There are questions that are too hard to think about without the aid of mathematics.

The other role of a model, of course, is to make predictions that can be tested. As I have repeatedly emphasized here, it is hard in biology to come up with quantitatively precise predictions, because the situations we are modelling are so complex. But I hope I have persuaded you that we can make some non-obvious qualitative predictions, and that sometimes these turn out to be true.[4]

The Best Possible Time to be Alive

The Logistic Map

Robert May

I

> It makes me so happy. To be at the beginning again, knowing almost noth-
> ing . . . A door like this has cracked open five or six times since we got up
> on our hind legs. It's the best possible time to be alive, when almost
> everything you knew is wrong.
>
> Tom Stoppard, Arcadia, Act 1, Scene 4

This passage from *Arcadia* is undoubtedly over the top, but it does capture
what it felt like – for me and others – in the early 1970s to be part of the scientific revolution caused by chaos theory.

Stoppard's play is a brilliant interweaving of three themes: landscape
gardening, Byron scholarship and chaos.[1] Always a stickler for accuracy, he
asked me to look over the text to see whether there were any scientific
errors (he didn't need much help). I embraced this opportunity with shame-
less pleasure, attending rehearsals in early 1993 as a groupie. I also wrote an
article for the play's programme note, which now is, I suspect, my most-
read publication, although it gains me no citations, being outside the usual
apparatus of scholarship.

In *Arcadia*, Stoppard makes the point – correctly, in my view – that all
the essentially new insights of chaos theory could have been made two
hundred years ago, long before the invention of electronic computers.

Contrary to the suggestion you often hear that chaos theory was a computer-generated discovery, all that's needed is a paper, pencil and a lot of patience; the computer simply increases the speed with which we can do the calculations, albeit dramatically. It was with these quintessentially low-tech materials that I began my work on chaos, and in those days all I ever had to hand was the early desktop machines, which seem antediluvian by today's standards. In science, chaos refers to the idea that the behaviour of something can be to all intents and purposes unpredictable even though it is described by a very simple 'deterministic' equation; by deterministic we mean that the equations, and all the parameters in them are completely known, with no statistical or uncertain elements. Such an equation appears to predict with certainty the future of something, given its state at some initial time.

The existence of such chaotic behaviour in simple deterministic equations was quite a shock to scientists, imbued as they have been by the powerful vision of Newton and those who followed him in the Age of Enlightenment. The Newtonian world is orderly and predictable, governed by laws and rules that can best be expressed in mathematical form, as equations. If the circumstances are simple enough – a planet moving around a sun, for example – the system can behave in a simple and predictable way. Situations that are effectively unpredictable – a roulette ball whose fate, the winning number, is governed by a complex concatenation of the croupier's hand, the spinning wheel and so on – were thought to arise only because the rules were many and complicated.

Over the past thirty years or so, since the advent of modern chaos theory, this Newtonian vision has splintered and blurred. We now know that simple equations can generate behaviour as complicated as anything we can imagine. I shall be discussing here arguably the simplest such chaos-generating equation, an equation so simple that a child can understand it (indeed, my daughter Naomi came across it in a computer class at junior school in the United States).

What, then, is this equation? Think of a number between zero and one; multiply it by the difference between that number and one; then multiply the result by a fixed constant, which we can call anything we like; let's call it a. The result is another number. In mathematical terms, if we call the initial number $x_{initial}$, the number x_{next} we have generated in this process is expressed economically by the equation

$$x_{next} = a \, x_{initial} \, (1 - x_{initial}).$$

This equation is the hero of this essay. It is very easy to apply; for example, if $x_{initial}$ is 0.25 and a is 10, the equation says that x_{next} is $10 \times 0.25 \times 0.75 = 1.875$. Mathematicians usually call this type of thing not an equation but a 'map', because it describes the 'mapping' of one number ($x_{initial}$) into another (x_{next}). This particular example is usually dubbed the Logistic Map.

The Logistic Map is fascinating to mathematicians because of its astonishing complexity. When people first come across this map, it seems to be the essence of simplicity – if you plug in some initial number, you generate another number; if you change the initial number by a tiny bit, you expect the number you generate always to be slightly and predictably different. At least, that was what I expected when I first came across the map in the early 1970s. But I soon found that this isn't always the case – for some values of a, the mapping gives results that look completely random and unpredictable. As I eventually came to understand, and as I shall describe, this happens when the map describes chaos.

What makes the Logistic Map fascinating to scientists is the observation that it can be applied successfully to ecology, the branch of biology that deals with the relations of organisms to one another and their environments. In particular, the map gives remarkable insights into changes in animal populations that take place over time. The spawning of salmon stocks, the number of ants crawling around their hill, even the fluctuations of grouse populations on moorland – the problem that Tom Stoppard's character Valentine is studying for his PhD in *Arcadia* – all obey the dicta of the map. As we shall see, the realization that chaos can sometimes underlie the large-scale behaviour of animal populations revolutionized ecologists' understanding of their field.

Chaos has come a long way since its early days. Scientists know that without it, we couldn't understand a huge range of scientific subjects. The electrical activity in our brains and hearts, the dripping of taps, the clustering of cars on expressways, even the intricate behaviour of the hydrogen atom – they all involve chaos. The very word 'chaos', in its scientific sense, is now part of many people's vocabulary. Who today hasn't at least heard of the 'butterfly effect', the classic illustration of how the ideas of chaos apply to weather forecasting?

More of that later. First, I want to look at how theoretical biologists were thinking about animal populations until the late 1960s. Then I want to describe how insights from the Logistic Map revolutionized the subject and taught us a new way of looking at nature. Stoppard was right – for those involved in this adventure, the 1970s were 'the best possible time to be alive'.

II

I became a theoretical biologist only after a series of professional transmutations. Back in Sydney in the late 1950s, I started out as a student of chemical engineering, then became a physicist, gaining a PhD with a thesis on superconductivity. I then spent a couple of years at Harvard University, in their division of engineering and applied physics. In the early 1960s I returned to Sydney University to teach theoretical physics, and later became professor of that subject. In the late 1960s my involvement as a founding member of Societal Responsibility in Science in Australia drew me, quite by accident, to an interest in the relation between complexity (in the sense of numbers of species, or the richness of the web of interactions among them) and stability (in the sense of ability to withstand or recover from disturbance) in ecosystems and, shortly after, to Princeton University as professor of biology. I was lucky to stumble into the nascent subject of theoretical ecology in its 'romantic phase', akin to theoretical physics in the 1920s and 1930s, when simple questions were being framed in appropriate mathematical ways and when surprising answers were consequently emerging.

Ecology is a young science. Arguably the first ecological text is the clergyman Gilbert White's *Natural History of Selborne*, published in 1789. This work goes beyond the fascination with descriptive natural history which characterized earlier such books, to begin to frame analytical questions about, for instance, what governs the abundances of the town's swifts and wasps. The century that followed saw the huge advances made possible by the advent of Darwin and Wallace's theory of evolution by natural selection, for my money the most important advance in humanity's intellectual history. In describing the 'struggle for existence' that underlies evolution, Darwin reached for metaphors of wedges in a barrel to illuminate a discussion of what we might call 'competition for niche space among species'. But he never quantified these ideas, and ecological studies lagged behind evolutionary ones at that time.

The British Ecological Society, the oldest society of its kind, was founded in 1913, much later than most other scientific societies in Britain. The Ecological Society of America followed soon after, in 1915. Up to around the middle of the twentieth century, both societies' publications were predominantly descriptive and classificatory, with much of the focus on plant communities. But by the middle of the century, animal ecologists were posing some theoretical questions; for example, why do some populations of some boreal mammals so often change periodically, alternately increasing and decreasing with time? More mathematical studies gave hints at the beginnings of oversimplified answers to these questions: for example, the populations of communities that have a single predator and a single prey have an inherent intensity to go in cycles. In the third quarter of the century, some theoretical biologists, including the hugely influential Robert MacArthur, speeded up the progress of the advance by combining empirical observations with analytical approaches (often explicitly mathematical) to frame clear lines of attack on ecological questions.

This is where I came in. It struck me that the equations that ecologists were using were in some important ways different from the more familiar ones of physics. The differences are not mainly in the technical nature of the equations, but rather that the equations of physics purport to give an exact account of whatever they are describing. For example, Einstein intended his equation of general relativity to describe deflection of a ray of light by the Sun to any accuracy you like – the more precise the information you put into the equations (for the mass of the Sun, the energy of the light ray, and so on), the more precise will be the equation's prediction for the deflection of the light. In population biology, things are often very different. There, the equations commonly refer to models of living systems that are always much too complicated to be amenable to the representational equations of the type beloved of physicists.

The models of biological communities tend rather to be of a very general, strategic kind – they are caricatures of reality. Just as a good caricature catches the essential truth behind the thing it is trying to depict but is for-givably vague about the unimportant details, so the most we can expect of the equations of population biology is that they capture the key points of the situation they are describing. So, for biologists studying animal populations, their equations are cartoons of reality, not the perfect mirror images sought by physicists. That's not to say that these biological equations are not vital to our understanding of nature. As the British mathematical biologist John Maynard Smith has noted, 'Mathematics without natural history is sterile, but natural history without mathematics is muddled.'

The ecologists out in the field had, for example, collected data that showed that animal populations in isolated communities generally stayed roughly constant or, as Maynard Smith commented in his 1968 classic *The Mathematical Ideas in Biology*, they fluctuate 'with a rather regular perio-dicity'. But what was the underlying reason for these population changes? If the mathematical models were any good, they would answer that question.

Roughly speaking, there were two schools of thought about how animal populations behaved. On one side, the Australian Charles Birch believed that most natural populations are driven by external effects so that they fluctuate wildly, driven by changes in the environment. Birch and his colleagues tended to draw their examples from insect populations that do just that. On the other side of the debate, another Australian, John Nicholson, took the view that populations are regulated by effects that depend primarily not on the environment but on the density of the population – the number of animals living in a given space. This latter picture indicated that the populations tend to increase when their densities are low and decrease when they are high, and that they therefore tended on average to be relatively steady. Nicholson and his confrères drew examples from relatively steady populations.

It seemed to some that only one of these two pictures could be right. But, as often happens in science when two opposing views of a problem both seem to be partially right yet ultimately irreconcilable, many of the protagonists were looking at the problem in too narrow a way. The problem, it turned out, can be understood much more simply in terms of a different way of thinking, of a different paradigm. The virtue of the Logistic Map was that it gave a clear and easy-to-grasp idea of this new and extraordinarily productive way of thinking, as I was soon to find out.

III

What she's doing is, every time she works out a value for *y*, she's using *that* as her next value for *x*. And so on. Like a feedback. She's feeding the solution back into the equation, and then solving it again. Iteration, you see.

Tom Stoppard, *Arcadia*, Act 1, Scene 4

Imagine a pond swimming with goldfish. During their isolated aquatic lives, these fish will eat, mate, possibly suffer diseases and unpredictable traumas such as a visit from a cat. One question that population ecologists are interested in is: how will the number of the fish change from one generation to the next?

A kind of answer to this question is given by the Logistic Map. To see how, let's think about the number of goldfish in the pond in terms of the fraction of the maximum total number of them that could possibly live in that environment. And let's call that fraction x. For example, if the maximum number of goldfish whose lives the pond could support is 1,000, then if there were 250 of them in the pond when we first counted them, x would be $250/1,000 = 0.25$.

The assumption at the heart of many simple mathematical descriptions of situations like this one is that the population $x_{initial}$ for one generation uniquely determines the population x_{next} for the next generation. But how, mathematically speaking, does x_{next} depend on $x_{initial}$? If we accept the simplest picture of the type pioneered by the English economist and clergyman Thomas Malthus (1766–1834), we might suppose that the population increases by a small fraction every year, provided the goldfish have unlimited food and reproduce freely, without restraint. So we might expect the connection between x_{next} and $x_{initial}$ be an equation like $x_{next} = 1.05 \, x_{initial}$, where, in this case, there is a 5 per cent growth in population each year. This would mean that if x at first has the value 0.25, its value in the next generation is $1.05 \times 0.25 = 0.2625$, and its value a generation later is $1.05 \times 0.2625 = 0.275625$, and so on. So the population gradually increases.

But life isn't like that. If the goldfish population were very large, the fish would soon run out of food and they would be fighting for it, disease would spread more easily and the community would be juicier prey for predators. The upshot would be that the population's rate of growth would soon fall. If, on the other hand, there were only a few fish enjoying themselves in the pond, with plenty of room to move about, their population would quickly burgeon. So how can the 'Malthusian' map $x_{next} = a \, x_{initial}$ (with a some constant) be modified so that it's more realistic? One answer is the Logistic Map $x_{next} = a \, x_{initial} (1 - x_{initial})$, which first became popular in the 1950s with population ecologists who were studying fish and insect populations. The quantity a represents the rate of growth, whose value is characteristic of the pond's environment. The new factor $1 - x_{initial}$ ensures that x_{next} does not grow too quickly, because as $x_{initial}$ rises, $1 - x_{initial}$ falls, keeping the next generation's population x_{next} in check (if x ever exceeds 1, the population is extinguished).

So what does the Logistic Map predict for the dynamical behaviour of the goldfish population (and of other phenomena to which it might apply)? In the 1950s, population experts applied the equation not only to communities of fish but to insects and other organisms. These experts made the common mistake of being blinkered by fashion – they looked for and found situations in which the populations settled to a steady value, in equilibrium. They even enquired about the values of the constant a that would guarantee such stability of the population. They did not enquire what happened when the constant *a* had values *outside* the range corresponding to the population settling to a stable, constant value. As the scientific world was soon to find out, the simplicity of the Logistic Map is profoundly deceptive.

Let's see how x_{next} changes each time we calculate it, or, in technical language, with each iteration. And let's choose three values for *a* (you'll see why I chose them in a moment): 2.4, 3.4 and 3.99. Take a look at Figure 1, where I have plotted how x_{next} evolves, in these three cases, beginning in each case with the initial value 0.01. In the first case ($a = 2.4$), x_{next} quickly settles down to a stable value – in the context of our population of goldfish, this means that the population of the fish in the pond becomes constant. In

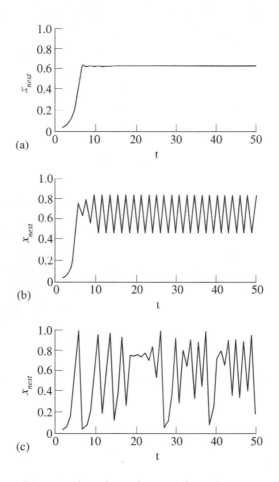

Figure 1 (a) The evolution of x_{next} for $a = 2.4$; (b) for $a = 3.4$; (c) $a = 3.99$.

the next case ($a = 3.4$), x_{next} continually jumps up and down between a high and a low value. The population of the goldfish keeps returning to the same value (it's said to be periodic) and this happens every two generations. The final case ($a = 3.99$) is bizarre: x_{next} jumps up and down, all over the place. This is 'chaos' – the goldfish population is fluctuating, seemingly without rhyme or reason, completely unpredictably.

In the early 1970s, soon after I became interested in understanding animal populations, I became fascinated with the Logistic Map. I wanted to try to understand mathematically how it evolves for *any* initial value and for *any* value of the constant a. It was tough going and progress was slow.

In the late autumn of 1973, soon after I arrived in Princeton to take up a permanent post, I drove down to the University of Maryland to give a seminar. I took with me some of the work I'd done on the Logistic Map, and several unanswered questions. At the seminar, I met Jim Yorke, a mathematician who was to become a friend and with whom I was to collaborate in first understanding the Logistic Map.

The story is easiest to tell if I begin with the conclusion. Have a look at Figure 2, which shows the map in all its complex glory. Along the bottom, on the horizontal axis, I've plotted values of the constant a; along the vertical axis, the value $x_{settled}$ to which x_{next} settles down after a few thousand

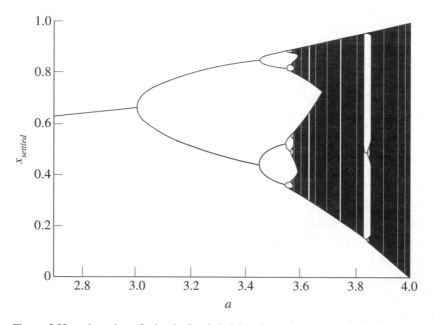

Figure 2 How the value of a in the Logistic Map determines the settled value of x.

iterations, for each value of the constant. When the constant is less than about three, x_{next} settles down to a unique value (just as it did in Figure 1(a)). However, as the value of the constant rises above 3, the settled value of x has not one but *two* possible values – there are stable cycles in the value of $x_{settled}$ (just as there were in Figure 1(b)). As the value of the constant increases, we see the emergence of a series of periodic behaviours, which I dubbed 'a cascade of period doubling'. Finally, when the constant is between 3.57 and 4, the map behaves bizarrely. This is the domain of chaos, where the settled value is so sensitive to the slightest change in the initial value of x that the final distribution of x can be regarded as random (you saw this in Figure 1(c)).

Let's pause for a moment to reflect on what this means for the pond's goldfish population. The constant influences how much the population changes from one generation to the next. Its value, different from one situation to another, depends on the goldfish (their fertility, appetite for food, aggression, visits from hungry cats and so on) and the environment in the pond (the food present, the salubriousness of the climate for the fish and so on). If the constant is low (below 3), the population will be stable. If it's between 3 and 3.57, the population will periodically alternate between high and low values. And if it's between 3.57 and 4, the population fluctuates wildly, and it's impossible to make long-term predictions even though we have a simple equation, that is fully determined.

When I arrived that Maryland seminar, I already understood the left-hand side of Figure 2 – the areas of stability and the cascade of period doublings. When I got to the point in the talk when I said that I didn't understand what was going on once a got bigger that 3.57, Jim Yorke interrupted. 'I know what comes next,' he said. With his colleague Tien-Yien Li, he had recently been investigating maps like the Logistic Map and had uncovered their chaotic behaviour. In fact, they even coined the word 'chaos' in its mathematical sense in their paper 'Period three implies chaos', which was later published in 1975.[2] Several of their colleagues advised them to choose a more sober word than 'chaos', but they went ahead and gave the field its arresting name. Yorke and Li had not looked at values of a below 3.57, and so had not appreciated the cascade of period doublings which characterizes the route to chaos. (An almost magical plethora of other period doublings – cycles with, for example, period 11 doubling to 22, 44, 88 and so on – lurk within the deeper complexities of chaos in the Logistic Map and its relatives.)

Putting together our separate understandings of bits of the puzzle, Jim

Yorke and I realized immediately that we had discovered something important. The behaviour of the Logistic Map was not just a mathematical quirk, but had much wider implications for predictions in simple mathematical models. We soon learned, to our surprise, that others had already trodden the same ground, beginning with the Finnish mathematician Pekka Myrberg almost twenty years before, in 1958. But Myrberg – and later pioneers in Russia, France and the United States – were looking at the patterns in the spirit of uncovering a fascinating mathematical phenomenon, with no explicit feeling for the wider implications for descriptions of the natural world. In contrast, Jim Yorke and I were looking at the Logistic Map in the specific context of down-to-earth ecological problems, trying to understand the origin of population fluctuations not just in goldfish ponds, but in other situations that could be described by the map. So although others had probed the map and uncovered something about its behaviour, York and I were the first to grasp the larger significance and consequences of these results.

But we certainly didn't understand all there was to know about the map. In one important detail, we missed a trick and someone else saw deeper than we did. To understand this point, look again at the big picture of the map's behaviour, Figure 2, and at the cascade of period doublings. It looks a bit like a remarkably symmetric tree lying on its side, with its branchings getting increasingly close together. In fact, as the number of iterations increases, the distance between each successive branching – as measured by the distance between the values of the constant a at which the branching takes place – is equal to a fixed number. The mathematician George Oster, of the University of California at Berkeley, first noticed this in 1976, and we derived an approximate formula for the ratio between successive branching values of the constant that predicted its value to be about 4.83.[3] For us, this was merely a mathematical detail and we forgot about it pretty much as soon as we'd finished the work.

But the American mathematician Mitchell Feigenbaum, working at the Los Alamos Laboratory in New Mexico, independently made this observation, on the basis of numerical studies rather than mathematical analysis. He, however, saw much further. First, his numerical studies of course gave a more accurate value for the ratio (about 4.6692) and showed that this number cropped up in many places where systems make transitions from stable to chaotic behaviour. More important, he boldly suggested that if the transition from smooth flow patterns to turbulence in fluids is, in its essentials, such a transition to chaos, then it should be possible to see the cascade of period doublings and measure the ratio he had predicted. Experimenters soon observed just this phenomenon. Feigenbaum deservedly won wide acclaim for his insight. This time it was my turn to be among those who discover something first mathematically, and for someone else to appreciate its wider importance, to 'discover it last', as it were.

These studies of the Logistic Map revolutionized ecologists' understanding of the fluctuations of animal populations. Remember, ecologists had long been arguing over whether the fluctuations were due to external effects (Birch's view) or to the density of the population of organisms (Nicholson's view). With the insights of the Logistic Map, it was clear that the Nicholson–Birch controversy was misconceived. Both parties had missed the point: population-density effects can, if sufficiently strong (as in Figure 1(c)), look identical to the effect of external disturbances. The problem is not to decide whether populations are regulated by density-dependent effects (and therefore steady) or whether they are governed by external noise (and therefore fluctuate). It's not a question of either/or. Rather, when ecologists observe a fluctuating population, they have to find out whether the fluctuations are caused by external environmental events (for example, erratic changes in temperature or rainfall), or by its own inherent chaotic dynamics, as expressed in the underlying deterministic equation that governs the population's development.

Ecologists had to think about populations in a new way – some would say a new Kuhnian paradigm – that rendered old ones obsolete. It's at times like this, with past assumptions being overturned and replaced with new ones, that science is at its most exciting for those lucky enough to be in the thick of it.

But this revolution did not happen overnight. Many colleagues took a lot of persuading and did not easily concede the profound implications that chaos theory has on science. In early 1976, I decided to write an evangelical paper that would suggest the wider significance of chaos. I wrote a review article on chaos theory, taking the Logistic Map as my exemplar of how chaos can figure in some of the simplest equations, hoping to persuade other scientists to look at where chaos figures in their work. I wrote the paper in a deliberately messianic style and submitted it to the leading UK science journal *Nature*. The editorial staff of the journal were sceptical, feeling, among other things, that the review was much too mathematical and unlikely to be of general interest. But one of the senior editors, Miranda Robertson, was persuaded, and she sent the manuscript to John Maynard Smith for review. His review was most generous ('it reads as if your mother wrote it', Miranda said), and so *Nature* did publish the paper in June 1976.[4] The paper had its intended effect of bringing chaos to a wide audience of scientists, and it now has several thousand citations.

More important than the good opinion of scientists is the good opinion of nature, as manifest by agreement with experiment. Since the early papers on the Logistic Map, several experimenters have demonstrated the felicity of the map's description of the dynamical behaviour of animal populations – not just goldfish, but insects and mammals, too. In one example, data on the population changes of the vole *Clethrionomys rufocanus* from 1922 to 1995 in Hokkaido, the northernmost island of Japan, spectacularly showed many of the features expected from studies of the Logistic Map.[5] Likewise, classic data on the Canadian lynx and the snowshoe hare in the wild have been illuminated by subsequent developments of these ideas. Experimenters have also looked at insect populations under controlled conditions in the laboratory and found the sort of behaviour one would expect on the basis of equations similar to the Logistic Map.

The applications of chaos extend way beyond ecology. Over the past ten years, it's become clear that chaos is just about everywhere in science and technology.[6] Mechanical engineers use its ideas to reduce the amount of the noise we hear in brakes and in the squealing wheels of railway carriages. Ship designers use it to avoid producing ships that keel over in storms. Electrical engineers use it to encode secure messages, to extract information from noisy signals and to help avoid power cuts. Astronomers use it to understand the distribution of asteroids in the solar system. Physicists use it to understand and predict the motion of fluids. Plainly, chaos is going to be an important part of twenty-first-century science.

IV

The unpredictable and the predetermined unfold together to make every-thing the way it is. It's how nature creates itself, on every scale, the snowflake and the snowstorm.

Tom Stoppard, *Arcadia*, Act 1, Scene 4

Why did it take scientists so long to recognize chaos? Newton created modern mathematical science at the end of the seventeenth century, so why didn't he or one of his successors study simple things like the Logistic Map and uncover its rich structure?

I think the answer is that, since Newton's day, mathematical studies of change focused almost exclusively on the dynamics of systems that change continuously with time. From Newton's laws of motion, through the later advances in understanding dynamical systems made by brilliant mathematicians such as Joseph Lagrange and Sir William Hamilton, the focus was on *differential* equations – equations with continuously changing variables (they feature, for example, distances that change smoothly, not in discrete jumps like graduations on a ruler). If any of these mathematical thinkers had happened upon the Logistic Map and spent some time working on it, I think it's a fair bet that they would have discovered chaos.

But they didn't. It was only at the very end of the nineteenth century that chaos was first glimpsed, by the great French mathematician Henri Poincaré when he was studying differential equations. Towards the end of the nineteenth century King Oscar of Sweden offered a prize for the first person who could show once and for all that the solar system as a whole (the sun, the planets, the asteroids and so on) was completely stable. It was in pursuit of this prize that Poincaré studied the 'three-body problem' of three gravitationally interacting objects (for example, the Sun, the Earth and the Moon), treated approximately as if they existed only at points. He demonstrated that the resulting system of differential equations could give rise to orbits of 'indescribable complexity'. He concluded that King Oscar's problem was insoluble, at least by using the techniques that he had to hand. He was right and he had been the first to see chaos, although few recognized this at the time. Pleasingly and appropriately, he was awarded the prize.[7]

The study of chaos scarcely progressed at all in the first half of the twentieth century, although in retrospect there are examples of scientists happening upon the shadows of chaos without fully appreciating what they had found. For example, the mathematicians Mary Cartwright and John Littlewood mentioned in a paper in the 1930s that they had found examples of relatively simple differential equations that showed amazingly complex behaviour, which we today would call chaotic. But the consensus was that each such example was complex and intractable in its own peculiar and particular way. Best put back in the cupboard. Not the sort of thing around which to organize a neatly structured syllabus.

Modern chaos theory actually began with a set of equations relating to weather forecasting, published in 1963, a year notable not only for modern science but also, as Philip Larkin has noted, for the Beatles and the sexual revolution. The equations were the work of the great meteorologist Edward Lorenz at the Massachusetts Institute of Technology. He had long been fascinated by the vagaries of the weather and, like many others, wanted to see if it would, one day, be possible to predict the weather with the same accuracy that we can foretell the motion of Halley's comet. At that time, the prediction of local weather, beyond a few days into the future, was widely seen as essentially one that would be solved using more powerful computers and with better information about initial weather conditions, to come from satellites.

Lorenz worked with some 'toy' meteorological equations, which caricatured the evolution of the weather – equations that specified how three quantities relating to the weather would change as time progressed. He was astonished to find that his equations had a remarkable property: their solutions were extremely sensitive to initial conditions.[8] If one set of initial conditions led to a certain outcome, another set – even if they differed from the original ones by the merest whisker, smaller than atomic dimensions – would after a short time lead to a completely different outcome. The reason was that his simple equations were behaving chaotically. They were actually caricaturing something we now know for sure about the local weather – that it is impossible to predict it for certain in the long term (in practice, longer than about seven to twenty days, depending upon the details of the present conditions).

The Lorenz equations are 'differential' equations, featuring *changes* with respect to time that progresses continuously, like the hand on a watch that turns smoothly, not one that jumps from one second to the next. This is in stark contrast to the way time is treated in the Logistic Map, which looks at snapshots taken at discrete times. Scientists are much less familiar with discrete maps than they are with differential equations, which until then were believed to be well-behaved and entirely predictable. So it was quite a shock to many scientists that even some simple differential equations, ones routinely thought to be understood, behaved with bizarre unpredictability.

For almost a decade, Lorenz's work had made next to no impression outside the relatively small group of scientists interested in weather. One of the reasons for this is that it requires a good deal of mathematical ability to deal with these equations, and great computational skill to represent their behaviour in pictorial terms. Some mathematicians even disputed that the Lorenz equations really did exhibit true chaos, suggesting that the apparent chaos might be an artefact of the numerical approximations that were used to study the equations. It was only in 1999 that the PhD student Warwick Tucker at the University of Uppsala in Sweden demonstrated rigorously that the Lorenz equations are definitely chaotic.[9]

In my view, the Lorenz equations might well have remained a piece of meteorological arcana if it had not been for the proselytizing of Jim Yorke and others in the early 1970s. Their task of persuading their fellow scientists of the importance of chaos was made much easier after the behaviour of the Logistic Map had been laid bare – everyone could appreciate the importance of chaos in terms of this trivially simple example, even if it was a discrete map of the sort that most scientists rarely used. But Lorenz's work demonstrated that the phenomenon of chaos existed in *differential* equations, featuring quantities that change smoothly and continuously, and it was this that made the majority of the scientific community sit up and take note. By the early 1980s, it was commonplace for scientists to be looking at their work with new eyes and, with their increasingly powerful computers, checking to see whether they had previously overlooked chaos in the phenomena they had been studying.

For most scientists, the unpredictability of some of the equations in classical science was a revelation. Since the 1920s, scientists had known that unpredictability is a crucial component of quantum theory, which describes the atomic and subatomic worlds. Quantum theorists know that it's possible only to predict the probabilities of the behaviour of an atomic electron. Few scientists expected that unpredictability was lurking beneath the simple equations that they and their predecessors had been using for two hundred years.

This revelation brought some fascinating insights. Early on in the chaos saga, my friend Henry Horn, a Princeton University ecologist, suggested that here at last is the reconciliation between free will and the foreordained nature of human fates as seen by Calvinism and some other religions. The Creator has placed us in a world of deterministic chaos, obeying defined rules with no random elements, but S/He alone can know the exact initial conditions that determine how the future unfolds. For us, the system's sensitivity to initial conditions means that it is unpredictable, and we interpret this as free will. Horn initially suggested this as a jocular aside, but it now enjoys a modest scholarly literature![10]

V

It was not only scientists who were becoming switched on to chaos in the early 1980s – the public were also becoming interested. The 'butterfly effect' became a buzz phrase that seems to have been used first after a presentation Edward Lorenz gave in Washington, DC, in 1972, entitled 'Does the flap of a butterfly's wings in Brazil set off a tornado in Texas?' Lorenz's non-mathematical talk about the unpredictability of the Earth's weather drew attention to the hair-trigger sensitivity of the planet's climatic system which would only come to seem understandable in terms of the ideas of chaos. Yet the notion of a butterfly effect was, as Lorenz has noted, not new. It figured, for example, in Ray Bradbury's intriguing short story 'A Sound of Thunder', written in 1952, long before the DC meeting. In this story, the death of a prehistoric butterfly, and its consequent failure to reproduce, changes the outcome of a presidential election.

The popularity of the phrase 'the butterfly effect' is probably due to James Gleick's deservedly popular *Chaos*,[11] whose first chapter was named after the term. This terrific book, first published in 1988 and now a classic of science writing, brought chaos not only to the public but also to the many scientists who had not heard of it. In my view, Gleick's account has three great virtues. First, it gives a compellingly readable and accurate account of a new and difficult branch of science. Second, it uses a cast of colourful characters in an effective way, to give the story life. Third, and most impressive, it conveys a real sense of the nature of scientific advance, with all its existential, unplannable complexities. For me, *Chaos* gives a more serious and illuminating representation of science than you will find in all the formal philosophizing of Karl Popper.

The only weakness of Gleick's book is that the relative amounts of credit to the protagonists in his drama are not quite right. Jim Yorke, for example, is given too little credit, the colourful Santa Cruz kids too much; I think his description of my role in events is about right. These inaccuracies annoyed some of the experts, hence the book's rather curmudgeonly reception by some whose work it publicized so effectively. Of course 'the play's the thing' for most people, but the playbill and credits loom large to the actors. Rarely has a popular science book caught the public imagination so effectively with an unfamiliar idea. Soon after it was published, chaos became a topic of conversation, something opinion-formers had to know about. Even Al Gore, after running unsuccessfully for the US presidency in 1988, hired a mathematician to teach him the essentials of the theory.

Artists, too, became fascinated by the idea of chaos. It featured in numerous works of visual art and in many novels.[12] In too many instances in which the theory is cited in popular culture, it has been confused with the trivial and age-old observation that things are complicated, more complicated than you think. This is the chaos of Spielberg's *Jurassic Park* movie, whose script was adapted from Michael Crichton's novel; I could hardly wait for the dinosaurs to eat the silly 'chaoticist'.

But it would be wrong to give the impression that all references to chaos in the arts are shallow. For the last word, let me return to the wonderful *Arcadia*, which Stoppard wrote after Gleick's book had captured his interest. Here are Valentine's words on nature's fundamental unpredictability, one of the most profound lessons of the Logistic Map:

> We're better at predicting events at the edge of the galaxy or inside the nucleus of an atom than whether it'll rain on auntie's garden party three Sundays from now. . . . We can't even predict the next drip from a dripping tap when it gets irregular. Each drip sets up the conditions for the next, the smallest variation blows prediction apart, and the weather is unpredictable in the same way, will always be unpredictable. (Act 1, Scene 4)

An Environmental Fairy Tale

The Molina-Rowland Chemical Equations and the CFC Problem

Aisling Irwin

A photograph, snapped from an Apollo spacecraft, caught the mood of 1970s Earth. The image was taken in a fleeting, shadowless moment, revealing Earth's beauty as never before. It exposed the planet's isolation – an oasis of blue floating in the black unknown. Above all, Earth seemed as fragile as a bauble: from the perspective of space, its inhabitants seemed to share an overwhelming common interest in the care of their delicate planet.[1]

The image enhanced the sense born in the 1960s that humanity now had the power to destroy its environment and thereby destroy itself. The Apollo missions' message of universality inspired humans to see themselves as 'riders on the Earth together . . . brothers'.[2] The same sentiment drew tens of millions of people together on the first Earth Day, in 1970, in a mass protest against assaults on nature.

During this awakening of environmental awareness, a few short lines were published that were to have as profound an impact on our understanding of the Earth's environment as any cosmic image.[3] These lines were prose, but without words, written in the symbols of another language. They prophesied global calamity, confirming that humans were damaging one of the Earth's life-support systems. With sublime economy, these chemical equations described the destruction of the ozone layer.

These lines owe their origins in part to the mood of the times. In turn, they shaped that mood. Politically, they began the era when 'riders' from around the Earth have been forced to negotiate with each other to defend their habitat. Scientifically, they have stretched the boundaries of disciplines, ushered in international research enterprises that mix dozens of approaches in order to understand the most complex of natural cycles. Environmentally, they have supplied us with two symbols: of Earth's vulnerability under human stewardship and, conversely, of the human potential to avert technological catastrophe.

The story of these equations emerges over a period of about half a century, from 1930 to the mid-1980s. It is partly the unfolding tale of scientific understanding of the atmosphere – once thought to be simple and inert, now pictured as the unceasing turmoil of thousands of interacting substances. It was along the road to this understanding that questions about ozone began to emerge. The process of answering them helped forge the idea of the planet as a single system – long, proliferating chains of cause and effect linking everything from microbes in the soil to obscure gases in the stratosphere.

To understand how scientists have extracted their understanding of ozone from the muddle of the atmosphere one must have a feel for the work of the chemist. Historically the chemist's destiny has been to search among the confusing manifestations of matter for its essence, to traverse the changeable material world in search of what is unchangeable, to find permanence, predictability and rules about matter. Chemistry is often seen as the dull relative in the scientific family, erroneously regarded as a purely descriptive science that lacks the glamour of physics or biology. It seems far from the physicist's struggle with fundamental forces and particles, or the mathematician's pure and abstract thought. Yet it was chemists who possessed the tools with which to reach into the cauldron of atmospheric reactions and focus on the important one. Then they were able to express it using a simple, symbolic language that has been centuries in the making. Chemists were able to predict interactions occurring 50 kilometres above the Earth – without going there – and even to determine the speed of these reactions. The chemists revealed the full extent of their power, however, by combining with other disciplines to produce models of the atmosphere, and make predictions that have been validated over ensuing decades.

Few equations have been so revelatory about humans' relationship with their environment, or have had such a dramatic effect on the world. And no other chemists have produced, in so few lines, work now credited as having the potential to become our 'salvation from environmental catastrophe'. These were the words of the Nobel Prize committee that gave the 1995 chemistry award to Mario Molina, Sherry Rowland and Paul Crutzen for their study of ozone destruction. It was the first time a Nobel Prize recognized research into a man-made impact on the environment.

It is a modern idea that there is anything damageable in the atmosphere. Air and its immutability were always taken for granted. From ancient times, people thought air was inert: the chemistry of the world happened below. The idea that there was a third physical state in addition to solid and liquid – that around and above us there could be diverse gases interacting with each other and with Earth below – was an innovation of the eighteenth century. After this conceptual shift, scientists added increasing amounts of complexity to their vision of the atmosphere. Chemical equations in their modern form have provided one way of expressing this.

Individual atmospheric gases such as carbon dioxide, oxygen and nitrogen only began to make their personalities known after 1750. In the twentieth century the atmosphere yielded to new techniques to reveal its hidden reaches. We now visualize the atmosphere as a series of thick spheres of protective 'airs', each more tenuous than the last and each cosseting Earth against the depredations of cold, oxygen-free, radiation-filled space. The first sphere hosts most human activity – it is where we live, fly most of our aeroplanes and experience the weather. These first 10–15 kilometres are known as the troposphere. Above it, where supersonic jets make brief appearances, is the next layer, the stratosphere. The next spheres are virtually empty and peter out after a few hundred kilometres, at the nearest edges of space.

But to use such simple divisions is deceptive. Earth is in fact both audience and actor in an elaborate atmospheric performance. Thousands of different substances mill over the globe. They waft at the whim of heat and cold, day and night, increasing and subsiding pressures, fluctuating solar radiation, seasons and dynamics played out to daily, yearly and longer rhythms. Molecules meet each other and react according to the dictates of position, time, temperature, light, pressure and the presence or absence of other molecules – and many of these factors are to some degree unknown. In the early 1950s scientists knew of 14 atmospheric compounds. Now they can identify over 3,000.

Modern chemistry has a panoply of origins: in the arts of metallurgy and brewing; in the ancient philosophers' puzzles over the nature of brute matter and the distinction between substance and form; in the mystical obsessions of alchemists. To explain the basics of matter, the latter groped for underlying principles such as the four Aristotelian elements (earth, air, fire and water), the seven metals, the universal spirit and the philosopher's stone. They understood the natures of substances through their links with the planets, with mythological characters, with theology. They represented them using symbols, colours, pictures, secret names and codes.

The achievement of chemistry over the last two hundred years has been to destroy these romantic underpinnings. What was a 'rambling, puzzling and chaotic' subject[4] struggled to rebuild itself on less elusive fundamental principles. The rudiments of these emerged in a burst of discoveries and insights in the eighteenth century, a period now known as the chemical revolution. Mystery and obscurity were replaced with simplicity and transparency of expression. The interactions of matter were no longer conveyed with the nebulous imagery of animals, monarchs and maidens, but with simple equations that could reduce a chemical story to its basics: beginning, middle and end.

Today we classify matter into over a hundred basic elements, ranging from familiar ones such as carbon and gold to obscure elements such as unnilquadium and rutherfordium, forced into brief existences by capricious scientists. The idea that basic elements lay at the heart of things originated from that master of the Chemical Revolution and victim of the French one, Antoine Lavoisier. He was an ambitious Parisian intellectual whose shares in the Ferme Générale, the loathed tax-collection company of the *ancien régime*, financed his science but lost him his head during the Reign of Terror. He defined elements, pragmatically, as substances that cannot be decomposed into anything simpler. The concept behind them was one of immutability and purity, the idea that an element is always the same regardless of its origin or method of preparation. Each element should have a name, he said, and if two elements combined to form a more complicated substance, its name should reflect the two elements that went to make it. Names, argued the idealists of the Chemical Revolution, should be abstract, with no meaning in ordinary language, so that they 'recall no idea that indicate false resemblances'.[5] In practice some names still evoke images of their discoverer, of their colour or even of a nearby planet.[6]

There remained the question of what these basic elements were themselves composed of. It had been tempting from early times to suppose that everything was ultimately composed of that popular stuff, primary matter – originally considered by Plato and Aristotle to be a featureless substance onto which various qualities and properties could be impressed. John Dalton, a Manchester teacher who joins Lavoisier as one of the architects of modern chemistry, proposed that the fundamental elements were composed of atoms. The atoms of any one element were identical but they differed from the atoms of other elements. An atom consists of a positively charged nucleus surrounded by a cloud of negatively charged electrons. Its unique identity lies in the number of positively charged protons in its core. The chemist is mainly concerned with understanding how interactions between atoms are governed by a common market in each one's outermost electrons. Atoms can be seen as engaged in a constant quest to find the perfect partners with which they can bond to form stable entities by sharing and exchanging electrons.

For each type of atom, endowed with a different electron distribution, stability is achieved with a different number and combination of these partners. Some atoms, such as chlorine (represented, for ease of discussion, as Cl), are too reactive to exist as single atoms and so they are most commonly found as diatomic molecules (represented as Cl_2). It is a similar story for oxygen atoms (O), which exist most stably as O_2, common oxygen. But oxygen atoms can also exist in a less stable form, as three connected atoms, the form known as ozone (O_3). This distinction, between diatomic oxygen and triatomic ozone, makes the difference between a colourless, odourless gas essential for respiration and a light-blue, pungent gas – a component of smog, notorious for its toxicity.

Five billion tonnes of ozone float in the stratosphere, up to 50 kilometres up, sheltering life below from the less benign forms of ultraviolet light. Ozone allows unmolested passage to Earth to the mildest (longest-wavelength) form of ultraviolet, known as UVA, which serves useful purposes such as triggering the manufacture of vitamin D in human skin. But ozone blocks the passage of more aggressive forms of ultraviolet light, UVB and UVC, which would otherwise make life impossible. UVB and UVC can weaken the human immune system, leaving us less able to fight disease. They can attack the skin and the eyes, leaving cancer and cataracts in their wake. They will destroy the simple life form of phytoplankton, which lies at the bottom of the oceanic food chain and whose absence could therefore cause entire ecological systems to collapse. Green plants – and therefore agricultural crops – are also vulnerable to its rays. Indeed, life could not have emerged from the water to live on the land until there was sufficient ozone in the atmosphere – that was about 420 million years ago. Ozone built up as part of the gradual change in Earth's atmosphere from one rich in carbon dioxide to one rich in oxygen.

A billion years after ozone formed, humans evolved to a state where they became capable of destroying it. Luckily, this was almost exactly the same moment when they began to understand ozone. It took a number of conceptual steps to comprehend the natural formation and destruction of the ozone layer, and then to perceive its vulnerabilities.

The year in which to begin this story is 1930, for three reasons. First, scientists unveiled the delicate mechanism by which ozone is naturally produced and destroyed in the stratosphere. Second, the noted American chemical engineer Thomas Midgley announced his invention of the useful chemicals known as CFCs (chlorofluorocarbons). Third, the Nobel Prize-winning physicist Robert Millikan (discoverer of cosmic rays) observed that there was little chance that humanity could cause significant harm to anything so colossal as the Earth.[7] It would be another forty years before scientists linked the first two pronouncements to reveal the error of the third.

When ultraviolet light of middling energy (UVB) reaches the ozone layer it generally meets an ozone molecule. Ultraviolet light can rupture bonds in most molecules – it is just a question of finding the frequencies to which that bond is vulnerable. UVB can rupture ozone, splitting it into diatomic oxygen and a free oxygen atom. When it exerts its shattering effect, a phenomenon known as photolysis, the resulting free oxygen atom is left in a highly excited state, searching for a new partner. Ultraviolet light can also break the robust bond in an oxygen molecule but in this case it requires the highest energy form, UVC. This breaks the oxygen molecule into two oxygen atoms. In the first reaction, ozone is destroyed; in the second, common oxygen is destroyed. A third makes up the cycle. The free oxygen atoms produced in the first two reactions are aggressive creatures, seeking to form new bonds as soon as possible. As soon as a free oxygen atom meets a diatomic oxygen molecule, it will latch on to it to form ozone again. If it meets, instead, a molecule of ozone, it can rob it of one of its oxygen atoms, turning both itself and the ozone molecule into diatomic oxygen molecules.

This cycle can be portrayed with some simple equations, using the symbols described above: O to represent a single oxygen atom; O_2 for two such atoms that have bonded to make an ordinary oxygen molecule; and O_3 for ozone. Using an arrow to depict chemical change, we produce one equation for the breakdown of ozone –

$$O_3 \rightarrow O_2 + O$$

– and another for the breakdown of diatomic oxygen:

$$O_2 \rightarrow O + O$$

The cycle is completed by the production of ozone described above:

$$O + O_2 \rightarrow O_3$$

An arrow, rather than an equals sign, links the two sides of the chemical equation. This is because the two sides are not equal in every sense. There are different chemicals on either side, with different characteristics (the blue, poisonous ozone and the colourless, life-enhancing oxygen). The arrow represents passage through time during which chemical interactions occur to produce new entities. But the two sides *are* equal in the sense that the number of atoms is conserved – none can magically appear or disappear. There are three oxygen atoms on either side of the first equation (and the same is true for the second and third equations).

The cycle continues, producing and re-forming ozone, and all the while the rupture of bonds absorbs energy and the formation of new bonds releases energy as heat.

The description of this cycle, by the English scientist Sidney Chapman, has a sequel that did not occur until forty years later. Chapman's equations did not fully explain the natural production and destruction of ozone. Calculations based on his work, and on the rates of the various chemical reactions involved, implied that ozone should have been present at much higher levels in the stratosphere than were actually observed. Scientists knew that there must therefore be another mechanism at work helping to break down ozone as fast as it was produced, in keeping with the steady levels recorded by instruments. It took those four decades for the final major participant in the natural ozone cycle to be identified – and when it was found, it turned out to be back on Earth, in the soil.

The discoverer of this cycle – Paul Crutzen – has made manifold contributions to understanding the ozone layer, the first of which began when he was twenty-six and took a job in the meteorology department at Stockholm University. It was the late 1960s and Sweden was abuzz at the time with the discovery of acid rain, perhaps the first environmental problem to spread across entire regions and therefore an important prelude to ozone depletion. But Crutzen wanted to study natural processes, so when the opportunity for research arose, he chose to work on stratospheric ozone.

By 1970 Crutzen had found that the missing agent in natural ozone destruction lay tens of kilometres below the ozone layer. Bacteria in the soil produce a particular oxide of nitrogen (N_2O) in small amounts. Crutzen saw that this oxide diffuses up through the troposphere, gradually changing into other, more reactive nitrogen oxides as it goes. These gases ultimately rise as high as the ozone layer. Ozone, as we have seen, is easily broken apart. One of the nitrogen oxides is called nitric oxide (NO) and it can snatch an oxygen atom from an ozone molecule and later pass it on to a free oxygen atom, turning it into diatomic oxygen. The net result is that it removes ozone by converting it to ordinary oxygen molecules.[8] Crutzen had supplied the missing link in natural ozone-layer chemistry and ushered in two important concepts that scientists would employ later in the story: that stable molecules from Earth may be able to diffuse upwards into the stratosphere; and that up there they may break down ozone.

The protagonist in the second seminal event of 1930, Thomas Midgley, was an American chemical engineer, born into a family of inventors. By the end of his life Midgley was the holder of over a hundred patents and president of the American Chemical Society.[9] He had made his name in 1921 for discovering that lead added to petrol could reduce engine knocking. Later, having transferred to General Motors Research Corporation's Frigidaire division, he announced the invention of dichlorodifluoromethane, the first of the family of chemicals that became known as CFCs. For these two inventions, one environmental historian has awarded him the accolade of having had more destructive effect on the atmosphere than any other single organism on the planet.[10]

Midgley's invention was a chemical of extraordinary placidity. It did not burn, barely dissolved in water and was not toxic. Its architecture – a central carbon atom surrounded by fluorine and chlorine atoms – was extremely stable. In the atomic quest to bond with the ideal partners and, through sharing of electrons, achieve stability, Midgley had produced the supreme combination of atoms, a molecule uninterested in further interaction with the outside world. He demonstrated this impassivity in front of an audience of chemists by inhaling a lungful of the gas and then exhaling it over a flame, which it extinguished. The fact that he emerged from this demonstration unharmed and did not exhale plumes of fire, ensured that CFCs had a memorable entrance into the scientific world. Although it was some time before industry harnessed CFCs, they were regarded as wonder molecules, ideal refrigerants because they boiled between –40°C and 0°C (depending on the CFC), and were cheap to manufacture and easy to store. Above all, they were safe.

Chemical symbols can depict a CFC more succinctly than words, using

C for carbon atom, Cl for chlorine and F for fluorine. A simple CFC consisting of a carbon, three chlorines and a fluorine, for example, is represented as $CFCl_3$. CFCs came into full use after World War II as aerosol-spray propellants, as refrigerants, in air conditioners and as cleaning fluids for electronic components – emissions rising from about 20,000 tonnes annually in the 1950s to about 750,000 tonnes in 1970. Their inertness was the key both to their capacity for devastation up in the stratosphere and to the failure of scientists to recognize this destructive power.

It was not until the late 1960s that the suspicion arose that we might be capable of disrupting natural cycles on a global scale. This change in attitude was partly driven by developments in science but it also required a change of intellectual paradigms in which certain questions were newly askable. It required the 1960s – the era of social unrest – notably the anti-war, civil rights, feminist and environmental movements. Out of this social ferment and activism emerged a new environmentalism, one that was apocalyptic in nature and global in outlook, and characterized by Rachel Carson's *Silent Spring*. It was unlike the environmentalism that had reigned before, which had been a reaction against industrialization and a romantic hankering for a pre-industrial arcadia. In the United States, anger and fear about the pollution of the countryside with pesticides, the death of lakes and the poisoning of rivers led the government to set up the Environmental Protection Agency. Two years after it did so, the first international conference on the environment, in Stockholm, heard the first official complaints that pollution produced by some countries was being washed out of the sky by the rain and falling into other countries. North America and Europe were thus forced, through the recognition of acid rain, to embrace the idea that artificially produced chemicals could interact with natural atmospheric processes on a scale that ignored national boundaries.

People then started to question the global environmental impact of some proposed new technologies. The first that was relevant to the ozone layer was the plan for a fleet of supersonic aircraft. An Anglo-French group proposed (and eventually built) Concorde; and there were also Soviet plans for a similar plane. In the United States, Boeing had a less advanced scheme to fill the skies with 800 such planes between 1985 and 1990.

Eight hundred supersonic jet planes roaring through the stratosphere, puffy exhaust trails left in their wake. Their routes would be through the ozone layer, their exhaust plumes nitrogen oxides. It was just a year since

Crutzen had shown that natural nitrogen oxides destroy ozone. It did not take long for scientists to realize that human-made nitrogen oxides could do the same. One researcher calculated that 500 supersonic craft might deplete ozone by 10 per cent within two years. Crucially, scientists spoke of the link with cancer, predicting that a 1 per cent decrease in ozone concentration would cause an additional 5,000–10,000 new cases each year in the United States alone. In this way a public already anxious about environmental destruction became familiar with the idea of the ozone layer as a delicate sheath around the planet whose integrity was vital for their own welfare. Soon attention was drawn to plans put forward by the American space agency NASA for a fleet of shuttles making weekly trips to outer space, their exhaust plumes filled with compounds containing chlorine. Scientists, notably Richard Stolarski and Ralph Cicerone of the University of Michigan, found in chlorine the same aggression towards ozone that Crutzen had found in nitrogen oxides.

Chemically, the concepts were all there, waiting to be put together to deduce ozone destruction by CFCs. Philosophically, the Western world was waiting for signs that it was propelling Earth towards destruction. But there is a deep gulf between what scientists understood about ozone destruction up to this point, and the idea that CFCs too might be culprits. The insights so far related to the sensational and violent release of aggressive chemicals into the ozone layer by futuristic planes and shuttles. CFCs were quiet, unreactive, and had been around for two decades already. In the theatre of the atmosphere they remained mere members of the audience.

The minds that made the connection between CFCs and ozone destruction had undoubtedly been prepared by the events of the preceding decades. Unwittingly, Mario Molina and Sherry Rowland possessed virtually all the chemical concepts necessary to make the link. In addition, they were already concerned about the adverse impacts of technology. Student activism had forced Molina to consider the public's fear of the demons that might emerge from laboratories. In 1968, as a researcher at the University of California, Berkeley, he had faced protests against research work on lasers, a reaction against studies elsewhere into the use of high-power lasers as weapons. Molina says he was 'dismayed' at the weapons link: 'I wanted to be involved with research that was useful to society, but not for potentially harmful purposes.'

Scientific curiosity with a touch of environmental awareness is what Molina says drove him to make the link between CFCs and ozone depletion. 'There was some environmental sense but it was very vague. It was to me more the idea of humans changing the environment without being fully aware of what consequences that might have – we were feeling some responsibility to really assess the consequences.'[11]

Molina and Rowland began their work on CFCs via an unexpected route: a suggestion by an English scientist that CFCs might serve as useful research tools for atmospheric researchers. The idea was that the ability of CFCs to waft, eternal and unmolested, through the atmosphere could be exploited by meteorologists in the effort to track air currents. The widespread use of CFCs over the past few decades meant that humans might have distributed them around the globe as if in deliberate preparation for a scientific experiment.

It was one of those crazy, brilliant ideas that often come to nothing but occasionally strike gold. James Lovelock, a scientist who works outside the establishment from his country house in Devon, has had several such ideas in his lifetime.[12] One of these was the Gaia model, born in the 1970s, which proposes that life on Earth regulates its own environment to keep itself healthy. Gaia is Earth as a single, vast organism, which, through a variety of biological feedback processes, is always moving towards the preservation of itself. Examples of such regulation are now plentiful: for example, plants and bacteria are believed to help control the temperature of the planet by removing carbon dioxide from the atmosphere and depositing it in the soil. But the idea, with its connotations of a living, sentient planet, was disputed by many scientists, particularly for being incompatible with Darwinian natural selection. Lovelock, who is in fact a rigorous scientist, has been the target of numerous bitter verbal attacks throughout his life as a result.[13] His proposal to track CFCs was also rejected by the scientific establishment – one referee remarked that he could see no possible use for the results, even if the mission were a success. So Lovelock funded himself, and joined a British Antarctic Survey supply ship, the *Shackleton*. He took with him a highly sensitive instrument he had developed for measuring extremely low levels of gases in the atmosphere. The invention was a decade old, and had already been used to detect minute concentrations of pesticides and other pollutants. It enabled scientists to discover the extent to which DDT had spread through the natural world and thus helped to provoke the upwelling of environmentalism in the 1960s.

Lovelock journeyed through the north and south of the Atlantic Ocean measuring CFC levels and on his return he brought a priceless piece of information. He had managed to calculate, by extrapolation from his measurements, the quantity of CFCs loose in the atmosphere. When he calculated, on the basis of industry figures, the quantity of CFCs that had so far been released into the atmosphere, he found that the two figures were very similar. The conclusion? CFCs do not break down and would probably linger in the atmosphere for ever.

This crucial piece of information was picked up in an odd, sideways way by Sherry Rowland, who was by then a chemist of high repute at the University of California at Irvine. Rowland, at forty-three, was well into a brilliant career in radioactive chemistry. He had entered academia running, propelled by a bright family, and fast-tracked through school and university. As a postgraduate chemist at the University of Chicago, he had worked for the inventor of radioactive carbon-dating, Willard Libby. For the previous five years Rowland had chaired the chemistry department, a job he relinquished in 1970, whereupon he began the search for a new research topic. About then, he became concerned about the environment because of its public prominence and his family's interest in the subject. So his attention was caught by the prospect of a conference on the application of his own subject – radioactivity – to environmental issues. After attending the conference in Salzburg, Austria, he met a fellow delegate on the train and heard of a new series of workshops aimed at improving understanding of the atmosphere by encouraging chemists and meteorologists to talk to each other. Rowland had always harboured an interest in the atmosphere, because of his work as a youthful graduate student with Libby. So he went to one of the workshops, held in Fort Lauderdale, Florida, in 1972. There he heard a scientist describe Lovelock's findings, second-hand, and promote the idea that scientists could use the inert CFCs to trace atmospheric movements.

But Rowland was a chemist and he knew that no molecule lasts for ever in the atmosphere, if only because everything will eventually work its way upwards into the stratosphere, where solar UV radiation will break it down. He wondered what the eventual fate of CFCs would be, and this curiosity led him to the heart of the story. Rowland took the problem back to the University of California at Irvine, where, in 1973, he began work on it with Molina, a thirty-year-old Mexican who had just joined him as a postdoctoral fellow. Molina was a fanatical scientist, who had been doing chemistry experiments since the age of eleven in a bathroom his parents had converted for him. He jumped at the task, whose subject was far from any of his previous scientific experience. Within three months they had done the work that was to shock the industrialized world and change their lives for ever.

First, Molina worked rapidly through every possible fate for CFCs in the troposphere – the 'sinks', such as oxidation or dissolution in rainwater, by which the atmosphere disposes of most molecules. He found nothing that could prevent CFCs from rising as far as the ozone layer. Next, he calculated how long it would take for a single CFC molecule, newly emerged from the puff of an aerosol can, to soar high enough to be destroyed: it would take half a century. So, a puff of spray that enhanced the fragrance of Brigitte Bardot in 1970 has been circulating for over thirty years, and is still a couple of decades away from destruction.

Molina and Rowland realized that once a CFC wafted sufficiently high, it would swiftly fall victim to ultraviolet light of an energy that could break one of its bonds, with the release of an aggressive chlorine atom. They might have finished their research at this point, with the demise of the CFC, but they decided instead to follow through their work and discover the fate of those shattered fragments. First, they had to consider the molecular interactions, and after this they had to fit this busy chemical scene into the theatre of atmospheric dynamics. As they worked with the various substances that might chemically interact, they made use of earlier research in chemical kinetics, the study of the speed with which molecules interact and the way in which these reactions occur. Painstaking work by chemists had already shown that an experiment conducted in the laboratory can reveal how rapidly a particular reaction takes place, even if that reaction involves chlorine atoms cavorting somewhere inaccessible, such as the cold, rarefied regions of the stratosphere. Scientists had already performed many of these experiments and recorded the results, so what could have been a decade's work for Molina and Rowland was done in just a few days.

The two then fitted these insights into models of the dynamic processes that control ozone movement. Molina, who now works at the Massachusetts Institute of Technology, told me:

> The challenge is: how does the chemistry function in this natural system as opposed to in the laboratory? It is easy to get lost in the details. You have to synthesize the essential features of the workings of the system to really have confidence in the results. It is something we now understand very well: how does the atmosphere function? But as atmospheric scientists we were just emerging at that time.

Molina remembered this time as a solitary one – working away on his own on what were often quite mundane tasks, regularly reporting to Rowland for guidance, and all the while building towards a conclusion that looked increasingly exciting. Much of the work was with pen and paper – making calculations, or sketching the rudimentary representations of molecules that chemists play with when they explore the potential courses of reactions. He also carried out a few simple experiments to analyse in detail what happened to various CFCs in the presence of ultraviolet light.

Models of the atmosphere at the time dealt only with the movement of gases upwards and downwards, picturing them as if in a column. The models could not tell researchers anything about variations from latitude to latitude or from season to season. Yet, with skilful handling, these rudimentary models produced a striking answer: the reactive chlorine atom would behave much as Crutzen's naturally produced nitrogen oxides do, devouring ozone. First, it would snatch an oxygen atom from ozone, leaving it as ordinary diatomic oxygen. Then it would pass this oxygen atom on to another, free oxygen atom, creating a new molecule of diatomic oxygen. The chlorine would thus be mopping up O_3 and O, and producing O_2 in the process. Crucially, when the chlorine atom finished carrying out this transfer, it was free again to begin another round – and indeed it can go on and on, wiping out thousands of ozone molecules as it goes. It stops only when it encounters a different substance, one to which it can hold fast – and which is thus known as a termination molecule. Once it is in this stable form, the chlorine's rampages are at an end and it drifts back down through the atmosphere, where the rain washes it away.

On average, according to the model, a single chlorine atom would wipe out 100,000 ozone molecules before meeting this fate. Eventually the production and destruction of ozone would reach a new equilibrium, in which the presence of CFCs had caused an approximately 10 per cent drop in levels of stratospheric ozone. The consequence: UVB rays would penetrate Earth's defences and strike living things below, triggering tens of thousands of extra skin cancers each year and exposing animals to disease by disrupting their immune systems.

Using the language of chemistry, Molina and Rowland captured the essence of their disturbing message. The CFC breaks down, releasing chlorine:

$$CFCl_3 \rightarrow CFCl_2 + Cl$$

The chlorine attacks ozone and produces an oxygen molecule:

$$Cl + O_3 \rightarrow ClO + O_2$$

And then produces another:

$$ClO + O \rightarrow Cl + O_2$$

The equations are perhaps best followed by charting the story of just one of the actors – the chlorine depicted in the first equation. For fifty years it has existed in the form symbolized on the left of the first equation – as part of a stable CFC. It is rudely snatched from this comfortable situation on exposure to ultraviolet light – emerging as a lone and aggressive chlorine atom on the right of the first equation. By equation two it has sought out an ozone molecule. This second equation describes how the chlorine takes one of this ozone's oxygen atoms to form a brief liaison. The final step is when our chlorine loses this oxygen atom, emerging at the end of the third equation as a free, aggressive chlorine once again.

The chemists recognized the sensational nature of their discovery and the importance of communicating it swiftly to the outside world. They published in the leading science journal *Nature*, where their equations appeared, embedded in a description of their work and squeezed into less than three pages of the edition published on 28 June 1974. The authors waited for the public reaction. Nothing happened.

Their work went unnoticed by the public and by the science journalists who are supposed to interpret for them. The two scientists realized that their paper was inaccessible, their warnings masked by the language of scientific journals. Their paper, 'Stratospheric sink for chlorofluoromethanes: chlorine atom-catalysed destruction of ozone', made but brief reference, in the penultimate paragraph, to the likelihood that 'important consequences may result' from the reactions their equations described, with the 'possible onset of environmental problems'. A supporting editorial was cautious in tone.

So Molina and Rowland held a press conference in September 1974, during a meeting of the American Chemical Society in Atlantic City, at which they also presented two further papers on the fate of CFCs. Molina explains:

> We thought the only chance society had was to get the public involved. We did not feel very comfortable in terms of the activities of dealing with the press and industry but we thought it was important to take that additional step. It was not difficult to decide to hold press conferences; we made up our minds early on to bring our findings to

the attention of the government and the public, realizing, though, that not everybody in the scientific community would applaud that decision. There is a certain resentment when scientists begin to publish their ideas in the news media.

During the press conference they called for a complete ban on the further release of CFCs into the atmosphere. There was no going back after they took this astounding step out of the scientific world and into the public one. By the end of 1974 their work was big news in America, splashed on front pages, featured in television programmes and supported by papers published by other leaders in the field. The next two years were a 'maelstrom', they later said. Their comments were of course a threat to the American chemical industry, which was at the time responsible for the bulk of the world's CFC production. Chemical manufacturers in the United States took on the scientists, pointing out that their claims were only theory, and that such shaky foundations were hardly good reason for undermining an entire industry with economic consequences. Against this, the scientists argued that the threat was so serious that waiting for 'proof' to emerge was not an option.

'The scientific ideas were not easy to explain at the beginning and so it was not an easy thing when we had confrontations with industry or other experts. These were our challenges,' says Molina. He believes their actions helped to alter attitudes among scientists about their responsibility to tell the public of their findings.

For two years the 'Incredible Stratospheric Travelling Road Show and Debating Society'[14] toured America. Molina and Rowland gave numerous testimonies at legislative, federal and state-level hearings along with Ralph Cicerone, of the University of Michigan, who had done similar research. The chemical giant DuPont, meanwhile, led the scientific opposition. One of its witnesses was none other than Lovelock, who, in 1974, appeared at a US Congressional hearing as a witness for the company. Although history will credit him with several contributions to the elucidation of the CFC story, he was initially sceptical of Molina and Rowland's claims. Years later, he had this to say to *New Scientist* magazine in his defence: 'Some might say I was a bloody fool, but I think I just did what came naturally. I liked the people [in the industry], they seemed to be a very honourable, decent bunch of scientists.'[15] In fact, always sceptical about man's power to destroy Earth in the face of such a formidable force as Gaia, he was unconvinced that a slightly increased exposure to UV radiation was a cause for concern.

Meanwhile, chemical manufacturers flew another scientist, Richard Scorer, who had denounced the research as 'pompous claptrap', around America to press home his message. Red-haired, fair-skinned campaigners against CFCs were particularly vulnerable to accusations of partisanship, and the debate frequently became 'vigorous and personal', according to Molina and Roland. They did not regret their actions, though.

The debate focused at this stage almost exclusively on the use of CFCs in aerosols, and little mention was made of their use as refrigerants or as blowing agents in the manufacture of expanded plastics such as polystyrene. Two years later, in 1976, the National Academy of Sciences produced a review of the science, supported by new research. This recommended the drastic curtailment of non-essential uses of CFCs unless new findings emerged within two years to mitigate the threat. This conclusion tipped the balance, and by 1978 several bodies, including the Environmental Protection Agency, had issued regulations on the phase-out of CFCs. Canada, Norway and Sweden also took action. By the beginning of the 1980s, an international response was growing. The United Nations Environment Programme (launched after the first international conference on the environment, held in Stockholm in 1972) fostered discussions on CFCs. By 1985 it had overseen the signing of the Vienna Convention for the Protection of the Ozone Layer by twenty nations.

The most astonishing aspect of this story is that so much action was taken despite the total lack of empirical evidence that CFCs were destroying ozone. The scientists had collected no data from the skies to show that ozone was disappearing. Indeed, as they later pointed out themselves, there had been no measurements anywhere in the stratosphere of any chlorine-containing substance.[16] The Vienna Convention was the first time that nations had agreed in principle to tackle a global environmental problem before anyone had either felt or empirically demonstrated its effects.

The explanation for this is to be found partly in the environmental awareness of 1970s America, in which several new bodies such as the Environmental Protection Agency were flexing their powers. As soon as Ronald Reagan replaced Jimmy Carter as American president in 1981, these freedoms were curtailed and further plans to reduce CFC production were shelved.[17] Had Molina and Rowland presented their work just a few years later, the initial response might have been very different. There was also the sinister nature of the threat, which presented concepts terrifying to the public, such as deadly invisible rays and skin cancer. Finally, the solution was not, in the grand scheme of things, costly. It certainly did not involve disruptive changes to individual lifestyles, and there was little public sympathy for an industry challenged over the production of luxury goods such as deodorant sprays.

The articulation of this issue was an early exploration of the precautionary principle – the idea that we cannot always wait for 'proof' of harm when issues of human health and environmental degradation are concerned. A corollary is that the burden should be on the manufacturer to provide proof of safety, rather than on the public to provide proof of harm. The CFC manufacturers were arguing the opposite, that they should be presumed innocent until shown to be guilty.

'They wanted the same rights as individuals,' notes Molina. In fact, the theory of ozone depletion by CFCs was an ideal basis for application of the precautionary principle. There could be dire consequences for human health if the scientists' warnings went unheeded – so the cost of delaying action might be very high. The proposed solution was not, relatively speaking, expensive or disruptive. A close parallel to this campaign was the one over another Thomas Midgley invention: leaded petrol. In Britain, the decision to phase it out came despite the lack of proof that it could damage children. The high risk could be balanced against the fact that the technology existed to provide an alternative, and cars could be adapted at no great cost.

Chemical equations can depict only fragments of reality. Molina and Rowland's brilliance lay in isolating a significant fragment, distinguishing it from the mêlée. The equations did not pretend to be a full account of activities in the ozone layer. Rather, they claimed to describe the most significant activities, viewed from the human perspective. The incompleteness of the equations reflected the incompleteness of the chemical knowledge of the time, and was emphasized eleven years later, when three British scientists announced in 1985 their discovery of the ozone hole. This was the moment when ozone depletion caught the international imagination, when Europeans became inflamed about the CFC issue, and the second phase of campaigning began. Yet the discovery was a shock as much to scientists as to anyone else. No one had been searching for a vast hole in the ozone layer over Antarctica, where it was discovered: they were looking for milder erosion at middle latitudes.

The fundamental problem for chemistry – a problem that is vividly illustrated by Molina and Rowland's efforts – is that it attempts to describe the real, material world, a place that is infinitely complicated. Chemistry – and the equations that represent it – can never be correct in the full sense of the word. Equations cannot enumerate all the conditions necessary for the reaction to occur, the conditions that may prevent the reaction occurring, and so on. Their purity is thus achieved at the expense of completeness. An equation runs the risk of becoming untrue irrelevant when external circumstances change. This is what happened with the discovery of the ozone hole.

The three scientists who stumbled on the hole over the pole were in their offices in Cambridge at the time, working for the British Antarctic Survey.[18] One of their routine jobs was to process data arriving from Halley Bay, where ozone levels in the stratosphere had been monitored since the 1950s. Their measuring instrument – known as a Dobson spectrophotometer – began to surprise them in the early 1980s with persistently low readings for ozone. Brian Gardiner, Joe Farman and Jonathan Shanklin suspected an instrumental error. Between them they had extensive experience of the Antarctic. They had worked in unearthly landscapes of blizzards, white cliffs and perspectiveless plains with only the attentions of penguins and the noise of cracking ice for company. They knew about frozen nights in research tents, and about tramping for hours over the ice to collect the readings of some lonely instrument. They knew that the extreme cold and the technical difficulties of operating remote instruments could distort an instrument's readings, so vigilance was required with the data it produced.

'The Antarctic was logically the very last place on Earth that you would see ozone depletion,' Gardiner comments.[19] But as they tightened their work and improved the accuracy of the measurements, the depletion figures grew larger rather than smaller. Each year was the same: severe depletion appeared suddenly, in October, at the beginning of the Antarctic spring, and lasted for two months, after which levels gradually recovered. The Dobson spectrophotometer seemed indeed to be staring up at a massive void of ozone, looming out of the Antarctic wastes. To unnerve the BAS scientists further, their instrument's more technologically advanced rival, aboard NASA's Nimbus satellite, which orbited above, had noticed nothing amiss with the ozone layer.

After three years, Farman, Gardiner and Shanklin believed they had tested their readings with sufficient rigour. They could no longer deny what the data before them revealed: that every October a third of the Antarctic ozone went missing. Gardiner, who now heads the meteorology and ozone unit at the BAS, speaks of his 'awful realisation' that the effect was real, that they had found a problem of global significance. Balancing scientific caution against the responsibility to make a warning as soon as possible was a stressful experience.

They published in 1985, their paper describing a vast and sudden emptiness over Halley Bay, a hole far deeper than Mount Everest, stretching from 10 kilometres upwards to 24 kilometres. Each year the depleted ozone seemed gradually to replenish itself only to disappear again, a celestial Cinderella, at the first sign of the end of the long Antarctic night. NASA's Nimbus satellite, watching Earth from above with a variety of mechanical eyes, was soon persuaded to produce similar data. NASA scientists found they had transmitted their complacency about ozone levels to their computers in the form of an instruction to the Nimbus instruments to ignore very low ozone readings as they were likely to be errors. Once corrected, Nimbus began to produce compelling images of the globe in which a dark patch over the Antarctic grew ominously larger with each successive year. The vivid portrayal of the size of this man-made stain on the world shocked both the public and the research community.

'I remember it happening – the results coming in,' says Professor Alan O'Neill, now a leading atmospheric scientist at Reading University. 'No one had any idea what was going on.'

This was what Molina and Rowland had failed to predict. Chemistry is far messier than its equations can ever describe; apart from the factors they had considered, others were at play.

The subsequent efforts to reconcile the claims of Molina and Rowland with the findings of the British Antarctic Survey proved to be a unifying force in atmospheric science, revealing that the atmosphere was a subject greater than any one scientific discipline. A new kind of scientific mind was born among meteorologists, physicists, mathematicians and chemists who were used to remaining within their disciplinary pens. This mind could glide easily between disciplines, content with imperfect data, partial results, models with question marks in them, blanks where nobody knows what is going on.

The work over the next eighteen months was done under pressure because of the urgent need to fix the hole and the political need for scientists, representing 'truth' to the outside world, to achieve consensus within. The Molina–Rowland equations had of course given the work a great head start and the two scientists, together with Crutzen and other key players, were soon hard at work trying to find explanations. Various disciplines engaged in intellectual battles to explain what had caused the hole.[20] Everyone wanted the problem to lie in their own domain, according to Gardiner. The troposphere lobby argued that it was an effect at the bottom of the atmosphere that had worked its way upwards. Mesosphere scientists argued that it had descended from above. Stratosphere experts supported sideways effects, with depletion occurring elsewhere in the globe and shifting south to the Antarctic. Solar physicists looked to the fluctuating activity of the Sun for an answer. Meanwhile physicists invented swirling diagrams to explain that the issue was not one of depletion but of redistribution of ozone around the globe, driven by the temperature changes during the onset of Antarctic springtime. Chemists, on the other hand, played with the unusually cold circumstances of the Antarctic winter to see how this might prepare the surroundings for a burst of ozone-depleting chemistry at the beginning of spring. Some of the theories were obviously ludicrous from the start but perhaps it was best to explore and discard them at the time, so they could not re-emerge later with any credibility. The scientists reached consensus with amazing speed and many commentators now regard this work as a model scientific response to an environmental problem.

The chemists won. Several teams suggested that the answer lay in the chemistry of the bitter Antarctic winter. From April to August the Sun eludes the Antarctic and a strong wind whips around the South Pole, isolating its air from the rest of the atmosphere. In this freezing, segregated world, vast wreathlike clouds form high in the sky. Antarctic explorers have long spoken of these ghostly clouds, which retain the glow of the Sun long after it has ceased to shine on the ground below. The clouds are full of ice crystals, finely dispersed solids that lie at the heart of the reason for the rapid ozone destruction. No one is quite sure of the mechanism, but it is known that on the surface of these solid ice particles reactions can take place that would not otherwise happen. The crystals act like a stratospheric dating agency, where gases that would otherwise have been ships passing in the Antarctic night can pause, meet each other and interact.

The scientists were able to show that the molecules that pause here to interact are the same 'termination molecules' (mainly hydrochloric acid and chlorine nitrate) that Molina and Rowland had already described. As we have already seen, the rampaging chlorine atom, after engaging in thousands of reactions with ozone, is finally captured as a more stable molecule, at which point its mischief is at an end and it descends to Earth. But in the presence of ice crystals these termination molecules dally and, on the crystals' solid surfaces, react to form new substances, sometimes the chlorine molecule itself (Cl_2). As soon as sunlight warms the Antarctic again, its ultraviolet rays set about shattering the chlorine-chlorine bonds, breaking them into active chlorine atoms, which thus return to their destructive activities. The Antarctic winter, therefore, provides unique conditions for a further episode of ozone destruction.

When scientists revealed the Antarctic ozone hole – swiftly made vivid and comprehensible by the Nimbus images – it entered the public consciousness outside America. Here, for once, was an imminent environmental disaster that ordinary people could do something to avert. All they had to do was alter their purchasing habits in minor ways. This link between the apocalyptic and the everyday caught the public imagination, launching the notion of environmental consumer power and leading many to do their bit by reading the small print on aerosol cans. People were further inspired by Prince Charles's announcement that even Princess Diana would confine herself to ethical sprays.

The discovery of the hole invigorated international negotiations. Today there are over ninety signatories to the latest amendment of the Montreal Protocol, with commitments to phase out a variety of other substances also suspected of depleting ozone. Although there are fears that China's production of CFCs could wipe out progress, many people now dare to speak as if the problem has been solved. Most scientists predict that the hole will be replenished sometime between 2050 and 2075.[21] The Antarctic ozone hole is still growing, but this is to be expected because of the time lag before CFCs reach the stratosphere. Perhaps we will indeed be able to gaze up at an intact ozone layer later this century, and regard the solution to ozone depletion as a textbook environmental response. Molina believes that the greatest impact of his work has been to create the sense of the world as a single system, which in turn strengthened the idea that human activities could affect the atmosphere of the entire planet.

'In some senses that became the first example – the science was really clear that these things can happen. But it also became a very important precedent in that these types of things could be solved.' Gardiner is more specific, arguing that we should learn from the ozone negotiations in tackling such issues as global warming[22] – we should 'attack the bigger problem of climate change by achieving intergovernmental agreement to limit the burning of fossil fuels before the consequences of that become a disaster'.[23]

Ozone has thus bewitched us, leading us to believe that because we have averted the first global environmental calamity, we can avoid further trouble in the same way. Just as nations managed to restrain their production of CFCs, surely they can cut back on the burning of fossil fuels to mitigate global warming?

But the next chapter of environmental problems is too complex to admit solutions as simple as those from the ozone story. Ozone was clean and simple; it was also urgent and overwhelming. Today's most notorious problems – global warming, deforestation, loss of biodiversity – are messy, insidious and slow to manifest themselves. In the case of global warming, fossil fuels are at the foundation of lifestyles and cultures. As Rowland himself told Bill Clinton, then American president in 1997, 'In the case of the CFCs these are gases that were manufactured by about twenty companies globally and all of them science-based, and the uses were almost entirely in affluent aspects of society. In the case of fossil fuels, everybody uses energy from fossil fuels nearly every day in every activity that they do.'[24] The ozone story was an extraordinary problem with an extraordinary solution.[25] Above all, it was the simplicity of the ozone equations that allowed them to exert their influence and that may have caused the world to become complacent about the ease with which solutions can be found to other impending environmental disasters.[26]

Two visions of Earth have dominated this story. Earth the vulnerable paradise, perfect in its delicate, blue-white beauty. And Earth the violated, stained at the pole by our industrial effluvia. Chemistry links the two and chemical equations express the essence of that link. Our story began with defenceless Earth, as helpless as the princess in the tower. Then came the environmental warriors who prepared to defend her virtue. They were called to action by wizard scientists who risked censure to proclaim their message of doom. In the final chapter, nations pulled together and averted the tragedy.

The Molina–Rowland equations lie at the heart of that fairy tale.

Afterword

How Great Equations Survive

Steven Weinberg

It is terribly difficult for us to put ourselves in the frame of mind of those who lived in past centuries, but many of their artefacts – buildings, roads, works of art – have survived, and some are even still in use. In the same way, though it is often hard for us to understand the thinking of past scientists who did not know what we know, the great equations that bear their names – Maxwell's equations for the electromagnetic field, Einstein's equations for the gravitational field, the Schrödinger equation for the wave function of quantum mechanics, and the other equations discussed in this book – are still with us, and are still useful. These equations are monuments of scientific progress, just as cathedrals are monuments to the spirit of the Middle Ages. Will there ever come a day when we do not teach these great equations to our students?

Although these equations are permanent parts of scientific knowledge, there have been profound changes in our understanding of the contexts in which they are valid, and of the reasons why they are valid in these contexts. We no longer think of Maxwell's equations as a description of tensions within the ether, as Maxwell did, or even as an exact description of electromagnetic fields, as his fellow physicist Oliver Heaviside did. We have known since the 1930s that the equations governing electromagnetic fields contain an infinite number of additional terms, proportional to higher and higher powers of the fields and the frequency with which the fields oscillate. These additional terms are tiny at the frequencies of visible light, but at much higher frequencies can lead to a scattering of light by light. Maxwell's theory is an *effective field theory*, a theory that is a good approximation only for fields that are sufficiently weak and slowly varying.

The additional terms that must be added to Maxwell's equations arise from the interaction of electromagnetic fields with pairs of charged particles and antiparticles that are continually being produced from empty space and then annihilate again. The calculations in the 1930s of these additional terms were carried out using quantum electrodynamics, the quantum theory of electromagnetism, electrons and anti-electrons. Quantum electrodynamics is itself not the final answer. It arises from the equations of a more fundamental theory, the modern Standard Model of elementary particles, in an approximation in which all energies are taken to be too small to create the quanta of the W and Z fields, the fields that appear in the Standard Model as the siblings of the electromagnetic field. And the Standard Model is not the final answer; we think it is only a low-energy approximation to a more fundamental theory whose equations may not involve electromagnetic fields or W or Z fields at all.

The equations of general relativity have undergone a similar reinterpretation. In deriving his equations, Einstein was guided by a fundamental insight, the principle of equivalence of gravitation and inertia, but he also introduced an ad hoc assumption of mathematical simplicity, that the equations should be of the type known as second-order partial differential equations. This means that the equations were assumed by Einstein to involve only rates of change of the fields (first derivatives) and rates of change of rates of change (second derivatives), but not rates of higher order. I don't know any place where Einstein explained the motivation for this assumption. In his 1916 paper on general relativity he claimed that there was 'a minimum of arbitrariness in the choice of these equations', because these were essentially the only possible second-order partial differential equations for gravitational fields that would be consistent with the principle of equivalence of gravitation and inertia, but at least in that article he made no attempt to explain why the equations had to be of second order. Perhaps he relied on the also unexplained fact that when Newton's theory of gravitation is written in terms of a gravitational field, the equation (Poisson's equation) obeyed by these fields is of second order. Or perhaps he felt that equations of such fundamental importance just have to be as simple as possible.

Today general relativity is widely (though not universally) regarded as another effective field theory, useful only for distances much larger than about 10^{-33} centimetres, and particle energies much less than an energy equivalent to the rest mass of 10^{19} protons. No one today would (or at least no one should) take seriously any consequence of general relativity for shorter distances or larger energies.

The more important an equation is, the more we have to be alert to changes in its significance. Nowhere have these changes been more dramatic than for the Dirac equation. Here we have seen not just a change in our view of why an equation is valid and of the conditions under which it is valid, but there has also been a radical change in our understanding of what the equation is about.

Paul Dirac set out in 1928 to find a version of the Schrödinger equation of quantum mechanics that would be consistent with the principles of special relativity. The Schrödinger equation governs the quantum-mechanical wave function, a numerical quantity that depends on time and on position in space, and whose square at any position and time gives the probability at that time of finding a particle at that position. The Schrödinger equation does not treat space and time symmetrically, as would be required by special relativity. Rather, the rate of change of the wave function with time is related to the second derivative of the wave function with respect to position (that is, the rate of change with position of the rate of change of the wave function with position). Dirac noted that the relativistic version of the Schrödinger equation for a particle without spin (the Klein–Gordon equation) is not consistent with the conservation of probability, the principle that the total probability of finding the particle somewhere must always be 100 per cent.

Dirac was able to construct a relativistic version of the Schrödinger equation consistent with the conservation of probability, known ever since as the Dirac equation, but it described a particle with a spin equal to one-half (in units of Planck's constant), not zero. This was regarded as a great triumph, for the electron was already known to have spin one-half, through the interpretation of atomic spectra a few years earlier. Moreover, by studying the effect of an external electromagnetic field on his equation, Dirac was able to show that the electron is a magnet with just the magnetic strength that the Dutch experimenters Samuel Goudsmit and George Uhlenbeck had inferred from spectroscopic data, and he was able to calculate the 'fine structure' of hydrogen, the tiny differences in energy between states that differ only in their total angular momentum. When Dirac died in 1984, one of his obituaries credited him with explaining why the electron must have spin one-half.

The trouble with all this is that there is no relativistic quantum theory of the sort for which Dirac was looking. The combination of relativity and quantum mechanics inevitably leads to theories with unlimited numbers of particles. In such theories the true dynamical variables on which the wave function depends are not the position of one particle or several particles, but *fields*, like the electromagnetic field of Maxwell. Particles are quanta – bundles of the energy and momentum – of these fields. A photon is a quantum of the electromagnetic field, with spin one, and an electron is a quantum of the electron field, with spin one-half.

After all, if Dirac's arguments were correct, they would have to apply to any sort of elementary particle. Nothing in Dirac's analysis made use of the special properties of electrons that distinguish them from other particles – the facts, for instance, that electrons are particles with a tiny mass, and that they are the particles found in orbit around the nucleus in all ordinary atoms. But contrary to what Dirac thought, quantum mechanics and relativity do not forbid the existence of elementary particles with spins different from one-half, and such particles are known to exist. There is not only the photon, with spin one, but massive particles of spin one, the W and Z particles, that seem just as elementary as the electron. There is not even anything in relativistic quantum mechanics that forbids the existence of elementary particles of spin zero. Indeed, such particles appear in our present theories of elementary-particle interactions, and much effort is being expended by experimental physicists to find these spinless particles.

Dirac's theory claimed as its greatest triumph the successful prediction of the antiparticle of the electron, the positron, discovered in cosmic rays a few years later. Dirac had observed that his equation had solutions of negative energy. To avoid a collapse of all atomic electrons into the negative energy states, he supposed that these states were almost all full, so that the Pauli exclusion principle (which forbids two electrons from occupying the same state) would preserve the stability of ordinary electrons of positive energy. The occasional unfilled negative energy state would be interpreted as a particle of positive energy and electric charge opposite to that of the electron, that is, as an anti-electron.

But from the perspective of quantum field theory there is no reason why a particle of spin one-half must have a distinct antiparticle. Particles of spin one-half that are their own antiparticle appear in some of our theories, though none has been detected yet. Of course, quantum field theory tells us that an electrically charged particle must have a distinct antiparticle, but this is just as true for particles of spin zero or one (which do not obey the Pauli exclusion principle) as for particles of spin one-half, and such antiparticles of integer spin particles are well known experimentally.

Confusion on this point continued for many years after Dirac's work, and may even continue today. In the 1950s a new accelerator was planned at Berkeley that for the first time would have enough energy to produce antiprotons. The objection was raised that everyone knew that the proton had to have an antiparticle, so why design an accelerator to target this particular discovery? One answer that was made at the time was that the proton did not seem to satisfy the Dirac equation, since it has a magnetic field considerably stronger than Dirac's theory would predict, and if it did not satisfy the Dirac equation, then there was no reason to expect it to have a distinct antiparticle. It was still not understood that Dirac's equation has nothing to do with the necessity for antiparticles.

So why did Dirac's equation work so well in predicting the fine structure of the hydrogen atom and the strength of the electron's magnetic field? It happens that the fusion of quantum mechanics with special relativity requires that a field whose quanta have spin one-half and interact only with a classical external electromagnetic field must satisfy an equation that is mathematically identical to the Dirac equation, though it has a quite different interpretation. The field is not a wave function – it is not a numerical quantity, like the Schrödinger wave function, but a quantum mechanical operator, and it has no direct interpretation in terms of the probabilities of finding the particle at different positions. By considering the action of this operator on states containing a single electron, one can calculate the particle's magnetic strength and the energies of the states of this particle in atoms. Because the equation for the electron field operator is mathematically the same as Dirac's equation for his wave function, the results of this calculation turn out to be the same as Dirac's.

All this is only approximate. The electron also interacts with the quantum fluctuations in the electromagnetic field, so its magnetic field and its energies in atomic states are not precisely equal to those calculated by Dirac, and it also has non-electromagnetic weak interactions with the atomic nucleus. But these are small effects in ordinary atoms. Although calculations of atomic structure based on Dirac's equation are only approximate, they are very good approximations, and continue to be useful.

So it goes. When an equation is as successful as Dirac's, it is never simply a mistake. It may not be valid for the reason supposed by its author, it may break down in new contexts, and it may not even mean what its author thought it meant. We must continually be open to reinterpretations of these equations. But the great equations of modern physics are a permanent part of scientific knowledge, which may outlast even the beautiful cathedrals of earlier ages.

Notes and Further Reading

Foreword
It Must be Beautiful

1 There is an analogy to be made here between the scientific equation and the poem. In Seamus Heaney's introduction to his version of *Beowulf* he notes that this great poem 'perfectly answers the early modern conception of a work of creative imagination as one in which conflicting realities find accommodation within a new order' (Faber and Faber, 1999 p. xvii). Something similar might be said of a great scientific equation: it accommodates within its 'new order' the apparently different quantities that feature within it.

2 A brief and perceptive analysis of Einstein's iconic status (including some remarks on $E = mc^2$) is given in Roland Barthes's essay 'The brain of Einstein' in *Mythologies* (Vintage, 1993), pp. 68–70.

3 For more on this, see the essay 'Poetry and science' by the late Czech poet and immunologist Miroslav Holub in *The Dimension of the Present Moment*, London (Faber & Faber, 1990), pp. 132–133. Also worth reading are the personal reflections of the poet Lavinia Greenlaw 'Unstable regions: poetry and science' in *Cultural Babbage*, ed. Francis Spufford and Jenny Uglow, London (Faber and Faber, 1996).

4 Richard Feynman, *The Character of Physical Law*, London (Penguin, 1992), pp. 35–36.

5 Eugene Wigner, 'The unreasonable effectiveness of mathematics', *Communications on Pure and Applied Mathematics*, vol. 13, 1960, pp. 1–14.

6 This vexed question is clearly discussed in John Barrow's *The Universe that Discovered Itself* (Oxford University Press, 2000), Chapter 5.

7 Several great theoretical physicists have written on the importance of beauty in their subject. See Chapter 6 of Steven Weinberg's *Dreams of a Final Theory* (Random House, 1993); Subrahmanyan Chandrasekhar's *Truth and Beauty: Aesthetics and*

Motivations in Science (University of Chicago Press, 1987); Chen Ning Yang's contribution 'Beauty in theoretical physics' in *The Aesthetic Dimension of Science*, ed. Deane W. Curtin, New York, Philosophical Library, 1980.

8 But 'Beauty is back', declares the post-modernist architectural critic Charles Jencks in his wide-ranging review 'What is beauty?' in *Prospect*, August/September 2001, pp. 22–27.

9 A thorough discussion of the concept of beauty in science is given in James W. McAllister's *Beauty and Revolution in Science* (Cornell University Press, 1996).

10 Dirac sets out his aesthetic beliefs and credentials in 'Pretty mathematics', *International Journal of Theoretical Physics*, vol. 21, 8/9 (1982), pp. 603–05.

11 In his *Opere Il Sattiatore* (1656), he wrote: '[The Universe] cannot be read until we have learnt the language and become familiar with the characters in which it is written. It is written in mathematical language.' As usual, Plato had been there first: he had said that 'the world was God's epistle written to mankind' and that 'it was written in mathematical letters'. It is worth noting that Galileo and the other founders of modern physics did not write equations, but ratios. These were, however, equivalent in some respects to equations. It was only a few decades later that equations became the preferred form of mathematical expression. See I. Bernard Cohen, *Revolution in Science* (Harvard University Press, 1985), pp. 139–140.

A Revolution with no Revolutionaries
The Planck–Einstein Equation for the Energy of a Quantum

1 For a well-informed discussion of modernism in a wide cultural context, see Thomas Vargish and Delo Mook, *Inside Modernism: Relativity Theory, Cubism, Narrative* (Yale University Press, 1999).

2 Ronald Taylor, *Berlin and its Culture* (Yale University Press, 1997).

3 The place of women in German university life at this time is described in Chapter 2 of *Lise Meitner: A Life in Physics* by Ruth Lewin Sime (University of California Press, 1996).

4 A comprehensive history of the beginnings of modern theoretical physics is given by Christa Jungnickel and Russell McCormmach, *Intellectual Mastery of Nature* vols. 1 and 2 (University of Chicago Press, 1986).

5 John Heilbron, *The Dilemmas of an Upright Man* (Harvard University Press, 2000). This is a comprehensive and wonderfully sensitive account of Planck's life, especially strong on his politics. This edition contains a valuable afterword that usefully clarifies the conclusions to which Heilbron came in the original, 1986 edition.

6 If you heat up the inside of the cavity to sufficiently high temperatures, the emerging radiation will appear not black but dull red, then orange, then yellow. That is why it is rather misleading to call the radiation 'black-body radiation' and better to call it cavity radiation.

7 The definitive history of the Reichsanstalt is David Cahan's *An Institute for an Empire* (Cambridge University Press, 1989).

8 Contrary to widespread belief among today's physicists, Planck began this work unaware that the cavity-radiation data constituted a crisis for classical physics. This point is comprehensively addressed in Martin Klein's *Max Planck and the Beginnings of Quantum Theory* (Archives for History of Exact Science, 1, 1962) pp. 459–79.

9　This is the view of Martin Klein, explained in his article 'Thermodynamics and quanta in Planck's work' (*Physics Today*, Vol. 19, no. 11), pp. 23–32. English translations of Planck's original papers are available in *Planck's Original Papers in Quantum Physics*, annotated by Hans Kangro (London, Taylor and Francis, 1972).

10　Thomas Kuhn gives his views most clearly in 'Revisiting Planck', (*Historical Studies in the Physical Sciences*, vol. 14, 2, 1984) pp. 231–52. He gave a more detailed but more opaque presentation of his case in *Blackbody Theory and the Quantum Discontinuity 1894–1912* (Clarendon Press, 1978).

11　A well-balanced assessment of the controversy concerning Kuhn's interpretation of Planck's work is given by Peter Galison in 'Kuhn and the quantum controversy' (*British Journal for the Philosophy of Science*, vol. 32, part 1, 1981), pp. 71–85.

12　The values that Planck obtained for the constants h and k were within a few per cent of their currently accepted values. In modern units, these values are $h = 6.63 \times 10^{-34}$ J s and $k = 1.38 \times 10^{-16}$ JK^{-1}, where J is short for joule, the modern unit of energy, s is short for second and K short for kelvin, the unit of temperature on the absolute scale. On this scale the lowest possible temperature is zero degrees and the freezing point of water under ordinary conditions is 273.16 degrees.

13　The energy of the atom in any substance – solid, liquid or gas – is roughly equal to Boltzmann's constant multiplied by the substance's temperature, with the temperature measured on the absolute scale.

14　The unique values of mass, length and time are now named after Planck. His formulae for their values in terms of his constant h, Newton's gravitational constant G and the speed of light in a vacuum c are: Planck mass = $\sqrt{(hc/G)}$, Planck length = $\sqrt{(hG/(c^3))}$, Planck time = $\sqrt{(hG/(c^5))}$. These quantities are now important to astrophysicists studying the beginnings of the Universe, as explained by Joseph Silk in *A Short History of the Universe* (W. H. Freeman and Co, 1997), pp. 74–76.

15　An especially fine biography of Einstein is Albrecht Fölsing's *Albert Einstein* (Penguin Books, 1997). For a comprehensive and authoritative account of Einstein's contributions to physics, see Abraham Pais's *Subtle is the Lord . . .* (Oxford University Press, 1982).

16　A compact and elegant introduction to the great papers that Einstein wrote in 1905, together with an annotated English translation of each one, is provided by *Einstein's Miraculous Year*, ed. John Stachel (Princeton University Press, 1998).

17　John L. Heilbron and Thomas S. Kuhn, 'The genesis of the Bohr atom' (*Historical Studies in the Physical Sciences*, vol. 1, 1969), pp. 211–90. For an account of Bohr's work on the atom see Abraham Pais's biography *Niels Bohr's Times: in Physics, Philosophy and Polity* (Clarendon Press, 1991), pp. 132–59.

18　Gerald Holton, 'R. A. Millikan's struggle with the meaning of Planck's constant' (*Physics in Perspective*, vol. 1, 1999), pp. 231–37.

19　Roger Stuewer, *The Compton Effect* (New York: Science History Publications, 1975). For a readable account of the work of Compton, and of the other American experimenters whose work figures in the $E = hf$ story, see Daniel J. Kevles, *The Physicists: The History of the Scientific Community in Modern America* (Harvard University Press, 1997).

20　This part of the $E = hf$ story is comprehensively covered in Max Jammer, *The Conceptual Development of Quantum Mechanics* (McGraw-Hill, 1966).

21 De Broglie's formula says that the wavelength of a quantum, whether of radiation or free matter, is given by Planck's constant divided by the size of the quantum's momentum.

22 K. K. Darrow, 'The scientific work of C. J. Davisson' (*The Bell System Technical Journal*, vol. 30, 1951), pp. 786–97.

23 George P. Thomson 'Early work on electron diffraction' (*American Journal of Physics*, vol. 29, 1961), pp. 821–25.

24 The concept of a field theory is explained in Steven Weinberg, *Dreams of a Final Theory* (London; Hutchinson, 1993), pp. 18–19.

The Sextant Equation
$E = mc^2$

Notes

1 Specifically: In the Russian spaceship frame, the flashes appear to come not along the horizontal, but because of the aberration effect, the flashes arrive at a small angle, $\alpha = v/c$, with respect to the horizontal. It is useful to think of the light's momentum as having a horizontal and a verticle component. The horizontal components of that momentum cancel out, because there is an equal component of this horizontal momentum coming from the left light source and from the right light source. But both flashes have a small component in the up direction that is $(E/2c)$ sin α. Because α is small, we can take sin $\alpha = \alpha$, so the upward component of momentum is $(E/2c)$ α, which (by the aberration effect) means that the vertical momentum is given by:

vertical component of one light beam $= (E/2c)$ $(v/c) = Ev/2c^2$.

2 The best single source on Lise Meitner is the excellent biography of her by Ruth Lewin Sime, *Lise Meitner: A Life in Physics* (University of California Press, 1996), see especially pp. 233–37.

3 See www.bizjournals.com/stlouis/stories/1998/11/09/smallb1.html.

Further Reading

The most comprehensive and scholarly edition of Einstein's work is, by far, *Einstein's Collected Papers* (Princeton University Press); for the period around 1905, see volume 2 and the English translation of that volume by Anna Beck.

The best scientific biography of Einstein is A. Pais, *'Subtle is the Lord . . .': Science and the Life of Albert Einstein* (Oxford University Press, 1982).

Excellent essays on the cultural and intellectual history of Einstein's work: Gerald Holton, *Thematic Origins of Scientific Thought* (Harvard University Press, 1973).

A general biography that is very useful: Albrecht Fölsing, *Albert Einstein: A Biography* (New York: Viking, 1997).

A collection of excellent essays from a variety of historical perspectives on Einstein: Peter Galison, Michael Gordin and David Kaiser, *Einstein's Relativities* (forthcoming, Routledge).

Technical, very helpful history of the special theory of relativity: A. I. Miller, *Albert Einstein's Special Theory of Relativity: Emergence (1905) and Early Interpretation (1905–1911)* (Reading, MA: Addison-Wesley, 1981).

On Wheeler and the H-bomb, Peter Galison, *Image and Logic: A Material Culture of Microphysics* (University of Chicago Press, 1997).

Situating Einstein within the tradition of electrodynamics, see Olivier Darrigol, *Electrodynamics from Ampère to Einstein* (forthcoming).

A short but very useful volume with translations of the five papers of 1905 is: John Stachel, *Einstein's Miraculous Year: Five Papers that Changed the Face of Physics* (Princeton University Press, 1998).

Einstein's own popularization of his theory is obviously also to be valued: Albert Einstein, *Relativity: The Special and the General Theory* (New York: Crown Publishers, 1961).

Einstein's essays on everything from politics to the philosophy of physics: Einstein, *Ideas and Opinions* (New York: Bonanza Books, 1954).

And my favourite elementary textbook on special is still, after all these years: A. P. French, *Special Relativity*, the M.I.T. Introductory Physics Series (New York: W. W. Norton, 1968).

The Rediscovery of Gravity
The Einstein Equation of General Relativity

Notes

1 It is a common modern practice to refer to this equation in the singular rather than the plural, as had been usual originally, because it is better to think of it as a single equation on the entire tensors that are involved (see the section on curved space-time geometry below), rather than on the family of components of those tensors.

2 Technically this number is called the *intrinsic* curvature, or *Gaussian* curvature, of the surface. We shall be coming to the notion of 'intrinsic' a little more fully later on.

3 The full mathematical theory of the type of four-dimensional geometry that is involved for Newtonian theory was first worked out by the outstanding French mathematician Élie Cartan in 1923/24.

4 The previous Euclidean expression for dl^2 now generalizes to the well-known form $ds^2 = g_{ab}dx^a dx^b$ (our 'dl' being written 'ds' in most literature). For an n-dimensional space we need n independent coordinates, here denoted $x^1, x^2, \ldots, x^n$. This may be a bit confusing, because the notation 'x^2' does *not* now stand for 'x squared', nor does 'x^3' stand for 'x cubed', etc. The notation 'x^a' (or 'x^b', etc.) is a generic symbol for one of these coordinates. Similarly, 'g_{ab}' is a generic symbol for one of the quantities $g_{11}, g_{12}, \ldots, g_{nm}$, these being $n(n+1)/2$ independent functions because $g_{ab} = g_{ba}$. The *Einstein summation convention* is being adopted here, according to which repeated indices get summed over. Hence the expression '$g_{ab}dx^a dx^b$' stands for '$g_{11}dx^1 dx^1 + g_{12}dx^1 dx^2 + \ldots + g_{nm}dx^n dx^n$'. In our two-dimensional case, $g_{11} = A$, $g_{12} = g_{21} = B$ and $g_{22} = C$ are functions of the two coordinates u and v, where $x^1 = u$ and $x^2 = v$.

5 There are explicit but complicated expressions telling us how to calculate the R_{abcd} from the g_{ab} and their first and second partial derivatives, with respect to the coordinates x^a.

6 Using Einstein's summation convention we can define, R_{ab} and R by the relations $R_{ab} = R_{acbd}g^{cd}$ and $R = R_{ab}g^{ab}$, where g^{ab} is the *inverse* of g_{ab}, in the sense of matrix algebra.

7 In fact Einstein later showed that this postulate can be *deduced* from his field equation, together with some other reasonable assumptions.

8 The vanishing of the 'covariant divergence' of T_{ab}.

9 The mathematician David Hilbert also came upon this equation at a similar time to Einstein, but by a different route, in the autumn of 1915. This has resulted in an uncomfortable priority dispute. But Hilbert's contribution, though technically important, does not really undermine Einstein's fundamental priority in the matter. See, in particular, J. Stachel (1999), *New Light on the Einstein–Hilbert Priority Question* in Journal of Astrophysics and Astronomy, Volume 20, Numbers 3 and 4, December 1999, 91–101.

10 There was a curious 'scare' in 1966 when Robert Dicke claimed that careful observations of solar oblateness by himself and Goldenberg showed that the Sun possessed a quadrupole moment of such a magnitude that it would thoroughly spoil the agreement of Mercury's perihelion advance with general relativity. Fortunately, subsequent observations and theoretical considerations showed that Dicke's conclusion was wrong.

11 Interestingly, although he did not have the basic ideas of general relativity, Poincaré had already predicted the existence of gravitational waves in 1905, based on analogies with Maxwell's theory of electromagnetism.

12 I have proposed a technically difficult but apparently feasible experiment, one version of which would have to be performed in outer space, for testing whether or not this proposal is correct.

Further Reading

W. Rindler, *Relativity: Special, General and Cosmological* (Oxford University Press, 2001).

W. Rindler, *Essential Relativity* (New York: Springer-Verlag, 1997).

L. A. Steen (ed.) *The Geometry of the Universe*, in Mathematics Today: Twelve Informal Essays, (New York: Springer-Verlag, 1978).

K. Thorne and C. W. W. Norton, *Black Holes and Time Warps: Einstein's Outrageous Legacy*, (New York, 1994).

A. Einstein, *Relativity: The Special and the General Theory* (Reprinted by Three Rivers Press, California, 1995).

Erotica, Aesthetics and Schrödinger's Wave Equation

1 For biographical details, see W. Moore, *Schrödinger: Life and Thought* (Cambridge University Press, 1989).

2 See D. Cassidy, *Uncertainty: The Life and Science of Werner Heisenberg* (New York: Freeman, 1992).

3 Ibid, p. 137, as recalled by Max Born.

4 For biographical details, see A. Pais, *Niels Bohr's Times: In Physics, Philosophy and Polity* (Oxford University Press, 1991).

5 E. Schrödinger, 'Über das Verhältnis der Heisenberg–Born–Jordanschen Quantenmechanik zu der meinen', *Annalen der Physik*, 70, 734–56 (1926), p. 735. This is known as the third communication.

6 The quotes in this paragraph are from E. Schrödinger, 'Quantisierung als Eigenwertproblem', *Annalen der Physik*, 80, (1926) pp. 437–90, which is the second communication.

7 For the case of determining the energy levels of an atomic electron Schrödinger's equation is written as:

$$\hat{H}\psi = E\psi$$

where $\hat{H}$ is called the Hamiltonian and is a mathematical representation of the total energy of the system (the sum of its energy of motion and its energy of position). ψ (psi) is the wave function that describes the system's characteristics such as its location in space at any time; and E is one of many numerical values for the system's possible energy values. So, on the left-hand side is a mathematical function in which ψ is operated on, while on the right-hand E is the energy of a particular energy level and ψ is the wave function that corresponds to it.

8 A note for aficionados. This form of the Schrödinger equation is actually a special case, though extremely important. This form does not involve time. The most general form of the equation is rather more complex:

$$\hat{H}\Psi = \frac{ih}{2\pi}\frac{\partial\Psi}{\partial t}$$

which says that when the Hamiltonian operates on the wave function, its result is equal to the rate of change of the wave function with time (other variables being held constant) multiplied by the square root of −1 and Planck's constant, divided by twice π.

9 Schrödinger never published his deliberations on a relativistic wave equation. The equation he dealt with soon became known as the Klein–Gordon equation and was revived in 1936 by Wolfgang Pauli and Victor Weisskopf, 'Über die Quantisierung der skalaren relativistischen Wellengleichung', *Helvetica Physica Acta,* 7, 709–31 (1934). For discussion of this equation, see A. I. Miller, *Early Quantum Electrodynamics: A Source Book* (Cambridge University Press, 1994).

10 Paul Dirac, 'The evolution of the physicist's picture of nature,' *Scientific American,* 208, 45-53 (1963), p. 47.

11 The second communication in note 2, p. 750.

12 See W. Heisenberg, 'Über den anschaulichen Inhalt der quantentheoretischen Kinematik und Mechanik', *Zeitschrift für Physik,* 43, 172–98 (1927), esp. pp. 184–85. Heisenberg points out that Schrödinger (in his second communication) made the rather trivial mathematical mistake of expanding the electron's wave function on the basis of harmonic oscillator wave functions which have the unique property of maintaining a localized wave packet. But this is not the case generally.

13 This, of course, should have been clear to everyone. The reason is that the Schrödinger equation is of the Sturm-Liouville genre which is an eigenvalue problem and so is immediately equivalent to an eigenvalue calculation from matrix algebra, which is the mathematics in Heisenberg's quantum mechanics. It is interesting that until Schrödinger's equation appeared, no one knew what to do with the eigenfunctions in quantum mechanics.

14 W. Heisenberg, 'Mehrkörperproblem und Resonanz in der Quantenmechanik', *Zeitschrift für Physik,* 38, (1926), pp. 411–26.

15 *Archive for History of Quantum Physics,* interview with W. Heisenberg, 22 February 1963, p. 30.

16 Ibid., p. 3. The paper is W. Heisenberg, 'Schwankungserscheinungen und Quantenmechanik', *Zeitschrift für Physik,* 40, pp. 501–06.

17 See note 12.

18 The origin of Planck's constant is described in the essay 'A Revolution with no Revolutionaries', in this book.

19 W. Heisenberg, *Physics and Beyond: Encounters and Conversations*, translated by A. J. Pomerans (New York: Harper & Row, 1971), p. 73.

20 Ibid., p. 74.

21 Ibid., p. 75.

22 Ibid., p. 76.

23 Roughly this is because an object's wave properties become less important the heavier it is. Consequently, and luckily, we are localized.

24 See E. Schrödinger, 'Die gegenwärtige Situation in der Quantenmechanik', *Die Naturwissenschaften*, 23, 807–12, 823–28, 844–49 (1935). Reprinted in English in W. Zurek and J. Wheeler (eds.), *Quantum Theory and Measurement* (Princeton University Press, 1983), pp. 152–67.

25 Quoted from W. Moore, *Schrödinger: Life and Thought* (Cambridge University Press, 1989), p. 314.

26 For a masterly survey of recent work on Schrödinger's non-relativistic wave mechanics, see J. S. Bell, 'Are there quantum jumps', in J. S. Bell, *Speakable and Unspeakable in Quantum Mechanics* (Cambridge University Press, 1987), pp. 201–12.

27 See A. I. Miller, *Einstein, Picasso: Space, Time and the Beauty that Causes Havoc* (New York: Basic Books, 2001).

28 See J. Bates, *The Genius of Shakespeare* (Oxford University Press, 1998) pp. 311–16.

29 Ibid.

30 Quoted from ibid., p. 316.

31 Quoted from ibid., p. 314.

32 See J. S. Bell, Against 'Measurement',' in A. I. Miller (ed.), *Sixty-Two Years of Uncertainty: Historical, Philosophical, and Physical Inquiries into the Foundations of Quantum Mechanics* (New York: Plenum Press, 1990), pp. 17-31.

33 R. P. Feynman, *The Character of Physical Law* (London: Penguin Books, 1992), p. 129.

A Piece of Magic
The Dirac Equation

Notes

1 That is, to predict the motion of a particle you need to know its charge and its mass: no more, no less. The value of the charge can be zero; then the particle will have only gravitational interactions.

2 In quantum mechanics, only certain values of the discrete spin are allowed. This is closely related to the restriction on allowed Bohr orbitals.

3 An interesting case is the photon, which is its own antiparticle. This is not possible for a charged particle, but the photon is electrically neutral.

4 In fact these particles obey wave equations that do have negative-energy solutions.

5 There is also a closely related object, the so-called Hermitean conjugate, that creates electrons and destroys positrons.

6 Seminal contributions were also made by the slightly older theorists Kramers and Bethe, and by the theorist-turned-experimentalist Lamb.

7 'Leptons' is just a generic word covering electrons, muons, the so-called tau particles, and their antiparticles. These particles all have very similar properties, including equal spin and charge. They differ in mass.

8 Up to a couple of profound but well-posed and solvable problems, as I'll shortly discuss.

9 Much later, in the 1960s, Heisenberg recalled, 'Up till that time [1928] I had the impression that, in quantum theory, we had come into the harbour, into the port. Dirac's paper threw us out into the sea again.'

10 This is a consequence of Gödel's completeness theorem for first-order predicate logic. Sophisticated readers may wonder how this result, that all valid theorems can be proved in mechanical fashion, can be consistent with Gödel's famous incompleteness theorem. (It's not a misprint: Gödel proved both completeness and incompleteness theorems. Clever fellow!) To make a long story short, Gödel's incompleteness theorem shows that in any rich mathematical system you will be able to formulate meaningful statements such that neither the statement nor its denial is a theorem. Such 'incompleteness' does not contradict the possibility of systematically enumerating all the theorems.

Further Reading

For background material on atomic physics and quantum theory, including excerpts from important original sources, I highly recommend H. Boorse and L. Motz, *The World of the Atom* (Basic Books, 1966). Of course, some of its more 'timely' parts appear somewhat dated today.

Dirac's classic is *The Principles of Quantum Mechanics*, 4th edn (Cambridge University Press, 1958).

A demanding but honest and beautiful treatment of the principles of quantum electrodynamics, with no mathematical prerequisites, is R. P. Feynman, *QED: The Strange Theory of Light and Matter* (Princeton University Press, 1985).

For a brief account of QCD, easily accessible after Feynman's book, with no mathematical prerequisites, see F. Wilczek, 'QCD made simple', *Physics Today*, vol. 53N8 (2000), pp. 22–28.

I'm at work on a full account, to be called simply *QCD* (Princeton).

For a conceptual review of quantum field theory, see my article 'Quantum field theory' in the American Physical Society Centenary issue of *Review of Modern Physics*, vol. 71, (1999), pp. S85–S95. This issue is also published as *More Things in Heaven and Earth – A Celebration of Physics at the Millennium*, B. Bederson, ed. (New York: Springer-Verlag, 1999). It contains several other reflective articles that touch on many of our themes.

Appendix

I quoted Dirac's equation in the form

$$\left[\gamma^{\mu}\left(i\frac{\partial}{\partial x^{\mu}} - eA_{\mu}(x)\right) + m\right]\psi(x) = 0.$$

Now let's identify the players. The wave function $\psi(x)$ is the object whose behaviour is being described. It has four components: $\psi_{e\uparrow}(x)$, $\psi_{e\downarrow}(x)$, $\psi_{p\uparrow}(x)$, $\psi_{p\downarrow}(x)$. Each is a function whose value depends on space and time, as indicated by the argument (x). For Dirac, these values were complex numbers, the square of whose magnitude (very roughly speaking) gives the probability for finding the corresponding particle type: electron with spin up, electron with spin down, positron with spin down, or positron with spin up, at the given space-time point. In the modern interpretation, the values are operators that create electrons or destroy positrons.

As usual in relativistic theories, the Einstein summation convention is in force. The superscripts and subscripts μ are supposed to take the values 0, 1, 2, 3, representing time and the three spatial directions, and sum up the contributions from all four values. The derivative operator $\partial/\partial x^0$ measures how fast the wave function is changing in time, the other derivative operators measure how fast it is changing in the different spatial directions. The $A(x)$ fields, with various subscripts, are the electromagnetic potentials. They specify the electric and magnetic fields that the electron feels. The electron charge is $-e$. It specifies the strength of its response to those fields. The electron mass is m.

Dirac's most characteristic technical innovation was to introduce the γ-matrices

$$\gamma^0 = \begin{pmatrix} 1 & 0 & 0 & 0 \\ 0 & 1 & 0 & 0 \\ 0 & 0 & -1 & 0 \\ 0 & 0 & 0 & -1 \end{pmatrix}$$

$$\gamma^1 = \begin{pmatrix} 0 & 0 & 0 & -1 \\ 0 & 0 & -1 & 0 \\ 0 & 1 & 0 & 0 \\ 1 & 0 & 0 & 0 \end{pmatrix}$$

$$\gamma^2 = \begin{pmatrix} 0 & 0 & 0 & i \\ 0 & 0 & -i & 0 \\ 0 & -i & 0 & 0 \\ i & 0 & 0 & 0 \end{pmatrix}$$

$$\gamma^3 = \begin{pmatrix} 0 & 0 & -1 & 0 \\ 0 & 0 & 0 & 1 \\ 1 & 0 & 0 & 0 \\ 0 & -1 & 0 & 0 \end{pmatrix}.$$

All the other elements in his equation – wave function, derivative, electromagnetic potentials, charge and mass – already appeared in the Schrödinger equation. The γ matrices were something quite new. They allowed Dirac to formulate an equation in which space and time appeared on an equal footing, while forcing him to introduce a wave function with four components.

Spelling it out more fully, the Dirac equation reads

$$
\begin{pmatrix}
i\partial_0 - eA_0 + m & 0 & -i(\partial_1 + \partial_3) + e(A_1 + A_3) & \partial_2 + ieA_2 \\
0 & i\partial_0 - eA_0 + m & -\partial_2 - ieA_2 & -i(\partial_1 - \partial_3) + e(A_1 - A_3) \\
i(\partial_1 - \partial_3) - e(A_1 - A_3) & -\partial_3 - ieA_2 & -i\partial_0 - eA_0 + m & 0 \\
\partial_2 + ieA_2 & i(\partial_1 - \partial_3) - e(A_1 - A_3) & 0 & -i\partial_0 + eA_0 + m
\end{pmatrix}
\begin{pmatrix}
\psi_{e\uparrow}(x) \\
\psi_{e\downarrow}(x) \\
\psi_{p\uparrow}(x) \\
\psi_{p\downarrow}(x)
\end{pmatrix} = 0 .
$$

Understanding Information, bit by bit
Shannon's Equations

Notes

1 For reasons we do not need to question here, a channel's bandwidth can be specified either in hertz (that is, the number of waves passing a point every second) or bits per second. The latter is twice the former, suggesting that the up and then down of a wave constitutes two bits. So a line that transmits a maximum of 10,000 bits per second also transmits a maximum of 5,000 hertz.

2 One of many ways of doing this is by noticing that large areas may be of the same colour, and instead of transmitting the same signal over and over, the computer transmits just two numbers: the colour and intensity of the picture point and the number of following dots which have the same values. Say that there are 1,000 consecutive picture points for which the colour has a value of 72 (from a maximum of 128, say) and the intensity is 93 (out of 128 again). Anticipating the meaning of a 'bit' as an on/off digit and just using the fact that 7 bits can have 128 different values, the computer instead of transmitting the 14 bits 1,000 times simply transmits the 16 bits followed by a binary number representing the repetitions. Ten bits can have 1,024 values, so 10 bits suffice to indicate this number of repetitions. The total number of bits transmitted is then 16 + 10, instead of 16,000. This is what is called a compression ratio of 16,000 to 26 (or 615 to 1). Of course if the picture is very varied you may not be so lucky. On average, ratios of 20 to 1 are commonly achieved.

3 In this and the next section I shall be taking liberties with mathematical rigour.

4 The size of the voltage is a measure of the energy transferred to each electrical charge. So the encoder translates the numbers that represent the colour and intensity of the picture points into voltage values that are transmitted to the other end of the cable.

Further Reading

C. E. Shannon, 'A mathematical theory of communication', *Bell System Technical Journal*, vol. 27, July and October 1948, pp. 379–423 and 623–56.

S. Roman, *Introduction to Coding and Information Theory* (Dortmund: Springer Verlag, 1996).

G. Boole, *An Investigation of the Laws of Thought* (London: Dover Publications, 1995).

C. E. Shannon, 'A symbolic analysis of relay and switching circuits', *Transactions of the American Institute of Electrical Engineering*, vol. 57, 1938, pp. 713–32.

N. Wiener, *Cybernetics* (Cambridge, Mass.: MIT Press, 1948).

I. Aleksander, *Impossible Minds: My Neurons My Consciousness* (London: Imperial College Press, 1996).

I. Aleksander, *How to Build a Mind* (London: Weidenfeld and Nicholson, 2000).

Hidden Symmetry
The Yang–Mills Equation

Notes

1 C.N. Yang and R. L. Mills, 'Conservation of isotopic spin and isotopic gauge invariance', *Physics Review*, vol. 96, 1954, pp. 191–95.

Further Reading

Y. Nambu, 'Quarks' *World Scientific*, 1985. Provides an attractive guide to the key developments in understanding particles and forces.

G. 't Hooft, *In search of the ultimate building blocks* (Cambridge University Press, 1997). A recent Noble prize-winner's account of particle physics in his lifetime.

G. Fraser, (ed) *The Particle Century*, (Institute of Physics, 1998). Covers the history of particle physics and the development of the Standard Model of particles and forces during the twentieth century.

C. N. Yang, *Selected Papers with Commentary* (Freeman, 1983). Contains Yang's recollections on many papers, including his work with Mills.

L. O'Raifeartaigh, *The Dawning of Gauge Theory* (Princeton University Press, 1997). Contains all the important papers mentioned in this chapter, together with an interesting but technical commentary, Highly recommended for physicists.

A Mirror in the Sky
The Drake Equation

1 Cocconi and Morrison, 'Searching for interstellar communications', *Nature* 184 (1959), pp. 844ff; reprinted in *The Quest for Extraterrestrial Life: A Book of Readings*, ed. Donald Goldsmith (University Science Books, 1980), as are many of the classic early papers in the field.

2 *Is Anyone Out There: The Scientific Search for Extraterrestrial Intelligence*, by Frank Drake and Dava Sobel (Delacorte Press, 1992). This account of the Green Bank meeting largely follows Drake's account.

3 *Project Cyclops: A Design Study of a System for Detecting Extraterrestrial Intelligent Life*, the account of a summer workshop at Ames Research Center, CR114445 (1971).

4 *Carl Sagan: A Life* by Keay Davidson (Wiley, 1999) p. 348.

5 *We Are Not Alone* by Walter Sullivan (McGraw-Hill, 1964).

6 In the introduction to *Interstellar Communication*, ed. A. G. W. Cameron (Benjamin, 1963); quoted in *The Biological Universe*, Steven Dick, p. 508. Dick's book provides an excellent historical account both of SETI and of other, related disciplines.

7 *NASA/TREK: Popular Science and Sex in America* by Constance Penley (Verso, 1997).

8 *The Making of Star Trek* by Stephen E. Whitfield and Gene Roddenberry (Ballantine/Del Rey 1968), p. 112. In the same book Roddenberry describes how – presumably having heard of the Drake equation – he put a calculation of the number of habitable planets available for exploration at the top of his original pitch to the network. Unfortunately he had no research materials to hand and was reduced to making up both the formula, $Ff^2 (MgE) - c^1RI^1 \times M = L/So*$, and the enormous number of planets held to follow from it. None of the network executives called his bluff.

9 *Star Maker* by Olaf Stapledon (1937).

10 *Intelligent Life in the Universe* by Iosif S. Shklovskii and Carl Sagan (Holden-Day, 1966).

11 *Communication with Extraterrestrial Intelligence*, ed. Carl Sagan (MIT Press, 1973). Those not wishing to face the whole thing will find an excellent account in *Carl Sagan: A Life in the Cosmos* by William Poundstone (Henry Holt, 1999).

12 'The nonprevalence of humanoids', by George Gaylord Simpson, *Science,* 143 (1964) pp. 769ff; reprinted in Goldsmith, op. cit.

13 'SETI and the wisdom of Casey Stengel' in *The Flamingo's Smile* by Stephen Jay Gould (W. W. Norton, 1984).

14 'Alone in a crowded universe', by Jared Diamond, *Natural History* (June 1990); reprinted in *Extraterrestrials: Where are They?*, ed. Ben Zuckerman and Michael H. Hart, 2nd edn (Cambridge University Press, 1995).

15 'Behold the stars', by Alfred Adler, *Atlantic Monthly* 234 (1974), pp. 109ff; reprinted in Goldsmith, op. cit.

16 'Where is everybody', a letter to *Physics Today* (August 1985) by Eric Jones, brings together various recollections of the lunch from Teller and others, not to mention the instigating cartoon.

17 Von Hoerner, 'The general limits of space travel', *Science* 137 (1962), pp. 18ff; reprinted in Goldsmith, op. cit.

18 Interviewed in *SETI Pioneers* by David W. Swift (University of Arizona Press, 1990), an excellent oral-history source.

19 Bracewell in Swift, op. cit.

20 Hart, 'An explanation for the absence of extraterrestrials on Earth', *Quarterly Journal of the RAS*, 16 (1975), pp. 128ff; reprinted in Goldsmith, op. cit.

21 Tipler, 'Extraterrestrial intelligent beings do not exist', *Quarterly Journal of the RAS*, 21 (1981) pp. 267ff.

22 The proceedings are *Strategies for the Search for Life in the Universe*, ed. Michael Papgiannis (Reidel, 1980).

23 For example, $R^* = 20, f_p = 1/40, n_e = 1/1,000, f_l = 1/10, f_i = 1/10, f_c = 1/2, L = 10,000$ gives $N = 0.025$.

24 Perhaps the fullest and most systematic account of the possibilities is 'The "Great Silence": the controversy concerning extraterrestrial intelligent life', Brin, *Quarterly Journal of the RAS*, 24 (1983), pp. 283ff.

25 Dyson's letter to Sagan is in his personal papers; reproduced in *Captured by Aliens: The Search for Life and Truth in a Very Large Universe* by Joel Achenbach (Simon & Schuster, 1999) (no date for the Dyson letter is supplied). Achenbach's book is terrific fun.

Equations of Life
The Mathematics of Evolution

1 Because we do not want individuals to have a negative number of offspring, we assume that the number is equal to some constant plus the payoff.
2 J. W. Burgess, *Scientific American* 234 (1976) pp. 100–06.
3 B. Sinervo and C. M. Lively, *Nature* 380 (1996), pp. 240–43.
4 The topics discussed in this essay are treated in more detail, in a nontechnical way, by K. Sigmund, *Games of Life* (Penguin Books, 1995).

The Best Possible Time to be Alive
The Logistic Map

1 *Arcadia* in Tom Stoppard, *Plays 5* (Faber and Faber, 1999). Useful commentary on the play is given in *Tom Stoppard: A Faber Critical Guide* (Faber and Faber, 2000).
2 T. Y. Li and Jim Yorke, 'Period three implies chaos', *American Mathematical Monthly*, vol. 82, (1975), pp. 985–92. This paper proved that for many maps, like the Logistic Map, with period 3 (so that they repeat themselves every three cycles), there will exist orbits with every period, along with an infinite number of chaotic orbits – ones that are irregular, without a fixed period.
3 In fact, we predicted it to be $2(1 + \sqrt{2})$.
4 'Simple mathematical models with very complicated dynamics', Robert May, *Nature*, vol. 261, 10 June 1976.
5 'The voles of Hokkaido', Robert May, *Nature*, vol. 396, 3 December 1998.
6 *The Nature of Chaos*, ed. Tom Mullin (Oxford University Press, 1993). See also *Nonlinearity and Chaos in Engineering Dynamics*, eds. J. M. T. Thompson and S. R. Bishop (Wiley, 1994).
7 Ian Stewart, *Does God Play Dice?: The New Mathematics of Chaos* (Viking Penguin, 1997).
8 Edward Lorenz has written an accessible account of his work in *The Essence of Chaos* (University of Washington Press, 1993).
9 'The Lorenz attractor exists', Ian Stewart, in *Nature*, vol. 406, 31 August 2000, pp. 948–49.
10 A useful reference is 'Randomness and perceived randomness in evolutionary biology', W. C. Wimsatt, *Synthese*, vol. 43 (1980), pp. 287–329.
11 James Gleick, *Chaos* (Heinemann, 1988).
12 Harriett Hawkins, *Strange Attractors: Literature, Culture and Chaos Theory*, (Prentice Hall, 1995).

An Environmental Fairy Tale
The Molina–Rowland Chemical Equations and the CFC Problem

1 Denis Cosgrove explores the influence of this image, taken from the Apollo 17 spacecraft, in *Apollo's Eye: A Cartographic Genealogy of the Earth in the Western Imagination* (Johns Hopkins University Press, 2001).
2 The words of a *New York Times* leader of 25 December 1968 that went on to claim that the Apollo flights would transform our sense of our place in the universe as profoundly as did the Copernican revolution.

3 Mario J. Molina and F. S. Rowland, *Nature*, vol. 249, 28 June 1974, pp. 810–12.

4 William H. Brock, *The Fontana History of Chemistry* (Fontana, 1992).

5 Guyton de Morveau, in M. P. Crosland, *Historical Studies in the Language of Chemistry* (Dover, 1978).

6 The metal mercury is an example, though it also now possesses a Latin name, hydrogyrum. Chlorine, a greenish gas, derives its name from *khloros*, the Greek word for green. The International Union of Pure and Applied Chemistry eventually banned the naming of elements after people, as it led to unseemly, mainly nationalistic, disputes.

7 John McNeill, *Something New Under the Sun: An Environmental History of the Twentieth Century* (Allen Lane, 2000).

8 This cycle can be represented by the equations $NO + O_3 \rightarrow NO_2 + O_2$; $NO_2 + O \rightarrow NO + O_2$.

9 His death was the result of an invention too far – a system of ropes and pulleys by which he hoisted himself out of bed in the mornings, having become disabled by polio. One day he became entangled in the system and strangled himself.

10 McNeill, op. cit.

11 Personal interview, December 2000.

12 *Homage to Gaia: The Life of an Independent Scientist*, James Lovelock's autobiography (Oxford University Press, 2000).

13 Darwinian natural selection explains an organism's activity in terms of the individual's struggle for survival; Gaia theory implies that organisms may be acting communally for the general good. See Richard Dawkins, *Unweaving the Rainbow*, for criticism of Gaia.

14 Lydia Dotto and Harold Schiff, *The Ozone War* (Doubleday, 1978).

15 *New Scientist*, 9 September 2000.

16 By the late 1970s measurements had been carried out showing that CFCs did indeed reach the stratosphere, and later measurements showed the presence of chlorine-free radicals. Changes in the ozone levels were demonstrated only in the mid-1980s.

17 Donald Hodel, a Reagan interior secretary, argued that wearing hats and sunglasses was preferable to trying to prevent ozone destruction.

18 The British Antarctic Survey web site describes current research and gives a flavour of life in Antarctica: http://www.antarctica.ac.uk.

19 Personal interview, December 2000.

20 For more detail about the competing theories, and about the story of the discovery of stratospheric ozone depletion as far as 1987, see John Gribbin, *The Hole in the Sky: Man's Threat to the Ozone Layer* (Corgi, 1988).

21 A study presented at the Stratospheric Processes and their Role in Climate (SPARC) Second General Assembly in Buenos Aires in December 2000 was optimistic on this issue.

22 Global warming is manifest as the rise in Earth's temperature which may be the result of the excessive accumulation in the atmosphere of gases produced by burning fossil fuels, cutting down forests and increased agriculture.

23 BBC interview, December 2000.

24 East Room Roundtable on Global Climate Change, 24 July 1997.

25 McNeill, op. cit.

26 Philip Shabecoff, in *Earth Rising: American Environmentalism in the 21st Century* (Island Press, 2000), argues that the twentieth-century environmental movement reached the limits of its effectiveness for two reasons: more complex problems with less clear solutions; and more aggressive and sophisticated counter-campaigns.

Index

HAZARDS OF TIME TRAVEL

NOVELS BY JOYCE CAROL OATES

With Shuddering Fall (1964)

A Garden of Earthly Delights (1967)

Expensive People (1968)

them (1969)

Wonderland (1971)

Do with Me What You Will (1973)

The Assassins (1975)

Childwold (1976)

Son of the Morning (1978)

Unholy Loves (1979)

Bellefleur (1980)

Angel of Light (1981)

A Bloodsmoor Romance (1982)

Mysteries of Winterthurn (1984)

Solstice (1985)

Marya: A Life (1986)

You Must Remember This (1987)

American Appetites (1989)

Because It Is Bitter, and
Because It Is My Heart (1990)

Black Water (1992)

Foxfire: Confessions of a
Girl Gang (1993)

What I Lived For (1994)

Zombie (1995)

We Were the Mulvaneys (1996)

Man Crazy (1997)

My Heart Laid Bare (1998)

Broke Heart Blues (1999)

Blonde (2000)

Middle Age: A Romance (2001)

I'll Take You There (2002)

The Tattooed Girl (2003)

The Falls (2004)

Missing Mom (2005)

Black Girl / White Girl (2006)

The Gravedigger's Daughter
(2007)

My Sister, My Love (2008)

Little Bird of Heaven (2009)

Mudwoman (2012)

The Accursed (2013)

Carthage (2014)

The Sacrifice (2015)

A Book of American
Martyrs (2017)

HAZARDS
OF
TIME TRAVEL

JOYCE CAROL OATES

4th ESTATE • London

4th Estate
An imprint of HarperCollins*Publishers*
1 London Bridge Street
London SE1 9GF

www.4thEstate.co.uk

First published in Great Britain in 2018 by 4th Estate
First published in the United States by Ecco,
an imprint of HarperCollins*Publishers* in 2018

1

A catalogue record for this book is
available from the British Library

ISBN 978-0-00-829544-8 (hardback)
ISBN 978-0-00-829545-5 (trade paperback)

Designed by Michelle Crow

Printed and bound in Great Britain by
CPI Group (UK) Ltd, Croydon

MIX
Paper from
responsible sources
FSC™ C007454

This book is produced from independently certified FSC paper
to ensure responsible forest management.

For more information visit: www.harpercollins.co.uk/green

For Stig Björkman,
and for
Charlie Gross

A self is simply a device for representing a functionally unified system of responses.

B. F. Skinner, *Science and Human Behavior*

They would not have come for me, naïvely I drew their attention to me. Willingly I dared what I should not have dared.

Of my own free will misjudging. Or rather, not judging—not thinking. In vanity and stupidity and now I am lost.

Sometimes on my knees in a posture of prayer I am able to break through the "censor barrier"—to remember . . .

But my brain hurts so! It is a terrible effort like struggling against the gravity of Jupiter.

My Exile-status forbids me to speak to anyone here of my sentence or of my life before Exile and so I am doubly lonely.

Though rarely alone in this strange place I am very lonely and am not sure that I can persevere.

My sentence is "only" four years. It might have been "life."

Or, it might have been Deletion.

On my knees each night straining to remember, to recall my old, lost self, I try to be grateful, my sentence was not Deletion.

And I try to be grateful, no one in my family was arrested as a collaborator/facilitator of Treason, with me.

I

VALEDICTORIAN

THE INSTRUCTIONS

1. In the Restricted Zone, the Exiled Individual (EI) is allowed a ten-mile radius of movement the epicenter of which is the official residence of the EI. This residence can be changed only by appeal to the Homeland Security Exile Disciplinary Bureau (HSEDB).

2. The EI is forbidden to question, challenge, or disobey in any way any local Restricted Zone authority. The EI is forbidden to identify himself/herself except as established by HSEDB. The EI is forbidden to provide "future knowledge" in the Restricted Zone and to search for or in any way seek out "relatives."

3. The EI will be issued a new and non-negotiable name and an appropriate "birth certificate."

4. The EI is forbidden to enter into any "intimate" or "confidential" relationship with any other individual. The EI is forbidden to procreate.

5. The EI will be identified as "adopted" by "adoptive parents" who are "deceased." The EI will be identified as

having no other family. This information will be the EI's official record in his/her Restricted Zone.

6.   The EI will be monitored at any and all times during his/her exile. It is understood that HSEDB can revoke the term of Exile and Sentencing at any time.

7.   Violations of any of these instructions will insure that the EI will be immediately Deleted.

DELETION

DI—"Deleted Individual."

If you are Deleted, you cease to exist. You are "vaporized."

And if you are Deleted, all memories of you are Deleted also.

Your personal property/estate becomes the possession of the NAS (North American States).

Your family, even your children, if you have children, will be forbidden to speak of you or in any way remember you, once you cease to exist.

Because it is taboo, Deletion is not spoken of. Yet, it is understood that Deletion, the cruelest of punishments, is always imminent.

To be Deleted is not equivalent to being Executed.

Execution is a public-lesson matter: Execution is not a state secret.

A certain percentage of executions under the auspices of the Federal Execution Education Program (FEEP) are broadcast via TV to the populace, for purposes of moral education.

(In a prison execution chamber made to resemble a hospital surgery, the CI [Condemned Individual] is strapped to a gurney by prison guards; then, in the clinical-white uniforms of "medics," prison staffers administer the lethal dose of poison into the CI's veins as tens of millions of home TV viewers watch.)

(Except us. Though Dad was already of MI [Marked Individual] status, and his Caste Rank [CR] vulnerable, neither Dad nor Mom allowed our TV to be turned on at Execution Hours which were often several times a week. My older brother Roderick objected to this "censorship" when he was still in school on the grounds that, if his teachers discussed the educational aspect of an execution in class, he would not be able to participate and would stand out as "suspicious"—but this plea did not persuade our parents to turn on the TV at these times.)

Deletion is a different status altogether, for while Execution is intended to be openly discussed, even to allude to Deletion is a federal offense punishable as Treason-Speech.

My father Eric Strohl had been MI since a time before I was born. As a young resident M.D. in the Pennsboro Medical Center he'd been under observation as a scientifically-minded individual, for such individuals were assumed to be "thinking for themselves"—not a reputation anyone would have wished to have. In addition, Dad was charged with associating with a targeted SI—(Subversive Individual)—who was later arrested and tried for Treason; Dad hadn't done more than *sympathetically listen* to this man address a small gathering in a public park when he and the others were caught in a Homeland Security "sweep"—and Dad's life was changed forever.

He was demoted from his residency in the medical center. Though he had an M.D., with special training in pediatric oncology, he could find work only as a lowly-paid medical attendant in the center, where there had to be maintained a bias against him, that he might never be allowed to "practice" medicine again. Yet, Dad never (publicly) complained—he was lucky, he often (publicly) said, not to be imprisoned, and to be alive.

From time to time MIs were obliged to restate the terms of their crimes and punishments, and to (publicly) express gratitude for their exoneration and current employment. On such occasions Dad took a deep breath and, as he said, *bartered his soul another time.*

Poor Dad! He was so good-natured in our household, I don't think I realized how terrible he must have felt. How *broken.*

Within the family it was understood that we didn't discuss Dad's status *per se,* but we seemed to be allowed—that is, we were not expressly forbidden—to allude to his MI status in the way you might allude to a chronic condition in a family member like multiple sclerosis, or Tourette's, or a predilection for freak accidents. Being MI was something *shameful, embarrassing, potentially dangerous*—but since MI was a (relatively) minor criminal category compared with more serious criminal categories, it wasn't a treasonable offense to acknowledge it. But Dad took risks, even so.

For one of the memories that comes to me, strangely clear and self-contained, like a disturbing dream suddenly recalled in daylight, was how one day when no one was home except us Dad took me upstairs to an attic room that had been shut up for as long as I could remember, with a padlock; and in that room Dad retrieved from beneath a loose floorboard, beneath a worn carpet, a packet of photographs of a man who looked teasingly familiar to me, but whom I could not recall—"This is your uncle Tobias, who was Deleted when you were two years old."

At this time I was ten years old. My two-year-old self was lost and irretrievable. In a quavering voice Dad explained that his "beloved, reckless" younger brother Tobias had lived with us while going to medical school and that he'd drawn the attention of the F.B.E./F.B.I. (Federal Bureau of Examiners, Federal Bureau of Inquisitors) after helping organize a May Day free-speech demonstration. At the age of twenty-three "your uncle Toby" had been arrested in this very house, taken away, allegedly tried—and Deleted.

That is, "vaporized."

What is that, Daddy?—"vaporized." Though I knew the answer would be sad I had to ask.

"Just—gone, sweetie. Like a flame when it's been blown out."

I was too young to register the depth of loss in my father's eyes.

For Dad had often that look of loss in his face. Exhausted from his hospital job, and his skin ashy, and a limp in his right leg from some accident after which a bone had not mended correctly. Yet, Dad had a way of smiling that made everything seem all right.

Just us, kids! We're hanging in here.

Except right now Dad wasn't smiling. Turned a little from me so (maybe) I wouldn't notice him wiping tears from his eyes.

"We aren't supposed to 'recall' Tobias. Certainly not provide information to a child. Or look at pictures! I could be arrested if—anyone heard."

By *anyone* Dad meant the Government. Though you would not say that word—"Government." You would not say the words "State"—"Federal Leaders." It was forbidden to say such words and so, as Dad did, you spoke in a vague way, with a furtive look—*if anyone heard.*

Or, you might say *They.*

You could think of *anyone,* or *they,* as a glowering sky. A low-ceiling sky of those large dirigible clouds rumored to be surveillance devices, sculpted shapes like great ships, often bruised colored and iridescent from pollution, moving unpredictably but always *there.*

Downstairs, in the vicinity of our electronic devices, Dad would never speak so openly. Of course you would never trust your computer no matter how friendly and throaty-seductive its voice, or your cell phone or dicta-stylus, but also thermostats, dishwashers, microwaves, car keys and (self-driving) cars.

"But I miss Toby. All the time. Seeing medical students his age . . . I miss how he'd be a wonderful uncle to you, and to Rod."

It was confusing to me. I'd forgotten what Dad had said— *Vaporized? Deleted?*

But I knew not to ask Dad more questions right now, and make him sadder.

Exciting to see photographs of my lost "Uncle Toby" who looked like a younger version of my father. Uncle Toby had had a frowning-squinting kind of smile, like Dad. And his nose was long and thin like Dad's with a tiny bump in the bone. And his eyes!—dark brown with a glisten, like my own.

"Uncle Toby looks like he'd be *fun.*"

Was this a stupid thing to say? Right away I regretted it but Dad only just smiled sadly.

"Yes. Toby was *fun.*"

He'd tried to warn his brother about being involved in any sort of free speech or May Day demonstrations, Dad said. Even during what had appeared to be a season of (relative) relaxation on the part of the Homeland Security Public Dissemination Bureau; during such seasons, the Government eased up on public-security enforcement, yet, as Dad believed, continued to monitor and file away information about dissenters and potential SIs (Subversive Individuals), for future use. *Nothing is ever forgotten*—Dad warned.

At such times rumors would be circulated of a "thaw"—a "new era"—for always, as Dad said, people are eager to believe good news, and to forget bad news; people wish to be "optimists" and not "pessimists"; but "thaws" are factored into cycles and soon come to an end leaving incautious persons, especially the young and naïve, vulnerable to exposure and arrest and—what comes after arrest.

After Uncle Toby's disappearance (as it was called) law enforcement officers had raided the house and appropriated his medical textbooks, lab notebooks, personal computer and electronic devices, etc., and all pictures of him either digital or hard copy that they could find; but Dad had managed to hide away a few items, at great risk to his own safety.

Saying, "I'm not proud of myself, honey. But I knew it would be wisest to 'repudiate' my brother—formally. By that time he'd been Deleted, so there was no point in defending him, or protecting him. I guess I was pretty convincing—and your mother, too—swearing how we didn't realize we were harboring an SI—a 'traitor'—so they let us off with just a fine."

Dad drew his sleeve across his face. Wiping his face.

"A devastating fine, actually. But we had to be grateful the house wasn't razed, which sometimes happens when there's treason involved."

"Does Mom know?"

"'Know'—what?"

"About Uncle Toby's things here."

"No."

Dad explained: "Mom 'knows' that my brother was Deleted. She never speaks of him of course. She might have 'known' that I'd kept back a few personal items of Toby's at the time but she's certainly forgotten by now, as she has probably forgotten what Toby looked like. If you work hard enough to *not think* of something, and wall off your mind against it, and others around you are doing the same, you can 'forget'—to a degree."

Brashly I was thinking *Not me! I will not forget.*

Touching one of my lost uncle's sweaters, soft dark-wool riddled with moth holes. And there was a yellowed-white T-shirt with a stretched neck. And a biology lab notebook with half the pages empty. And a wristwatch with a stretch band and a blank dead face forever halted at 2:20 P.M. that Daddy tried to revive without success.

"Now you must promise, Adriane, never to speak of your lost uncle to anyone."

I nodded *yes, Daddy*.

"Not to Mommy, and not to Roddy. You must not speak of 'Uncle Toby.' You must not—even to me."

Seeing the perplexed look in my face Dad kissed me wetly on the nose.

Gathering up the outlawed things and returning them beneath the floorboards and the worn carpet.

"Our secret, Adriane. Promise?"

"Yes, Daddy. Promise!"

SO YES, I knew what Deletion was. I know what Deletion *is*.

I am not likely to emulate my uncle Toby. I am no longer interested in being "different"—in drawing attention to myself.

As I have sworn numerous times I determined to serve out my Exile without violating the Instructions. I am determined to be returned to my family one day.

I am determined not to be "vaporized"—and forgotten.

Wondering if beneath the floorboards in the attic there's a pathetic little cache of things of mine, gnarled toothbrush, kitten socks, math homework with red grade *91*, my parents hastily managed to hide away.

THE WARRANT

Hereby, entered on this 19th day June NAS-23 in the 16th Federal District, Eastern-Atlantic States, a warrant for the arrest, detention, reassignment and sentencing of STROHL, ADRIANE S., 17, daughter of ERIC and MADELEINE STROHL, 3911 N. 17th St., Pennsboro, N.J., on seven counts of Treason-Speech and Questioning of Authority in violation of Federal Statutes 2 and 7. Signed by order of Chief Justice H. R. Sedgwick, 16th Federal District.

"GOOD NEWS!"

Or so at first it seemed.

I'd been named valedictorian of my class at Pennsboro High School. And I'd been the only one at our school, of five students nominated, to be awarded a federally funded Patriot Democracy Scholarship.

My mother came running to hug me, and congratulate me. And my father, though more warily.

"That's our girl! We are so proud of you."

The principal of our high school had telephoned my parents with the good news. It was rare for a phone to ring in our house, for most messages came electronically and there was no choice about receiving them.

And my brother, Roderick, came to greet me with a strange expression on his face. He'd heard of Patriot Democracy Scholarships, Roddy said, but had never known anyone who'd gotten one. While he'd been at Pennsboro High he was sure that no one had ever been named a Patriot Scholar.

"Well. Congratulations, Addie."

"Thanks! I guess."

Roddy, who'd graduated from Pennsboro High three years before, and was now working as a barely paid intern in the Pennsboro branch of the NAS Media Dissemination Bureau (MDB), was grudgingly admiring. I thought—*He's jealous. He can't go to a real university.*

I never knew if I felt sorry for my hulking-tall brother who'd cultivated a wispy little sand-colored beard and mustache, and always wore the same dull-brown clothes, that were a sort of uniform for lower-division workers at MDB, or if—actually—I was afraid of him. Inside Roddy's smile there was a secret little smirk just for *me*.

When we were younger Roddy had often tormented me— "teasing" it was called (by Roddy). Both our parents worked ten-hour shifts and Roddy and I were home alone together much of the time. As Roddy was the older, it had been Roddy's task to *take care of your little sister*. What a joke! But a cruel joke, that doesn't make me smile.

Now we were older, and I was tall myself (for a girl of my age: five feet eight), Roddy didn't torment me quite as much. Mostly it was his expression—a sort of shifting, frowning, smirk-smiling, meant to convey that Roddy was thinking certain thoughts best kept secret.

That smirking little smile just for me—like an ice-sliver in the heart.

My parents had explained: it was difficult for Roddy, who hadn't done well enough in high school to merit a scholarship even to the local NAS state college, to see that I was doing much better than he'd done in the same school. Embarrassing to him to know that his younger sister earned higher grades than he had, from the very teachers he'd had at Pennsboro High. And Roddy had little chance of ever being admitted to a federally mandated four-year university, even if he took community college courses, and our parents could afford to send him.

Something had gone wrong during Roddy's last two years of high school. He'd become scared about things—maybe with reason. He'd never confided in me.

At Pennsboro High—as everywhere in our nation, I suppose—there was a fear of seeming "smart"—(which might be interpreted as "too smart")—which would result in calling unwanted attention to you. In a True Democracy all individuals are *equal*—no one is *better than anyone else*. It was OK to get B's, and an occasional A–; but A's were risky, and A+ was very risky. In his effort not to get A's on exams, though he was intelligent enough, and had done well in middle school, Roddy seriously missed, and wound up with D's.

Dad had explained: it's like you're a champion archer. And you have to shoot to miss the bull's-eye. And something willful in you assures that you don't just miss the bull's-eye but the entire damned target.

Dad had laughed, shaking his head. Something like this had happened to my brother.

Poor Roddy. And poor Adriane, since Roddy took out his disappointment on me.

It wasn't talked about openly at school. But we all knew. Many of the smartest kids held back in order not to call attention to themselves. HSPSO (Home Security Public Safety Oversight) was reputed to keep lists of potential dissenters/ MIs/ SIs, and these were said to contain the names of students with high grades and high I.Q. scores. Especially suspicious were students who were good at science—these were believed to be too "questioning" and "skeptic" about the guidelines for curriculum at the school, so experiments were no longer part of our science courses, only just "science facts" to be memorized ("gravity causes objects to fall," "water boils at 212 degrees F.," "cancer is caused by negative thoughts," "the average female I.Q. is 7.55 points lower than the average male I.Q., adjusting for ST status").

Of course it was just as much of a mistake to wind up with C's and D's—that meant that you were *dull-normal*, or it might mean that you'd deliberately sabotaged your high school career. Too obviously "holding back" was sometimes dangerous. After graduation you might wind up at a community college hoping to better yourself by taking courses and trying to transfer to a state school, but the fact was, once you entered the workforce in a low-level category, like Roddy at MDB, you were there forever.

Nothing is ever forgotten, no one is going anywhere they aren't already at. This was a saying no one was supposed to say aloud.

So, Dad was stuck forever as an ME2—medical technician, second rank—at the district medical clinic where staff physicians routinely consulted him on medical matters, especially pediatric oncology—physicians whose salaries were five times Dad's salary.

Dad's health benefits, like Mom's, were so poor Dad couldn't even get treatment at the clinic he worked in. We didn't want to think what it would mean if and when they needed serious medical treatment.

I hadn't been nearly as cautious in school as Roddy. I enjoyed school where I had (girl) friends close as sisters. I liked quizzes and tests—they were like games which, if you studied hard, and memorized what your teachers told you, you could do well.

But then, sometimes I tried harder than I needed to try.

Maybe it was risky. Some little spark of defiance provoked me.

But maybe also (some of us thought) school wasn't so risky for girls. There had been only a few DASTADs—Disciplinary Actions Securing Threats Against Democracy—taken against Pennsboro students in recent years, and these students had all been boys in category ST3 or below.

(The highest ST—SkinTone—category was 1: "Caucasian." Most residents of Pennsboro were ST1 or ST2 with a scattering of ST3's. There were ST4's in a neighboring district and of course dark-complected ST workers in all the districts. We knew they existed but most of us had never seen an actual ST10.)

It seems like the most pathetic vanity now, and foolishly naïve, but at our school I was one of those students who'd displayed some talent for writing, and for art; I was a "fast study" (my teachers said, not entirely approvingly), and could memorize passages of prose easily. I did not believe that I was the "outstanding" student in my class. That could not be possible! I had to work hard to understand math and science, I had to read and reread my homework assignments, and to rehearse quizzes and tests, while to certain of my classmates these subjects came naturally. (ST2's and 3's were likely to be Asians, a minority in our district, and these girls and boys were very smart, yet not aggressive in putting themselves forward, that's to say *at risk*.) Yet somehow it happened that Adriane Strohl wound up with the highest grade-point average in the Class of '23—4.3 out of 5.

My close friend Paige Connor had been warned by her parents to hold back—so Paige's average was only 4.1, well inside the safe range. And one of the obviously smartest boys, whose father was MI, like my Dad, a former math professor, had definitely held back—or maybe exams so traumatized him, Jonny had not done well without trying, and his average was a modest/safe 3.9.

Better to be a safe coward than a sorry hero. Why I'd thought such remarks were just stupid jokes kids made, I don't know.

Fact is, I had just not been thinking. Later in my life, or rather in my next life, as a university student, when I would be studying psychology, at least a primitive form of cognitive psychology, I would learn about the phenomenon of "attention"—

"attentiveness"—that is within consciousness but is the pointed, purposeful, focused aspect of consciousness. Just to have your eyes open is to be conscious only minimally; to *pay attention* is something further. In my schoolgirl life I was conscious, but I was not *paying attention*. Focused on tasks like homework, exams, friends to sit with in cafeteria and hang out with in gym class, I did not pick up more than a fraction of what hovered in the air about me, the warnings of teachers that were non-verbal, glances that should have alerted me to—something . . .

I would realize, in my later life, that virtually all of my life beforehand had been *minimally conscious*. I had questioned virtually nothing, I had scarcely tried to decipher the precise nature of what my parents were actually trying to communicate to me, apart from their words. For my dear parents were *accursed with attentiveness*. I had taken them for granted—I had taken my own bubble-life for granted . . .

So it happened, Adriane Strohl was named valedictorian of her graduating class. Good news! Congratulations!

Now I assume that no one else who might've been qualified wanted this "honor"—just as no one else wanted a Patriot Democracy Scholarship. Except there'd been some controversy, the school administration was said to favor another student for the honor of giving the valedictory address, not Adriane Strohl but a boy with a 4.2 average and also a varsity letter in football and a Good Democratic Citizenship Award, whose parents were allegedly of a higher caste than mine, and whose father was not MI but EE (a special distinction granted to Exiled persons who had served their terms of Exile and had been what was called 110 percent rehabilitated—Exile Elite).

I'd known about the controversy vaguely, as a school rumor. The EE father's son had not such high grades as I did, but it was believed that he would give a smoother and more entertaining valedictory address, since his course of study was TV Public Relations and not the mainstream curriculum. And maybe administrators were concerned that Adriane Strohl would not be entertaining but would say "unacceptable" things in her speech?

Somehow without realizing, over a period of years, I'd acquired a reputation among my teachers and classmates for saying "surprising" things—"unexpected" things—that other students would not have said. Impulsively I'd raised my hands and asked questions. I was not *doubtful* exactly—just curious, and wanting to know. For instance was a "science fact" always and inevitably a fact? Did water *always* boil at 212 degrees F., or did it depend upon how pure the water was? And were boy-students *always* smarter than girl students, judging from actual tests and grades in our school?

Some of the teachers (male) made jokes about me, so that the class laughed at my silly queries; other (female) teachers were annoyed, or maybe frightened. My voice was usually quiet and courteous but I might've come across as *willful*.

Sometimes the quizzical look in my face disconcerted my teachers, who took care always to compose their expressions when they stood in front of a classroom. There were approved ways of showing interest, surprise, (mild) disapproval, severity. (Our classrooms, like all public spaces and many private spaces, were "monitored for quality assurance" but adults were more keenly aware of surveillance than teenagers.)

Each class had its spies. We didn't know who they were, of course—it was said that if you thought you knew, you were surely mistaken, since the DCVSB (Democratic Citizens Volunteer Surveillance Bureau) chose spies so carefully, it was analogous to the camouflage wings of a certain species of moth that blends in *seamlessly* with the bark of a certain tree. As Dad said, *Your teachers can't help it. They can't deviate from the curriculum. The ideal is lockstep—each teacher in each classroom performing like a robot and never deviating from script under penalty of—you know what.*

Was this true? For years in our class—the Class of NAS-23—there'd been vague talk of a teacher—how long ago, we didn't know—maybe when we were in middle school?—who'd "deviated" from the script one day, began talking wildly, and laughing, and shaking his/her fist at the "eye" (in fact, there were probably numerous "eyes" in any classroom, and all invisible), and was arrested, and overnight Deleted—so a new teacher was hired to take his/her place; and soon no one remembered the teacher-who'd-been-Deleted. And after a while we couldn't even remember clearly that one of our teachers *had been* Deleted. (Or had there been more than one? Were certain classrooms in our school *haunted?*) In our brains where the memory of _____ should have been, there was just a blank.

Definitely, I was not aggressive in class. I don't think so. But compared with my mostly meek classmates, some of whom sat small in their desks like partially folded-up papier-mâché dolls, it is possible that Adriane Strohl stood out—in an unfortunate way.

In Patriot Democracy History, for instance, I'd questioned "facts" of history, sometimes. I'd asked questions about the subject no one ever questioned—the Great Terrorist Attacks of 9/11/01. But not in an arrogant way, really—just out of curiosity! I certainly didn't want to get any of my teachers in trouble with the EOB (Education Oversight Bureau) which could result in them being demoted or fired or—"vaporized."

I'd thought that, well—people liked me, mostly. I was the spiky-haired girl with the big glistening dark-brown eyes and a voice with a little catch in it and a habit of asking questions. Like a really young child with too much energy in kindergarten, you hope will run in circles and tire herself out. With a kind of naïve obliviousness I earned good grades so it was assumed that, despite my father being of MI caste, I would qualify for a federally mandated State Democracy University.

(That is, I was eligible for admission to one of the massive state universities. At these, a thousand students might attend a lecture, and many courses were online.)

Restricted universities were far smaller, prestigious and inaccessible to all but a fraction of the population; though not listed online or in any public directory, these universities were housed on "traditional" campuses in Cambridge, New Haven, Princeton, etc., in restricted districts. Not only did we not know precisely where these centers of learning were, but we also had not ever met anyone with degrees from them.

When in class I raised my hand to answer a teacher's question I often did notice classmates glancing at me—my friends, even— sort of uneasy, apprehensive—*What will Adriane say now? What is wrong with Adriane?*

There was nothing wrong with *me*! I was sure.

In fact, I was secretly proud of myself. Maybe just a little vain. Wanting to think *I am Eric Strohl's daughter.*

THE ARREST

The words were brisk, impersonal: "Strohl, Adriane. Hands behind your back."

It happened so fast. At graduation rehearsal.

So fast! I was too surprised—too scared—to think of resisting.

Except I guess that I did—try to "resist"—in childish desperation tried to duck and cringe away from the officers' rough hands on me—wrenching my arms behind my back with such force, I had to bite my lips to keep from screaming.

What was happening? I could not believe it—I was being *arrested*.

Yet even in my shock thinking *I will not scream. I will not beg for mercy.*

My wrists were handcuffed behind my back. Within seconds I was a captive of the Homeland Security.

I'd only just given my valedictorian's speech and had stepped away from the podium, to come down from the auditorium stage, when there came our principal Mr. Mackay with a peculiar expression on his face—muted anger, righteousness, but fear also—to point at me, as if the arresting officers needed him to point me out at close range.

"That is she—'Adriane Strohl.' That is the treasonous girl you seek."

Mr. Mackay's words were strangely stilted. He seemed very angry with me—but why? Because of my valedictorian's speech? But the speech had consisted entirely of questions—not answers, or accusations.

I'd known that Mr. Mackay didn't like me—didn't know me very well but knew of me from my teachers. But it was shocking to see in an adult's face a look of genuine *hatred*.

"She was warned. They are all warned. We did our best to educate her as a patriot, but—the girl is a born *provocateur*."

Provocateur! I knew what the term meant, but I'd never heard such a charge before, applied to me.

Later I would realize that the Arrest Warrant must have been drawn up for me before the rehearsal—of course. Mr. Mackay and his faculty advisers must have reported me to Youth Disciplinary before they'd even heard my valedictory speech—they'd *guessed* that it would be "treasonous" and that I couldn't be allowed to give it at the graduation ceremony. And the Patriot Democracy Scholarship—that must have been a cruel trick, as well.

As others stood staring at the front of the brightly lighted auditorium the Arrest Warrant was read to me by the female Arresting Officer. I was too stunned to hear most of it—only the accusing words *arrest, detention, reassignment, sentencing—Treason-Speech* and *Questioning of Authority.*

QUICKLY THEN, Mr. Mackay called for an "emergency assembly" of the senior class.

Murmuring and excited my classmates settled into the auditorium. There were 322 students in the class, and like wildfire news of my arrest had spread among them within minutes.

Gravely Mr. Mackay announced from a podium that Adriane Strohl, "formerly" valedictorian of the class, had been arrested by the State on charges of Treason and Questioning of Authority; and what was required now was a "vote of confidence" from her peers regarding this action.

That is, all members of the senior class (excepting Adriane Strohl) were to vote on whether to confirm the arrest, or to challenge it. "We will ask for a show of hands," Mr. Mackay said, voice quavering with the solemnity of the occasion, "in a full, fair, and unbiased demonstration of democracy."

At this time I was positioned, handcuffed, with a wet, streaked/guilty face, at the very edge of the stage. As if my classmates needed to be reminded who the arrestee Adriane Strohl was.

Gripping my upper arms were two husky Youth Disciplinary Officers from the Youth Disciplinary Division of Homeland Security. They were one man and one woman and they wore dark blue uniforms and were equipped with billy clubs, Tasers, Mace, and revolvers in heavy holsters around their waists. My classmates stared wide-eyed, both intimidated and thrilled. An arrest! At school! And a show-of-hands vote which was not a novelty in itself except on this exciting occasion.

"Boys and girls! Attention! All those in favor of Adriane Strohl being stripped of the honor of class valedictorian as a consequence of having committed treason and questioned authority, raise your hands—yes?" There was a brief stunned pause. Brief.

Hesitantly, a few hands were lifted. Then, a few more.

No doubt the presence of the uniformed Youth Disciplinary Officers glaring at them roused my classmates to action. Entire rows lifted their hands—*Yes!*

Here and there were individuals who shifted uneasily in their seats. They were not voting, yet. I caught the eye of my friend Carla whose face too appeared to be wet with tears. And there was Paige all but signaling to me—*I'm sorry, Adriane. I have no choice.*

As in a nightmare, at last a sea of hands were raised against me. If there were some not voting, clasping their hands in their laps, I could not see them.

"And all opposed—no?" Mr. Mackay's voice hovered dramatically as if he were counting raised hands; in fact, there was not a single hand, of all the rows of seniors, to be seen.

"I think, then, we have a stunning example of democracy in action, boys and girls. 'Majority rule—the truth is in the numbers.'"

The second vote was hardly more than a repeat of the first: "We, the Senior Class of Pennsboro High, confirm and support the arrest of the former valedictorian, Adriane Strohl, on charges of Treason and Questioning of Authority. All those in favor . . ."

By this time the arrestee had shut her teary eyes in shame, revulsion, dread. No need to see the show of hands another time.

The officers hauled me out of the school by a rear exit, paying absolutely no heed to my protests of being in pain from the tight handcuffs and their grip on my upper arms. Immediately I was forced into an unmarked police vehicle resembling a small tank with plow-like gratings that might be used to ram against and to flatten protesters.

Roughly I was thrown into the rear of the van. The door was shut and locked. Though I pleaded with the officers, who were seated in the front of the vehicle, on the other side of a barred, Plexiglas barrier, no one paid the slightest attention to me, as if I did not exist.

The officers appeared to be ST4 and ST5. It was possible that they were "foreign"-born/ indoctrinated NAS citizens who had not been allowed to learn English.

I thought—*Will anyone tell my parents where I am? Will they let me go home?*

Panicked I thought—*Will they "vaporize" me?*

Heralded by a blaring siren I was taken to a fortresslike building in the city center of Pennsboro, the local headquarters of Homeland Security Interrogation. This was a building with blank bricked-up windows that was said to have once been a post office, before the Reconstitution of the United States into the North American States and the privatizing and gradual extinction of the postal service. (Many buildings from the old States remained, now utilized for very different purposes. The building to which my mother had gone for grade school had been converted to a Children's Diagnostic and Surgical Repair Facility, for instance; the residence hall in which my father had lived, as a young medical student, in the years before he'd been reclassified as MI, was now a Youth Detention and Re-education Facility. The Media Dissemination Bureau, where my brother Roddy worked, was in an old brownstone building formerly the Pennsboro Public Library, in the days when "books" existed to be held in the hand— and read!) In this drafty place I was brought to an interrogation room in the Youth Disciplinary Division, forcibly seated in an uncomfortable chair with a blinding light shining in my face, and a camera aimed at me, and interrogated by strangers whom I could barely see.

Repeatedly I was asked—"Who wrote that speech for you?"

No one, I said. No one wrote my speech, or helped me write it—I'd written it myself.

"Did your father Eric Strohl write that speech for you?"

No! My father did not.

"Did your father tell you what to write? Influence you? Are these questions your father's questions?"

No! My own questions.

"Did either of your parents help you write your speech? Influence you? Are these questions their questions?"

No, no, *no*.

"Are these treasonous thoughts their thoughts?"

I was terrified that my father, or both my parents, had been arrested, and were being interrogated too, somewhere else in this awful place. I was terrified that my father would be reclassified no longer MI but SI (Subversive Individual) or AT (Active Traitor)—that he might meet the same fate as Uncle Tobias.

My valedictorian speech was examined line by line, word by word, by the interrogators—though it was just two printed double-spaced sheets of paper with a few scrawled annotations. My computer had been seized from my locker and was being examined as well.

And all my belongings from my locker—laptop, sketchbook, backpack, cell phone, granola bars, a soiled school sweatshirt, wadded tissues—were confiscated.

The interrogators were brisk and impersonal as machines. Almost, you'd have thought they might be robot-interrogators—until you saw one of them blink, or swallow, or glare at me in pity or disgust, or scratch at his nose.

(Even then, as Dad might have said, these figures could have been robots; for the most recent AI devices were being programmed to emulate idiosyncratic, "spontaneous" human mannerisms.)

Sometimes an interrogator would shift in his seat, away from the blinding light, and I would have a fleeting but clear view of a face—what was shocking was, the face appeared to be so *ordinary*, the face of someone you'd see on a bus, or a neighbor of ours.

My valedictorian address had been timed to be no more than eight minutes long. That was the tradition at our school—a short valedictorian address, and an even shorter salutatorian address. My English teacher Mrs. Dewson had been assigned to "advise" me—but I hadn't shown her what I'd been writing. (I hadn't shown Dad, or Mom, or any of my friends—I'd wanted to surprise them at graduation.) After a half-dozen failed starts I'd gotten desperate and had the bright idea of asking numbered questions—twelve, in all—of the kind my classmates might have asked if they'd had the nerve—(some of these the very questions I'd asked my teachers, who had never given satisfactory answers)—like *What came before the beginning of Time?*

And *What came before the Great Terrorist Attacks of 9/11?*

Our RNAS calendar dates from the time of that attack, which was before my birth, but not my parents' births, and so my parents could remember a pre-NAS time when the calendar was different—time wasn't measured in just a two-digit figure but a *four-digit figure*! (Under the old, now-outlawed calendar, my mother and father had been born in what had been called the twentieth century. It was against the law to compute birth dates under the old calendar, but Daddy had told me—I'd been born in what would have been called the twenty-first century if the calendar had not been reformed.)

NAS means North American States—more formally known as RNAS—Reconstituted North American States, which came into being some years after the Great Terrorist Attacks, as a direct consequence of the Attacks, as we were taught.

Following the Attacks there was an Interlude of Indecisiveness during which time issues of "rights"—(the Constitution, the Bill of Rights, Civil Rights law, etc.)—vs. the need for Patriot Vigilance in the War Against Terror were contested, with a victory, after the suspension of the Constitution and the Bill of Rights by executive order, for PVIWAT, or Patriot Vigilance. (Yes, it is hard to comprehend. As soon as you come to the end of such a sentence, you have forgotten the beginning!)

How strange it was to think there'd been a time when the regions known as (Reconstituted) Mexico and (Reconstituted) Canada had been separate political entities—separate from the States! On a map it seems clear, for instance, that the large state of Alaska should be connected with the mainland United States, and not separated by what was formerly "Canada." This too was hard to grasp and had never been clearly explained in any of our Patriot Democracy History classes, perhaps because our teachers were not certain of the facts.

The old, "outdated" (that is, "unpatriotic") history books had all been destroyed, my father said. Hunted down in the most remote outposts—obscure rural libraries in the Dakotas, below-ground stacks in great university libraries, microfilm in what had been the Library of Congress. "Outdated"/"unpatriotic" information was deleted from all computers and from all accessible memory— only reconstituted history and information were allowed, just as only the reconstituted calendar was allowed.

This was only logical, we were taught. There was no purpose to learning useless things, that would only clutter our brains like debris stuffed to overflowing in a trash bin.

But there must have been a time *before that time*—before the Reconstitution, and before the Attacks. That was what I was asking. Patriot Democracy History—which we'd had every year since fifth grade, an unchanging core of First Principles with ever-more detailed information—was only concerned with post-Terrorist events, mostly the relations of the NAS with its numerous Terrorist Enemies in other parts of the world, and an account of the "triumphs" of the NAS in numerous wars. So many wars! They were fought now at long-distance, and did not involve living soldiers, for the most part; robot-missiles were employed, and powerful bombs said to be nuclear, chemical, and biological. In our senior year of high school we were required to take a course titled "Wars of Freedom"—these included long-ago wars like the Revolutionary War, the Spanish-American War, World War I, World War II, the Korean War, the Vietnam War, and the more recent Afghanistan and Iraq wars—all of which our country had won—"decisively." We were not required to learn the dates or causes of these wars, if there were actual causes, but battle-places and names of high-ranking generals, political leaders, and presidents; these were provided in columns to be memorized for exams. The question of *Why?* was never asked—and so I'd asked it in class, and in my valedictory address. It had not occurred to me that this was *Treason-Speech*, or that I was *Questioning Authority*.

The harsh voices were taking a new approach: Was it one of my teachers who'd written the speech for me? One of my teachers who'd "influenced" me?

The thought came to me—*Mr. Mackay! I could blame him, he would be arrested . . .*

But I would never do such a thing, I thought. Even if the man hated me, and had me arrested for treason, I could not lie about him.

AFTER TWO HOURS of interrogation it was decided that I was an "uncooperative subject." In handcuffs I was taken by YD officers to another floor of Home Security which exuded the distressing air of a medical unit; there I was strapped down onto a movable platform and slid inside a cylindrical machine that made clanging and whirring noises close against my head; the cylinder was so small, the surface only an inch or so from my face, I had to shut my eyes tight to keep from panicking. The interrogators' voices, sounding distorted and inhuman, were channeled into the machine. This was a BIM (Brain-Image Maker)—I'd only heard of these—that would determine if I was telling the truth, or lying.

Did your father—or any adult—write your speech for you?

Did your father—or any adult—influence your speech for you?

Did your father—or any adult—infiltrate your mind with treasonous thoughts?

Barely I could answer, through parched lips—*No. No, no!*

Again and again these questions were repeated. No matter what answers I gave, the questions were repeated.

Yet more insidious were variants of these questions.

Your father Eric Strohl has just confessed to us, to "influencing" you—so you may as well confess, too. In what ways did he influence you?

This had to be a trick, I thought. I stammered—*In no ways. Not ever. Daddy did not.*

More harshly the voice continued.

Your mother Madeleine Strohl has confessed to us, both she and your father "influenced" you. In what ways did they influence you?

I was sobbing, protesting—*They didn't! They did not influence me . . .*

(Of course, this wasn't true. How could any parents fail to "influence" their children? My parents had influenced me through my entire life—not so much in their speech as in their personalities. They were good, loving parents. They had taught Roddy and me: *There is a soul within. There is "free will" within. If—without—the State is lacking a soul, and there is no "free will" that you can see. Trust the inner, not the outer. Trust the soul, not the State*. But I would not betray my parents by repeating these defiant words.)

At some point in the interrogation I must have passed out— for I was awakened by deafening noises, in a state of panic. Was this a form of torture? Noise-torture? Powerful enough to burst eardrums? To drive the subject insane? We'd all heard rumors of such torture-interrogations—though no one would speak openly about them. Shaken and excited Roddy would come home from his work at Media Dissemination to tell us about certain "experimental techniques" Homeland Security was developing, using laboratory primates—until Mom clamped her hands over her ears and asked him to *please stop*.

The deafening noises stopped abruptly. The interrogation resumed.

But it was soon decided then that I was too upset—my brain waves were too "agitated"—to accurately register truth or falsity, so I was removed from the cylindrical imaging machine, and an IV needle was jabbed into a vein in my arm, to inject me with a powerful "truth-serum" drug. And again the same several questions were asked, and I gave the same answers. Even in my exhausted and demoralized state I would not tell the interrogators what they wanted to hear: that my father, or maybe both my parents, had "influenced" me in my treasonous ways.

Or any of my teachers. Or even Mr. Mackay, my enemy.

I'd been taken out of the hateful BIM, and strapped to a chair. It was a thick squat "wired" chair—a kind of electric chair—that sent currents of shock through my body, painful as knife-stabs. Now I was crying, and lost control of my bladder.

The interrogation continued. Essentially it was the same question, always the same question, with a variant now and then to throw me off stride.

Who wrote your speech for you? Who "influenced" you? Who is your collaborator in Treason?

It was your brother Roderick who reported you. As a Treason-Monger and a Questioner of Authority, you have been denounced by your brother.

I began to cry harder. I had lost all hope. Of all the things the interrogators had told me, or wanted me to believe, it was only this—that Roddy had reported me—that seemed to me possible, and not so very surprising.

I could remember how, squeezing my hand when he'd congratulated me about my good news, Roddy had smiled—his special smirking-smile just for me.

Congratulations, Addie!

"DISCIPLINARY MEASURE"

Next morning I was taken from my cell and returned to Youth Interrogation.

Half-carried from my cell, handcuffed and my ankles shackled, I was very very tired, very sick, scarcely conscious.

It was my hope that my parents would be waiting for me—that they'd been summoned to come get me, and take me home. I would accept it that I'd be forbidden to attend graduation—forbidden even to graduate from high school; I would accept it that I might be sent to a Youth Rehabilitation camp, as it had been rumored the boys at Pennsboro High had been, who'd been arrested by Youth Disciplinary. All I wanted was to see my parents—to rush at them, and throw myself into their arms . . .

Some months ago my parents had celebrated my seventeenth birthday with me. It had been a happy time!—but seemed now a lost, childish time. I had not felt like seventeen and now, desperate for my parents, I scarcely felt like a teenager at all.

Of course my parents were nowhere near. Probably they did not know what had happened to me. And I did not dare ask about them.

Instead I was being sternly informed: several Patriot Scholars had been arrested yesterday afternoon, simultaneous with my arrest, in a Youth Disciplinary "sweep." After a season or more of relatively few such sweeps and seizes, YDDHS was "cracking down" on "potential subversives."

These Patriot Scholars were graduating seniors from other high schools in the area. Their names, too, had been handed over to YDDHS by the principals of their schools. The question was put bluntly to me: Was I, Adriane Strohl, a collaborator with these students? Was I a co-conspirator?

Their names were told to me: I'd never heard of any one of them.

Their faces appeared on three overhead TV screens: I'd never seen any one of them before.

A camera was turned on me, in a blaze of blinding light. I had to assume that my frightened face was being beamed into other interrogation rooms, where the arrested Patriot Scholars were being held.

Repeatedly I was asked: Was I a collaborator with any one, or more, of these individuals? Was I a *co-conspirator*?

All I could answer was *No*.

Weakly, hopelessly—*No*.

CHEN, MICHAEL was a very young-looking Asian-American boy with sleek black hair worn to his collar and widened, very dark and alarmed eyes. He, too, had been named valedictorian of his graduating class, at Roebuck High School. A glance at CHEN, MICHAEL and you knew that this was a smart boy, probably ST3.

PADURA, LAUREN was a thick-bodied girl with a strong-boned if rather ashen face and damp eyes, probably ST2. She was sitting as upright as possible, though handcuffed and shackled at the ankles; she was a student at East Lawrence High. A glance at PARUDA, LAUREN, and you knew that this was a girl who thought for herself and very likely, in her classes, asked questions as I had.

ZOLL, JOSEPH JAY was a tall lanky dark-blond boy with a blemished face, thick glasses, and a small mustache on his upper lip, ST1 like me. He had been named salutatorian of his class at Rumsfeld High and looked like a math/computer whiz—one glance at ZOLL, JOSEPH JAY and you knew that here was a boy you wanted for your friend, whose kindness, patience, and computer expertise would be invaluable.

The four of us, on the TV monitors, were not looking great. Our eyes were bloodshot. Our mouths were trembling. Whatever we'd done, we regretted it. We didn't look *innocent*. We didn't any longer resemble high school seniors—we looked much younger. *Just kids. Scared kids. Kids needing their mom and dad. Kids without a clue what was happening to them.*

A panicked thought came to me—what if one of the Patriot Scholars suddenly confessed to "collaborating" with the rest of us? Would we all be executed?

A brisk voice informed us that we had thirty seconds to compose our confessions. At the end of these thirty seconds, if no one had confessed, one Patriot Scholar would be suitably "disciplined" with a Domestic Drone Strike (DDS)—on camera.

We were terrified—paralyzed. No one spoke.

Sleek-black-haired CHEN, MICHAEL opened his mouth and tried to speak, but no words came out.

There were tears, and agitation, in PADURA, LAUREN'S face—but no words issued from her mouth.

Then, I heard myself pleading, in a thin wavering voice, that we were not "collaborators"—we didn't know one another, we'd never seen one another before, we didn't know one another's names . . .

Indifferently the off-screen voice was counting: eleven, sixteen, twenty-one . . . Twenty-seven, twenty-eight . . .

My heart was pounding so violently, I thought that it would burst. My eyes darted from one TV monitor to the others—as the Patriot Scholars trembled and cringed in their chairs, narrowing their eyes, yet unable to shut their eyes entirely.

On one of the screens there was a blinding flash. The boy with the mustache on his upper lip—ZOLL, JOSEPH JAY—was slammed sideways as if he'd been struck with a laser ray, that entered the side of his head like liquid fire, exploded and devoured his head, and then his torso and lower body, in less than three seconds.

What was left of ZOLL, JOSEPH JAY fell to the floor, slithering and phosphorescent, and by quick degrees this, too, vanished . . . I had a glimpse of the other Patriot Scholars staring in horror at their TV screens before all four screens went dark and a roaring in my ears became deafening.

When I woke from a sick, dead faint, I was being lifted from my chair. Yet so terrified still, I didn't dare to open my eyes.

EXILE: ZONE 9

"Adriane. I am your Youth Disciplinary Counsel."

She was a woman of about Mom's age. Her face was shiny and glaring, so that I could barely look at her. Or maybe my vision had become oversensitive from the previous day's interrogations in blazing light.

Her name was S. Platz. Her manner was almost jovial, as if she and I shared a joke.

"Try to lift your head and look at me, dear. As if you have nothing to hide. We are being 'surveyed' and videotaped—you must know."

After the terror I'd been feeling, and the hopelessness, S. Platz was so astonishing to me, I couldn't at first believe her; I was sure that she must be another torturer. Each time I closed my eyes I saw ZOLL, JOSEPH JAY struck down like an animal—or an "enemy" figure in a video game.

I would never forget that horrific sight, I thought.

I would never want to forget, for the executed boy's sake.

S. Platz, unlike the other interrogators, did not continue to ask me the same questions, and she did not speak in a brisk impersonal voice.

She asked one of the uniformed officers to unshackle me—both my wrists and my ankles. She asked if my wrists and ankles hurt, and if I was "very tired" and would like to "sleep an uninterrupted sleep" in a real bed—to help with "healing."

Almost inaudibly I said *Yes.*

(Wondering if "uninterrupted sleep" meant something terrible?)

But S. Platz seemed so kindly! Tears flooded my eyes, I was so grateful for her sympathy.

Yes thank you. Oh yes—I would like to sleep . . .

"I have good news for you, Adriane. Youth Disciplinary has ruled that the discipline for your violation of the federal statutes is—Exile."

Exile! I had heard of Exile—of course. This form of discipline was often confused with Deletion because, so far as anyone knew, including the families of Exiled Individuals, the Exiled Individual simply—vanished.

It was said to be highly experimental, and dangerous. Exiled Individuals were *teletransported*—every molecule in their bodies dissolved, to be *reconstituted* elsewhere. (No one who was left behind knew where. A colony on another planet? This was a prevailing rumor, according to Roddy. But which planet? If any had been colonized by the Government, ordinary citizens knew nothing about it.) But often, the teletransportation failed, and individuals were injured, incapacitated, killed, or, in effect, "vaporized"—and no one ever saw them again.

Only if the Exiled Individual reappeared, years later, after having served out his sentence, could it be assumed that he had been alive all the while, but in a remote place. EIs were generally allowed to live but had to submit to a process of "Re-education" and "Reconstitution."

Exile was considered to be the "humane"—"liberal"—disciplinary measure, appropriate for younger people who had not committed the most serious crimes—yet.

In our Patriot Social Studies class we were taught that the "Re-educated" and "Reconstituted" individual, successfully having completed his sentence of Exile and returned to the present time, designated EI1, was often an outstanding Citizen-Patriot; several notable EI1's had been named to federal posts in Homeland Security Public Safety Oversight and Epidemic Control; and the most renowned of EI1's had risen to a high-ranking executive post in the Capitol, as an assistant to the director of the Federal Bureau of Interrogation.

There were rumors that the president was himself an EI1—a former "traitor"-genius now totally converted to NAS and its democratic tradition.

S. Platz was saying: "Your case was carefully adjudicated, Adriane, after several of your teachers entered pleas for you. Their claim was that you are 'naïve'—'very young'—'not subversive'—and 'not radical'—and that, if you are separated from the influence of your MI-father and are allowed to Re-educate yourself, you would be of value to society. Therefore, we are transporting you to Zone Nine. There, you will attend an excellent four-year university to train yourself in a socially useful profession. Teaching is strongly suggested. Or, if your science grades are good, you would be allowed to apply to medical school. Zone Nine is not so 'urban' as our Eastern zones, nor is it so 'rural' as most of our North-Midwest zones. It is not on any NAS map—it is a place that 'exists' only by way of special access, for, in our present time, in the North-Midwest States that now encompass what was known as 'Wisconsin' in the era of Zone Nine, things are very different." Seeing that I was looking confused and frightened, S. Platz said, "You need not concern yourself with any of this, Adriane—you will simply be transported to the university in Zone Nine, where you will be a 'freshman.' You will be given a new, abbreviated identity. You will be, as you are, seventeen years old. And your name will be 'Mary Ellen Enright.' If necessary, when you return to us your training can be brought up-to-date. Everything that you need to know is explained in The Instructions, which I will give you now."

Though S. Platz spoke clearly I was having difficulty understanding. Badly wanting to ask *But can I see my parents again? Just once before I am sent away* . . .

S. Platz handed me a sheet of paper stiff as parchment. When I tried to read it, however, my eyes filled with moisture, and lost focus.

"So there is really no need to ask me questions—is there?"

S. Platz paused, smiling at me.

And now I saw that the counsel's steel-colored eyes were not smiling, but only just staring, and assessing.

The realization came to me—*If I don't react properly, I will be vaporized on the spot. The woman has this power.*

With a numb smile I managed to murmur *Thank you!*

And not a word about my parents. Or the life I would be leaving—the life to which I would be "lost."

II

ZONE 9

THE HAPPY PLACE

She was a strange girl. At first, we didn't like her.

She never smiled at us. Her face was like a mask. She prayed on her knees—we'd seen her! She cried herself to sleep every night like us, only worse.

We were all homesick, our first semester at Wainscotia. Missed our parents and our families so! But this girl was sad in a way we weren't—like her heart was broken. And she would not be comforted—which does not seem natural.

We were Christian girls, mostly Protestant. We attended chapel on Sundays. (She never did—we noticed that.) We believed in prayer— seriously!

We believed in helping one another. We believed in smiling-through-tears. You cried, and then you laughed—you opened a box of brownies your mom had sent to you, to share with your room-mates, or anyone who showed up in the room.

You cried, and wiped away your tears—and you were yourself again.

She was the girl who scorned our mothers' brownies, and scorned the opportunity to walk with us to class on the steep-hilly sidewalks leading into the campus, who almost never came with us to the din-

ing hall and sat with us. She'd gone to freshman orientation alone, and slipped away alone. The only girl in Acrady Cottage who didn't participate in Vesper Song.

Probably the only person in the freshman class who claimed she'd "lost" her green-and-purple freshman beanie. Who paid no heed to upperclassmen ordering her to "step out of the way, frosh"— stared through them as if she didn't see them and kept walking in her stiff-hunched way like a sleepwalker you pitied and would not want to waken.

The girl-with-no-name we called her. For if you called out a cheery Hi there, Mary Ellen!—she didn't seem to hear, or to register the name, even as she quickened her pace hurrying away.

We knew little about her. But we knew that she was a scholarship girl, like us.

Acrady Cottage was a residence for freshman scholarship girls— meaning that most of us couldn't have afforded Wainscotia State except for financial aid and part-time jobs on campus.

(She had a part-time job in the geology library.)

Scholarship girls were thrifty girls! Most of our textbooks were used—some of them pretty badly battered.

We wore hand-me-down clothes and clothes sewn by our mothers or grandmothers—or by us.

Quite a few of us were 4-H girls. We were three Wisconsin State Fair blue-ribbon winners, in Acrady Cottage.

Acrady wasn't a residence hall like the ivy-covered stone buildings elsewhere on campus—it was just a house, plain weathered-gray shingle board, where freshman girls lived who could not afford to live anywhere else.

But Acrady Cottage had spirit!

At Vesper Song, Acrady excelled though we were one of the smaller residences on campus—twenty-two girls.

Twenty-two girls including "Mary Ellen"—the girl who kept to herself like someone in quarantine.

First week at Wainscotia we were all homesick. But trying to be cheerful—"friendly."

Not her—"Mary Ellen Enright."

She'd avoid us if she could. Her own roommates!—that isn't easy.

Four of us in the third-floor room. Cramped space and just two (dormer) windows.

She'd taken the bed that was farthest, in a corner. And her desk jutted out from the wall, to partially hide the bed from the rest of the room. No window in that corner.

She'd shrink from us. Trying to smile—a weird bright smile that didn't reach her eyes.

And then, when she thought we were asleep and wouldn't see her, on her knees in the corner, praying.

And crying herself to sleep.

She looked like someone who has traveled a long journey and has not fully recovered.

We'd wondered—Is she foreign?

But what kind of foreign?

Even the way she spoke was strange. If you cornered her to say hello and ask how she was and she couldn't escape or avoid answering she would stammer a reply that was almost—but not quite— intelligible. We could recognize in her speech the rhythms and vowel-sounds of English, so if we didn't exactly know what she was saying, we could guess.

And she spoke so fast and nervous! Like none of the other girls of Acrady Cottage.

Of course, we were all from the Midwest. Most of us from

Wisconsin. Where Mary Ellen was from was—one of the eastern states, we'd been told. Evidently people talked faster there.

Our part of the U.S., it's generally known as the Happy Place. (Midwest!) And Wainscotia is a very special university at the heart of the Happy Place.

This girl Mary Ellen, we had to wonder: Was she a Christian? Was she—Jewish?

None of us from Wisconsin had ever seen a Jewish person, before coming to Wainscotia. But there were Jewish people here— professors, it was said, as well as students. A few.

There was even a Jewish fraternity, and there was a Jewish sorority. So they could live with their own kind. This was amazing to us!

In large cities in Wisconsin, like Milwaukee, and Madison, there were Jewish people—we knew that. But we were mostly from upstate Wisconsin, or rural counties. German, Scandinavian, Scots and Irish backgrounds—and English of course.

Also, "Mary Ellen" did not look like us. This was difficult to explain, but we all agreed. Her hair was darkish blond, spiky like it hadn't been combed or brushed properly; never curled or waved, and not washed often enough. It needed cutting, and trimming. And she went to bed without hairpins or rollers. Ever.

Didn't do anything for her hair. Didn't even seem to recognize what rollers were!

(Like she didn't seem to know how a phone worked—the way you dial with your forefinger. Of all of us in Acrady Cottage, only Mary Ellen would shrink from answering the phone when it rang near her, so we had to wonder—was it possible her family didn't own a telephone?)

(And she didn't smoke! And was forever coughing from our cigarette smoke—coughing in fits, until her eyes watered—though she never complained as you'd have expected a nonsmoker to complain.

You could see the misery in her face but it was what you'd call long-suffering misery.)

She might have been pretty—almost—except she never wore lip-stick. A guy looking at her would look right past her, there was so little to catch the male eye. (We all wore red—very red—lipstick!) She didn't even pluck her eyebrows which is about the least a girl can do, to make herself attractive.

She looked like a convalescent. Some wasting disease that left her skinny, and her skin ashy-pale and sort of grainy as if it would be rough to the touch like sandpaper. Her eyes would have been beautiful eyes—they were dark-brown, like liquid chocolate—with thick lashes—but they were likely to be narrowed and squinting as if she were looking into a bright, blinding light. And she did not look you in the eye—like a guilty person.

Along with not smoking, she also didn't eat potato chips, glazed doughnuts, Cheez-bits, M&M's, and those little cellophane bags of Planters peanuts that were our favorite snacks while we were studying, that left salt all over our fingers. She didn't eat any kind of chewy meat like beef or pork or even chicken, sometimes. She'd eat fish casseroles.

Which was why she was so skinny. Flat-chested, and her hips flat, like a guy's, she didn't have to struggle into girdles like the rest of us.

Didn't even know what a "girdle" was—just stared at Trishie's when she was getting dressed for a frat party, like she'd never seen anything so scary!

Scholarship girls in Acrady Cottage studied, hard. We worked long hours at our desks for we had to retain B averages or lose our schol-arships. But Mary Ellen Enright worked harder, and longer than any of us—so far as we knew, Mary Ellen did almost nothing else except her schoolwork, and her part-time job in the geology library!

At her desk with her back to the room, and to her roommates, she was unnaturally still, concentrated upon her work motionless as a mannequin but so tense, you could see the tension in her neck and shoulders.

As if she was holding back a scream, or a sob.

Yet she remained at her desk, with the crook-necked lamp shining a halo of light onto the desktop, for as long as she could endure it—if none of her roommates objected to the light.

(At first we did not object. We had no trouble sleeping with a single light on in the room. But as time passed, and we grew less patient with our misfit-roommate, we did object, and Mary Ellen took her work downstairs to the work-study lounge where she would disturb no one, and no one would disturb her. And sometimes she slept there, on a couch. And we did not have to hear her sob herself to sleep!)

It was hard to say what was wrong with Mary Ellen. The way she'd stare at us, and look away—like she'd seen something scary—made us think there was something wrong with us.

We hoped (sort of) that she would drop out of school. Or transfer. This was cruel of us, and not very Christian—we were not proud of such thoughts. But we were girls who'd only just graduated from high school a few months before and maybe Mary Ellen scared us, so close to the edge. Close to mental and physical collapse as we'd sometimes felt ourselves away from home for the first time and at the University of Wisconsin–Wainscotia where 9,400 students were enrolled.

A certain percentage of freshmen dropped out in the first several weeks of their first semester at Wainscotia. You'd hear of someone—mostly girls, but there were boys, too—who just "broke down"—"couldn't sleep"—"cried all the time"—"felt lost."

But Mary Ellen seemed determined not to be one of these. She was deeply unhappy and seemed to us—(to some of us, at least)—on the verge of a nervous collapse but there was something willful about her—like a cripple who doesn't seem to know she's a cripple, or a stutterer who doesn't seem to know that she's a stutterer.

Another strange thing: alone among the girls of Acrady Cottage Mary Ellen Enright received no mail.

And yet stranger: Mary Ellen Enright seemed to expect no mail, for she walked by the mailboxes without glancing at her own.

We asked our resident adviser Miss Steadman what she thought we could do to make Mary Ellen less lonely and Miss Steadman suggested leaving her alone for the time being, for "Mary Ellen" had traveled a long distance—from one of the eastern states like New York, New Jersey or Massachusetts—and was feeling a more acute homesickness than we were feeling, whose families lived in the state and whom we could visit on weekends by bus.

Is "Mary Ellen" Jewish? we asked.

Miss Steadman said she did not think so. For "Enright" was not known to be a Jewish name.

But she seems like she's from—somewhere else. Like she's not an American—sort of.

Miss Steadman frowned at this remark which clearly she did not like.

Miss Steadman would tell us only that Mary Ellen Enright was the sole resident of Acrady Cottage whose complete file hadn't been made available to her, as our resident adviser. She'd received a file from the office of the dean of women for Mary Ellen but it was very short, and some of it had been blotted out with black ink.

Hilda McIntosh described how in the first week of classes she'd come into her room on the third floor one afternoon—and there was her roommate Mary Ellen Enright standing in front of Hilda's desk staring at her typewriter.

This was Hilda's portable Remington typewriter, one of her proudest possessions which she'd brought to college. Not everyone in Acrady Cottage had a typewriter, and those who didn't were envious of Hilda!

And there was Mary Ellen staring at the typewriter with some look, Hilda said, "way beyond envy."

Like the girl had never seen a typewriter before! Like it was some new invention.

So Hilda said, speaking softly not wanting to startle Mary Ellen—(but startling her anyway, so that she jumped, and quivered, and her eyelids fluttered)—You can try it, Mary Ellen, if you want to. Here's a piece of paper!

Hilda inserted the paper into the typewriter. Hilda indicated to the girl how she should type—striking several keys in rapid succession.

The girl just stared blank-faced.

Like this, see? Of course, you have to memorize the keyboard. That comes with practice. I learned in high school—it isn't hard.

The girl touched one of the keys, lightly. Like she hadn't strength to press it down.

In a faint voice she said, It doesn't w-work . . .

Hilda laughed. Of course it works!

The girl peered at the back of the typewriter, as if searching for something missing.

In a faint voice saying, But—there's nothing connecting it to—to . . .

Hilda laughed. This was like introducing a rural relative to, well, an indoor toilet! Funny.

Look here, Hilda said.

Hilda sat down at her desk and the keys clattered away like machine-gun fire:

SEPTEMBER 23, 1959

ACRADY COTTAGE

WAINSCOTIA STATE UNIVERSITY

WAINSCOTIA FALLS, WISCONSIN

USA

UNIVERSE

The new girl Mary Ellen stared at this display of magic—the flying keys, the typed letters that became words and sentences— and could not seem to speak. As if her throat had shut up. As if the clattering typewriter frightened her. As if she couldn't bear the sight of—well, what was it? Hilda couldn't imagine.

Hilda encouraged Mary Ellen to try the typewriter again but Mary Ellen backed away as if it was all too much for her. And sud denly then, her eyes rolled back in her head, her skin went chalky- white—and she fell to the floor in a dead faint.

TYPEWRITER

"Mary Ellen?"

One of them was speaking to me. She'd come up behind me.

I was very frightened. I knew there were informers like my brother Roddy—of course. But I could not comprehend if I was behaving like a guilty person or whether, in my EI status, I was behaving in a manner appropriate to my circumstances.

The girl was the one called "Hilda McIntosh." She had a round bland moon-face and a very friendly smile. Her hair was a chestnut-colored "pageboy." I could not bring myself to look her in the face, still less in the eyes, for fear of what I would see.

The empty gaze. The iris in the eye the size of a seed.

I wondered: Was this person an informer? Did she know who *Mary Ellen Enright* really was? *Had she followed me?*

It was often the case, here in Zone 9, that individuals followed me. Yet in such practiced ways, I could not be certain that they were following me by design or by coincidence.

I'd had to escape from—wherever it was—the lecture hall in the building near the chapel—Hendrick Hall—as oxygen was being sucked from the room by the others—(I'd counted sixty-six student-figures seated in steep-banked rows)—and the professor-figure at the podium continued lecturing on the rudiments of logic. *Some Y is X. X is M. What is the relationship of M to Y?*

I did not cross the green. The open, vulnerable expanse of the green. Making my way like a wounded wild creature close beside buildings and through narrow passageways in order not to attract attention.

Not daring to glance up, to see who was "seeing" me.

The EI will be monitored at any and all times during his/her exile.

Violations of these Instructions will insure that the EI will be immediately Deleted.

In a sequence of hills, mostly downhill, approximately one mile to Acrady Cottage where I would hide in the third-floor room assigned to "Mary Ellen Enright."

It was the epicenter of the Restricted Zone 9. It was my imprisonment. Yet, I felt safe there.

I entered the cottage by a side door. Made my way up the back stairs hoping not to be detected by any girl-figures and I avoided the resident adviser's suite on the first floor where the door was always flung open in a way to suggest *Welcome!* but which I worried might be an informer's trick.

It had been a blade twisted in my heart, that my brother Roddy had informed on me. I could not recall much of those terrible hours of interrogation at YDDHS but I did recall this revelation and my shock and yet my unsurprise for *Of course, Roddy always hated me. He would wish me Exiled—or worse.*

I wanted to think that one day I would see Roddy again—and I would forgive Roddy. Tears stung my eyes for I could not bear to think that Roddy did not want my forgiveness.

But if I confronted Roddy, it would mean that I would see my parents again. Desperately I wanted to think this!

It was a time, early afternoon, when the freshman residents of Acrady Cottage were mostly out.

I was counting on being invisible. I could not breathe normally, if I knew myself *visible*.

In these early days in Zone 9 I did not think of those others who might be in Exile here—for surely there were others like myself. Like one trapped in a small cage who has no awareness of others similarly trapped, and in her desperation no sympathy to spare, I could only think of my own situation.

Hurrying up the stairs. Panting, and sweating. For it was a hot dry September in this place, and Acrady Cottage was not air-conditioned.

In Zone 9 very few buildings were air-conditioned. Evidently in this era air-conditioning was rare. And what air there was, to my dismay and disgust, was often polluted by cigarette smoke.

Astonishing to me, my roommates smoked! Each one of them. As if they knew nothing about cigarettes causing cancer, or did not care. Worse, I saw their annoyed expressions when I coughed and choked and yet—how could I help myself? Smoking had been banished in NAS-23 for as long as I'd been alive. (Intravenous nicotine was promoted in its place.)

Thinking *Is this my punishment? Secondary smoke inhalation.*

Had to wonder who my roommates really were. Why I'd been assigned to room 3C of Acrady Cottage, with these individuals. In NAS-23 it was said—"No accidents, only algorithms." I could not think that there were coincidences in the stratagems of Homeland Security. I could not think that at least one and perhaps all of my roommates were informants assigned to *Enright, Mary Ellen.*

Possibly, one of them was a robot. But which one?

SUCH RELIEF, when I was alone in our room.

(But was I ever "alone" in our room?)

That prevailing odor of cigarette smoke like body odor made my nostrils pinch.

Here was my opportunity to examine a clumsy black machine with a keyboard on the desk of one of my roommates—a "typewriter."

I'd heard of typewriters of course. I'd seen photographs of typewriters and my parents had spoken of owning a typewriter, I think. (Or had it been my grandparents?) But I had never seen an actual typewriter, from an era before computers.

In a kind of trance I stared at the strange machine. Something about it that made me uneasy.

So badly I missed my laptop computer, I felt almost faint. I missed my cell phone, that fitted so comfortably in the palm of my hand it was like a growth there, a rectangular luminous eye.

Could not comprehend the logic of this machine. Could it be so crude, only just—*typing*?

No Internet? No e-mail? No texting? Only just—*typing*?

There was nowhere to look in or about the typewriter! There was no *screen*.

Profound to think that this clumsy machine connected with no reality beyond itself. Just—a *machine*.

You were trapped in yourself, at a typewriter. You could not escape into cyberspace. In Zone 9, you could not access cyberspace.

Hard to comprehend: in 1959 *cyberspace did not exist*.

And yet, that was not possible—was it? For one of the great accomplishments of the twenty-first century—we'd been told and retold—was the establishment of cyberspace as an entity separate and distinct and (presumably) independent of human beings, thus independent of constraints of time and space.

Not that I'd ever understood this. To truly understand you'd have to know math, physics, astrophysics, the most advanced

computer science that was in fact, in NAS-23, classified information . . .

"Mary Ellen?"—the voice was close behind me.

The smiling girl had crept up silently behind me—"Hilda" was
her name. She'd scared me so, my heart leapt in my chest like a
shot bird.

Hilda was very friendly of course. They were all very friendly.

Their eyes eating at me, like hungry ants. Memorizing, assessing. Planning the words they would use in their reports to
Homeland Security.

In her flat midwestern voice that seemed mocking to my ears
Hilda was saying that the machine was her "almost-new Remington," of which she seemed to be proud.

"You can try it, if you want to. Here's a piece of paper!"

Hilda inserted a sheet of paper and rolled it into position.
She indicated to me how I should type, striking several keys in
succession—random keys, with her deft fingers.

But I just stared. I was feeling light-headed.

I would have tried to speak but my tongue felt like cotton
batting too big for my mouth.

Hilda said, "Like this, see? Of course, you have to memorize
the keyboard. So that your fingers type without you having to
think. That comes with practice. I learned in high school—it isn't
hard."

I touched one of the keys. Nothing happened.

"It doesn't w-work . . ."

Hilda laughed at me. Not maliciously but as an older sister
might laugh at a naïve younger sister.

"Of course it works, Mary Ellen! Like this."

Mary Ellen. Was this name uttered in a friendly way, or in a
mocking way?

I wanted to think that the girl-figure "Hilda" was actually a girl like myself, who was being sincerely friendly. I did not want to think that the girl-figure was a registered agent of Homeland Security or (possibly) a virtual representation of an undergraduate girl manipulated on a screen by a (distant) Homeland Security agent teasing me in my Exile in Zone 9.

It was distracting to me, that Hilda stood so close. Many of the girls of Acrady Cottage stood close to me, and caused me to back away. Our way of behaving with one another in NAS-23 was noticeably different: the unspoken rule was *do not come close.* Since the arrest, and the terrible sight on the TV monitor of the boy executed, I was in dread of strangers coming too close. My skin prickled with the danger.

Hilda was so very friendly, so *nice*, she seemed entirely oblivious of my wariness. It would be said of Hilda that she was a "pretty" girl—(as it would be said of me, I was sure, that I was a "plain" girl). Shorter than I was by at least two inches, and plumper. Where my body was lean almost like a boy's her body was shapely as a mature woman's. Like the other girls Hilda wore a sturdy brassiere—a "bra"—that might have stood by itself, made of firm, metallic-threaded fabric; beneath her clothes, this "bra" asserted itself like an extra appendage. Half-consciously I shrank away hoping that Hilda would not, seemingly unconsciously, brush against me with her sharp-pointed breasts.

Hilda sat at her desk with an exaggerated sort of perfect posture like a young woman in an advertisement and typed, rapidly and flawlessly, to demonstrate to me how easy "typing" was:

SEPTEMBER 23, 1959

ACRADY COTTAGE

WAINSCOTIA STATE UNIVERSITY

WAINSCOTIA FALLS, WISCONSIN

USA

UNIVERSE

"See? Now you try it, Mary Ellen."

September 23, 1959! It could not be true—could it?

This was Zone 9—of course. This was my Exile. I must accept my Exile, and I must adjust. Yet—

The horror swept over me: this was eighty years into the past, and more. I had not yet been born. My parents had not yet been born. There was no one in this world who loved me, no one who even knew me. No one who would claim me. I was utterly alone.

"Mary Ellen? What's wrong?"

With a look of genuine—sisterly—concern, Hilda reached out for me, even as I shrank away.

"Don't t-touch me! No . . ."

I was terrified, nauseated. Yet too weak to escape—a black pit opened at my feet, and sucked me down.

THE LOST ONE

Help me! Help me—Mom, Daddy . . .

I miss you so much . . .

It was a ravenous hunger in me, to return home. A yearning so strong, it seemed almost that a hand gripped the nape of my neck, urging me forward as in a desperate swooning plunge.

I am all alone here. I will die here.

THEY'D STRAPPED ME DOWN. Wrists, ankles, head—to prevent "self-injury."

A painful shunt in the soft flesh at the inside of my elbow, through which a chill liquid coursed into my vein. It was a mechanical procedure they'd done many times before.

In a flat voice the pronouncement: *subject going down.*

I saw myself as a diminishing light. A swirl of light, turning in upon itself and becoming ever smaller, more transparent.

Abruptly then—I was gone.

Dematerialization of the subject. Teletransportation of the subject's molecular components. Reconstitution in Zone 9.

"'**MARY ELLEN ENRIGHT.**' This is she?"

The question was put to someone not-me. Yet I could observe the lifeless body from a slightly elevated position and felt pity for it.

Like a zombie. Exiled.

I would wonder—*Does a zombie know that it is a zombie? How would a zombie comprehend.*

This was funny! But laughter caught in my throat like a clot of phlegm.

In this very cold place. Where blood coursed slow as liquid mercury.

I was very confused. I could not clear my head. My brain had been injured. I had heard them joking.

NSS it was called—*Neurosurgical Security Services.* Rumors had circulated in high school. The subject was taboo.

Before *teletransportation* they'd inserted a microchip into a particular part of my brain called the hippocampus, where memory is processed before being stored elsewhere in the brain. At least, I thought this must have happened. I did not think it had been a dream.

Part of my scalp had been shaved, a pie-shaped wedge of skull removed, the microchip installed. (Evidently) I felt no pain. A zombie does not feel pain. Even the sawed-out portion of the skull and the lacerated scalp were cold-numb and remote to me. And yet I felt such a powerful wave of gratitude, I could have wept—*They did not remove my parents from me. They left me my parents at least.*

For that part of my brain might have been removed, which contained all memory of my parents.

In Exile you cling to what you have, that has not (yet) been taken from you.

From this cold place I was carried, with others who'd been *tele-transported*, in a vehicle resembling an emergency medical van.

The vehicle did not move rapidly. There was no siren.

This was not an emergency but routine.

The vehicle made stops at several destinations, before mine. In my semiconscious state I had little awareness of what was happening. I was trying to speak to my parents whose faces were vivid to me in their concern for me. I was trying to say *In four years I will see you again. Don't forget me!*

I could not have said if I was seventeen years old, or seven years old.

I could not have said which year this was. I had no idea where I was.

We had left the lights of a city and were traveling now in a vast rural night. It was astonishing to me, stars in the night sky overhead were large and luminous as I had never seen stars before in my old, lost life.

The air was purer here, in Zone 9. So sharp to inhale! The night sky was not obscured by the scrim of pollution to which we were all accustomed in the old, lost life.

We who were being carried in the van in the night were strapped to stretchers and could not turn our heads to regard one another. We were very tired, for we'd come a long distance.

It might have been the case, not all of the *teletransported* had made the journey fully alive. It was not clear to me initially whether I was fully alive.

One of the other *teletransported* was hyperventilating in panic. Something must have gone wrong with his medication. I could not turn my head to see. Or, my head was strapped in place. I held myself very still and breathed calmly as Dad would instruct me in the presence of the Enemy. I thought—*They will vaporize him.* It was a desperate thought—*They will vaporize him, and not me.*

At the next stop, I was taken from the van.

Unstrapped from the stretcher, and made to stand.

"Use your legs, miss. There is nothing wrong with your legs. Your brain sends the signal—left leg, right leg. And your head— lift your head."

I was able to walk a few yards, before I collapsed.

In the morning I woke beneath a thin blanket, on a lumpy cot. The bandages were gone from my head. The straps were gone from my wrists and ankles. Most of the grogginess had faded.

It would be explained to me: I was a freshman student at Wainscotia State University in Wainscotia Falls, Wisconsin. I had arrived late the previous night, feverish. I had been brought to the university infirmary and not to my residence. And now, in the morning, since my fever had disappeared, I was to be discharged.

"Your things have been delivered to your residence, Miss Enright."

"Yes. Thank you."

"Your residence is Acrady Cottage, on South University Avenue."

"Thank you."

Acrady Cottage. South University Avenue. It was up to me to find this place, and I would do so.

I was feeling hopeful! Small gulping waves of wonder would rush over me from time to time, amid even the paralysis of fear.

For the crucial matter was: my parents were living, and I would return to them, in four years. My parents had not been "vaporized" even in my memory.

And the crucial matter was: "Mary Ellen Enright" was evidently a healthy specimen. She had not died in *teletransportation*. If her brain had been injured, it was not a major injury.

If it was a minor injury, maybe it would heal.

When I tried to rise from the cot, however, I felt faint, and would have lost my balance—but the strong-muscled young woman in the white nurse's uniform reached out to catch me.

"There you go, 'Mary Ellen'! On your way."

She laughed. Our eyes locked, for a fleeting second.

She had pinned-back blond hair, so pale it was almost white. Above her left breast, a little plastic name tag—IRMA KRAZINSKI.

She knows who I am. Yet, she is not an Enemy.

Later I would think—*Maybe she is one like me and will pity me.*

AT THE RESIDENCE a large cardboard box awaited M. E. ENRIGHT in the front foyer.

"You are—'Mary Ellen'? This just arrived."

The box measured approximately three by four feet. It was so crammed, one of its sides was nearly bursting.

And the box was badly battered, as if it had come a long distance, in rainy weather. Transparent tape covered it in intricate layers crisscrossing like a deranged cobweb. Even with a pair of shears provided by the resident adviser of Acrady Cottage it was very difficult to open.

"My! Someone took care that this box would not rip open in delivery!"

Inside were clothes: several skirts, blouses, sweaters, a pair of slacks, a navy-blue wool jumper, a fleece-lined jacket, flannel pajamas, white cotton underwear, white cotton socks, a pair of sneakers, and a pair of brown shoes identified by the resident adviser as "penny loafers." There were also "Bermuda shorts" and a "blazer"—clothes of a kind I had never seen before. And sheer, long "stockings"—I'd never seen before. All these items were secondhand, rumpled, and smelled musty.

I was staring inside the box. I felt dazed, dizzy. I thought—*These are castoff clothes of the dead.*

"Shall I help you carry these upstairs? It might be more practical just to leave the box here and take your things up in our arms . . ."

"No. I can take them by myself. Thank you."

The resident adviser, Miss Steadman, was being very kind. But I did not want even to look at her. I did not want to speak with the woman more than necessary and I did not want to be alone with her in the room to which I was assigned for even a few minutes.

I did not want her to see these clothes close up. I did not feel comfortable with her registering that, to me, some of these things were unfamiliar. Nor did I want her to smell the sour, stale odor that lifted from them, any more than she already had.

I did not want her to feel sorry for me. *That poor girl!—indeed, she is* poor.

Also, Miss Steadman's words, her manner of speech, were strange to me. It was clear that she was speaking English yet so slowly, with such odd nasal vowels, it made me anxious to listen to her.

At the bottom of the box was an envelope with M. E. ENRIGHT stamped on it. I would not open this envelope until I was alone in room 3C when I would discover that it contained five twenty-dollar bills that were crisp as if freshly printed, and a stiff sheet of paper headed THE INSTRUCTIONS.

There was no personal note. I felt a small stab of disappointment for I had thought—I mean, I'd wanted to think—that S. *Platz* had taken a personal liking to me.

In my arms I carried my new belongings upstairs to room 3C. I grew short of breath quickly for I had not recovered from my long journey. Miss Steadman watched me with concerned eyes but did not attempt to help me another time.

Freshmen would be arriving on the Wainscotia campus the next day. I'd been sent into Exile at the perfect time and I did think that S. Platz must have had something to do with this timing.

Room 3C was at the rear of the cottage. A large room with two dormer windows and a slanted ceiling. Bare floorboards, bare walls with scattered holes for picture hanging and small nails.

Four beds, four desks: four roommates!

It was surprising to me, I would be rooming with three other girls and not alone.

But a relief, the room was ordinary. Except for the slanting ceiling that, if I wasn't alert, would bump against my head.

Quickly my eyes glanced about. It would be an involuntary reaction in Zone 9: establishing that a new space held no (evident) danger. Nothing in it (that I could see) to frighten, threaten, or disorient.

Nothing unique to Zone 9. Rather, a room that could be *anywhere*.

I took the bed in the farthest corner, beneath the slanted ceiling. I would leave the windows, the better-positioned beds, and the largest closets for my roommates for I did not want them to dislike me.

"'Mary Ellen'! Are you sure, you want that bed way off in a corner?"—so my roommates asked when they arrived, with evident sincerity.

These were nice girls. (Were they?) Staring at me with curious eyes but they were not rude, or did not mean to be rude.

Though they were enough alike to be three sisters they were strangers to one another. "White" girls—ST1. All were from rural Wisconsin and had gone to Wisconsin high schools. Their broad flat northern-midwestern accents were identical. Their names were immediately confused in my head like a buzzing of insects.

I thought—*One of them may be my executioner.*

"When did you arrive, Mary Ellen? Last night?"

"Where're you from, Mary Ellen?"

"Did your parents bring you? Are they still here?"

"Sorry, Mary Ellen! We're taking up a lot of room, I guess . . ."

Much of the day the room was crowded with parents, relatives, young children, helping my roommates move in.

I went away to hide. The sounds of strangers' voices, loud, assured, happy-seeming, those broad flat vowels, were oppressive to me. But I did not cry.

At evening I returned to the room at the top of the stairs for I had nowhere else to go. Acrady Cottage was my home now.

EVENTUALLY, WHEN I BEGAN to wear the clothes that had come in the box, I would discover that only a few items fitted me.

Some things were too small, too short, too tight—most were too large.

Faint half-moons of stains beneath the armpits of sweaters. Loose buttons, missing buttons. Broken zippers. Dark-smeared something, possibly food, I hoped not blood, on a skirt.

The girls of Acrady Cottage would whisper among themselves, to see me so badly dressed—like a pauper, with clothes from Goodwill—but I never minded for I was grateful for what had been given to me.

My favorites were a pleated Black Watch plaid skirt (as it was identified for me by a roommate) with an oversized ornamental brass safety pin holding the skirt together—ingeniously; a dark-rose turtleneck sweater that reminded me of a sweater back home, though this sweater was much larger; a long-sleeved white blouse with a "lace" collar that fitted me, and gave me a serious, somber look, that I particularly liked because it seemed to suggest *This is a good girl, a nice girl, a shy girl, a girl who would never, ever be subversive or raise her voice. Please be kind to this girl thank you!*

In my old, lost life I had never worn blouses. I had never worn "lace"—or known anyone who had.

I had never worn skirts, dresses. I had worn only jeans. In fact, just two or three pairs of jeans, that had not cost much and that I wore all the time without needing to think.

In Zone 9 girls wore skirts to classes, sometimes dresses. They wore "sweater sets"—cardigans over matching short-sleeved sweaters. Sometimes they encased their legs in *nylon stockings* which I did not think I could manage without tearing, though I would try.

How my friends would laugh, to see me in a lacy blouse. In nylon stockings. In the Black Watch plaid skirt with the big brass safety pin holding the pleated material together—*Oh God what has happened to Addie. Is that even* her?

NOT A PLEASANT SIGHT. Electrodes in my roommates' heads.

Well, not *electrodes.* I knew better.

"Like this, Mary Ellen. I can't believe you've never 'set' your hair!"

Laughing at me. Not unkindly. (I wanted to think.)

But I could not manage it: putting my hair in plastic "rollers" before going to bed.

First, you wetted your hair with some special smelly setting-solution. Brushed and combed your hair. Separated your hair into numerous strands, and rolled these strands onto "rollers" (three sizes: largest pink, medium blue, smallest mint green) which were secured to the scalp as tightly as possible with bobby pins.

Yes, your scalp might hurt from the pins and from having to lie with your head on a pillow, on rollers.

You might even get a little headache! But it was worth it, for the effect of the smooth glossy pageboy the next day.

Tried just once. Awake half the night. Drifting off to sleep and waking in a nightmare sweat of electrodes in my brain. And in the morning most of the rollers had come out, and when I brushed out my hair it was as limp and straggling as ever, or almost.

Hilda said, "Next time, Mary Ellen, I'll set your hair in rollers. Don't you dare say *no*."

COED

So lonely! It was as if my body had been gutted from within.

As if, in the place where my heart had been, there was an emptiness that nothing could fill.

Other freshmen were homesick, and other girls in my residence cried from time to time in this new place. But their homesickness was a kind of exquisite torture, a way of measuring their love for their families. They called home, often on Sunday evenings (when the rates were lower) and received calls from home. They wrote home, and received letters from home. Their mothers sent them baked goods to share with their roommates and friends. And their homes were accessible to them—just a few hours away by car.

I began to be ashamed, as well as despondent, that I received no mail from home—no calls, no packages. I could not bear it, the girls of Acrady Cottage pitied me and spoke of me wonderingly behind my back.

Yet, there must be genuine orphans in the world, with no families and no relatives. The category into which *Enright, Mary Ellen* had been placed could not have been so empty of inhabitants.

I wondered if I would discover someone like myself? Or— someone like myself would discover *me*?

A "coed" at the State University at Wainscotia, Wisconsin, enrolled in the College of Liberal Arts with the likelihood of an education major, Class of 1963.

If this was Exile, it was not the cruelest Exile.

I knew this: the cruelest Exile would be death.

Badly wishing I could let my parents know that I was (still) alive. (But was I alive? Often I wasn't sure.)

Wishing that I could know that my parents were (still) alive in NAS-23 and that we would be together again, in four years.

WAINSCOTIA STATE UNIVERSITY covered many acres of land in semi-rural Wainscotia Falls in northeastern Wisconsin, a day's drive from the city of Milwaukee in the south. Most of its nine thousand students were from small Wisconsin towns or farms. One of the largest colleges was Agriculture and Animal Husbandry.

Other prominent colleges were the School of Education, the School of Business, the School of Nursing, and the School of Engineering.

So many thousands of people linked by a single sprawling campus! Though much of the time the campus looked calm—in the morning, on the hilly paths, students hurrying to classes, in small groups, in pairs, alone—as the chapel bell sounded sonorously. *You must move along. You must take your place. You have your name, your identity. You have no choice.*

There was a thrill in this! There was the solace of the impersonal.

I was enrolled in five courses, three of them "introductions"—to English literature, to psychology, and to philosophy. These were large lecture courses with quiz sections that met once weekly. It was possible to be invisible in large lecture halls and to imagine that no one was observing me.

If I saw, on campus, or in one of my classes, a girl from Acrady Cottage, my vision blurred and didn't register what I saw. If a girl waved to me, or smiled at me—I did not seem to see. Wherever it was possible, I was invisible.

Eventually, they would leave me alone, I believed. I would learn in Intro to Psychology the phenomenon of *operant extinction*: *when reinforcement is no longer forthcoming, a response becomes less and less frequent.*

All my determination was to survive. To get through the challenge of the first weeks, the first semester, and the first year; to get through four years; to complete my Exile, and be *teletransported* back home.

I did not want to think that absolute obedience to The Instructions might not save me. I did not want to think about the future except in the most elemental terms: *behavior, reward.*

Almost it seemed to me, S. Platz had personally promised me—my Exile would end, one day.

In the meantime I was obliged to obey The Instructions. I had immediately memorized them though I did not quite understand what was meant by the admonition against providing "future knowledge" in the Restricted Zone.

So far as I had been informed the microchip in my brain blocked many memories of my past (that were, of course, in 1959, anticipations of the "future"). I could not confidently "foresee," still less predict. When I tried to recall classic discoveries of the intervening decades—the discovery of DNA, for instance—the development of molecular genetics—brain "imaging"—any modern history apart from Patriot History—it was like trying to peer through a frosted glass window.

You can see shadowy shapes beyond, maybe. But you cannot *see.*

How ironic it was, I'd been, for those few cruel days, *valedictorian of my high school class*!

AS FOR THE ADMONITION against seeking out "relatives"—I would not have known how to begin.

In the Cultural Relocation Campaign that swept the country when I was in middle school, hundreds of thousands—millions?—of individuals were evicted from their homes, to be settled in relatively depopulated areas which the Government wanted to "reconstitute"; among these were Mom's and Dad's parents—Roddy's and my grandparents—who were evacuated to western Nebraska and northern Maine, respectively; but I had no idea where they'd lived previously, still less where they might have been living in 1959.

And I would not have dared leave the ten-mile radius of the "epicenter."

I would be the ideal student—the ideal "coed." I would attract no unwanted attention. I would never betray or even feel the mildest curiosity. I would never enter into any "intimate" relationship with anyone—hardly! This was my resolve.

During the first week of classes instructors took attendance by reading off names. It was exciting to hear a cascade of strangers' names in which there was secreted, like a cuckoo egg in a nest of unsuspecting birds, the name to which I'd been assigned: *Enright, Mary Ellen*. This seemed to me so wholly a fictitious name, so totally unconvincing a name, I steeled myself as shyly I raised my hand and murmured "Here." But no one took any notice. Most of the names were, like "Enright," immediately forgettable. Though scattered amid these were odd nasal-sounding names that turned out to be, as I would later discover, Swedish, Norwegian, Germanic.

Instructors were not likely to linger over *Enright, Mary Ellen*.

Virtually all of my fellow students, indeed all of my professors, were ST1: "Caucasian." Exceptions were staff workers (cafeteria, janitorial, grounds) who were likely to be ST5 or darker.

I wondered if in any of my classes there were other Exiled Individuals. Could we recognize one another? Did we dare recognize one another? But how? At what risk?

We would all be in disguise. In secondhand clothes, with secondhand names, determined to survive.

I took a kind of grim pride in my own disguise: sweaters and skirts from out of the battered cardboard box, fleece-lined jacket in cold weather, sneakers, even "penny loafers" with white cotton socks darned at the toes, but not darned very substantially. My textbooks and anthologies were all secondhand, their spines stamped USED. In the campus bookstore I'd bought a spiral notebook whose cover was speckled black and white, and on the pristine lined pages of this notebook I took fanatic notes in a kind of miniature handwriting, to save on space. In lecture halls I was the faceless girl in the first row hunched over her notebook, rapidly writing as the professor spoke, scarcely daring to look up.

How strange to be *writing*. This was a so-called hand-eye coordination skill that had nearly vanished by the time I was born—though Mom and Dad had insisted upon teaching Roddy and me how to *write*. And strange to be *reading* in a book, on paper "pages" which you had to turn with your fingers, and which, if you wanted, you could tear out; but you needed no "power" to maintain the book, and no electronic medium.

Strangest of all, the university library—a vast brownstone building of numerous floors descending even into the earth, filled with row upon row of "stacks" containing "books" to be touched, and opened, *by hand*. And in reading rooms, high ceilings, myriad lights, and polished floors—and students!

Just to climb the stone steps to this building, like some ancient temple, left me dazed, and apprehensive.

Often in Zone 9 I found myself out of breath. My heart beat

erratically as it had beat—(I remembered this amid so much that had been wiped from my memory)—when I'd witnessed the Domestic Drone execution of—(but what was the boy's name? *Zoll-* was the first syllable)—in a time that seemed to me now long ago, and fading. My head ached somewhere behind my eyes, where the microchip had been inserted. If I tried to think of—(was the word *home*? *parents*?)—there emerged a barrier like Plexiglas. Against this barrier I would strain and strain—like a trapped creature trying to press through a wall.

Yet, if I gave up this effort, and turned my concentration upon my work, reading passages of print in textbooks and anthologies, underlining, taking notes, making outlines, writing first drafts of papers—in the way of a "normal" undergraduate in Zone 9—the pressure in my brain relaxed, and my breathing grew calmer.

This is your life now. What you must be, now.

LOST FRIENDS

Sometimes I woke with a sob, and it wasn't my parents I was missing but my friends.

And it came to me in a rush of emotion—I loved my friends! I'd taken them for granted sometimes, for which I was sorry now.

Trying to recall their names: Carla, and Melanie, and Deborah, and . . . was it Paige?

Like trying to peer through something clotted. A thick gauze. Strength was required to remember. My eyes ached in their sockets with the effort of trying to see my friends' faces that were beginning to fade.

We'd been together since middle school. We'd become closer in high school under certain stresses and pressures. *To be like everyone else. But—how is that possible? No one is like anyone else.*

At Pennsboro High there'd been a vicious social hierarchy about which no one was supposed to speak: the sons and daughters of Government officials at the top, and the rest of us spread out beneath. With my father an MI, my social caste had been determined from the first day my mother had brought me to preschool.

There'd been a time in tenth grade when Carla had been deeply sad and anorexic, and her friends had helped her through

the crisis; there'd been a time in our junior year when Glenna
had been in a perpetual state of anxiety, when her research-
scientist father had been removed from his laboratory on a
charge of *Science-Treason*. And there was the terrible ten-month
period when Deborah's father was in custody at Homeland Security Interrogation, and no one knew if he'd ever be released.
(Eventually Mr. Albright was released, but in a perpetual benign
zombie-state, so it was as if Deborah had lost her father after all.)
Using a text-code we'd invented for ourselves, which we were
shrewd enough to change every few weeks, we'd texted one another continuously, day after day, for years. We were closer in
some essential ways than we were with our parents who did not
speak openly to us, and certainly we were closer with each other
than with any of our male classmates. ("Female" and "male"
were expected to distrust each other. It was not considered good
behavior to have close friends of the opposite sex before high
school graduation.)

Somewhere in my damaged memory was the vague knowledge that my friends had tried to warn me about the valedictorian
address—Paige had advised me to work with our English teacher
on the speech, which, she thought, should have been modeled
after previous valedictorian addresses, that had not antagonized
the Youth Disciplinary censor, but I'd ignored her suggestion
totally—in fact, I'd been insulted and hurt by it.

How stupid I'd been! Paige had wanted to protect me, and I'd
been heedless.

I wondered what graduation at my high school had been like.
The boy-athlete whom Mr. Mackay had wanted to give the valedictory address would have given it, in my place. Someone else
would have been named salutatorian. No one except my friends
would miss me.

Wasn't someone else supposed to be valedictorian?

Someone else? Who?

That girl . . . What was her name . . .

What girl? I don't remember any girl.

You know, the one with—I guess brown hair . . .

Oh gosh! I remember her—sort of . . .

She was arrested for Treason. She's gone now.

Gone—where?

Gone.

HE, HIM

And then, in this place of utter loneliness, I fell in love.

He was not the first person who looked kindly upon me in Zone 9. Or protectively. Or even curiously.

He was the first person who *knew*. Staring at me, in an instant he knew who I was. What I was.

And I thought—*There are two of us now. Him, and—Mary Ellen.*

WOLFMAN

"'Enright, Mary Ellen'"—the eyes moved onto me curiously, coolly.

I could scarcely breathe. Digging my nails into the palms of my hands.

Our instructor was returning midterm exams. He was smiling, though not with his eyes. His eyes moved restlessly over us, impersonal and detached, calculating.

I wondered: were we "subjects" to him? Wolfman was a research psychologist, an assistant professor in psychology.

After several weeks of class he knew the names of certain individuals in our class but not, until now, mine.

Hesitantly I'd risen from my seat to take the little blue booklet from Wolfman. He had not known me before. I had not distinguished myself in the quiz section by raising my hand, answering his questions or asking questions of my own, as a scattering of other students had, more aggressive than Mary Ellen Enright. But now, Wolfman was looking at Mary Ellen Enright.

"Good work, Miss Enright. You've been reading Skinner outside the textbook?"

"Y-Yes."

Continuing to stare at me, just a moment too long.

In class Wolfman was brusque and sunny and sharp-witted—a knife-blade that has been honed razor-sharp, you would not carelessly reach out to touch. Now his ironic smile faded. His expression seemed one of surprise—genuine surprise. As if about to ask me something further but changed his mind, quickly looking away and calling the next name.

A sensation of faintness enveloped me. But I managed to return to my desk, clutching the blue book without daring to open it.

Does he know?

Does he—somehow—recognize me?

Is he in Exile too—like me?

For the remainder of the hour I remained in a state of something like suspended animation. I could not bring myself to look at the quiz section instructor at the front of the room, casually sitting at the edge of a table; meekly I leaned over my notebook, to take notes. Nor did Wolfman glance at me. Through a roaring in my ears I could barely hear his voice expounding upon some principle of psychology and by the end of the period, when the bell sounded, I remained benumbed in my desk as others trooped out of the room. When at last I dared to raise my eyes Ira Wolfman had departed.

Only when I was safely alone did I open the blue book. Extravagant, exclamatory, the grade was in red ink—99%.

And in red ink the witty scrawl below—*No one is perfect.*

The EI is forbidden to identify himself/herself except as established by HSEDB.
The EI is monitored at all times.

A soul in Exile. I was convinced, I'd come into contact with a kindred soul.

Those of us who'd been assigned to Wolfman's quiz section were considered both lucky (Wolfman was *fun*) and not-so-lucky (Wolfman was *tough*). He could be intimidating as well as entertaining. He was, to midwestern eyes and ears, "different"—even "foreign."

Wolfman spoke more fluently, more rapidly and on the whole less patiently than most other adults in Zone 9 whom I'd encountered. His hair was thick and dark and lifted from his forehead in wayward curves that reminded me of the thick swirly feathers of an owl. (Thinking of the handsome great horned owl specimen in the Natural History collection.) Though Wolfman was clean-shaven you could see the shadow of a beard darkening his jaws and beneath his jaws, on his throat. Most days for our class he wore a sport coat, dark trousers and a white or pale blue shirt. Often, not always, a necktie. His eyes were dark, of the hue of wet slate, quizzical and restless. You did think that, at times, Wolfman's attention—his deepest and most profound attention—was elsewhere.

I am here but not truly here. You have no idea who I am.

It was Wolfman's custom to fling out questions as one might fling out a handful of coins to underlings—some students responded alertly, others held back suspecting a trick. Others were confused, disoriented. We were likely to be cautious of Ira Wolfman despite his genial and inviting smile.

There were twenty-five students in the quiz section of whom three were female. Rarely had I tried to answer the instructor's questions for I feared his possible sarcasm, the stroke of a razor-sharp blade across the knuckle of an unsuspecting undergraduate.

The week's subject was "neurobehavioral analysis"—a Skinnerian postulate that the nervous system is inactive most of the time and roused only by stimuli from the environment, that result in "reflexes." It was a behavioral premise that an animal is, essentially, a machine. A human being is an animal and is therefore, essentially, a machine. Individual, group, and mass behavior can be programmed, conditioned, predicted and controlled. Radical behaviorism is a science: you can graph the consequences of experiments. A human being is not *inward* but *outward*—the sum of (measurable) behavior. A human being *is* behavior, to be observed and charted by others. What the environment can't control is the genetic self. But this too is determined. Zombies come in all shapes, sizes, and types.

I did not want to think that this was true. But in Zone 9, in the psychology department of Wainscotia State University, such "truths" prevailed.

The lecture course taught by Professor A. J. Axel—Introduction to Twentieth-Century Psychology—was one of the most popular and revered in the College of Liberal Arts, with an enrollment of more than two hundred students. Many were pre-med, for this was a requirement. Wolfman's quiz section met on Friday mornings when it was his task to explicate Professor Axel's often abstruse lecture for us, to shine a beacon of clarity into the dark patches of our ignorance. You could see that the youthful Wolfman took pleasure in befuddling us further, before enlightening us. He took pleasure in scribbling graphs—"learning curves"—on the blackboard, to illustrate the professor's argument. It was a pleasurable task to explain experiments in detail to which Professor Axel only alluded.

Occasionally, always very subtly, Wolfman corrected Professor Axel. *As Professor Axel meant to say . . .*

A. J. Axel was claimed to be one of the most distinguished academic psychologists in the United States, a former Harvard collaborator with the greatest experimental psychologist of the twentieth century, B. F. Skinner. He had been a protégé of the famed Dr. Walter Freeman as well, assisting Freeman in performing a number of lobotomies in the Midwest, in the early 1950s. He was the director of the Wainscotia Center for Social Engineering. Tall, white-haired, gentlemanly, invariably dressed in a tweed coat, white shirt and tie, Axel was a figure of dignity and erudition. Yet, Axel was not always comprehensible to his audiences for his vocabulary was highly specialized, as in a secret code. Such mysterious terms as *operant conditioning, schedules of reinforcement, law of effect, reinforcers, punishers, escape learning, avoidance learning* were predominant.

To undergraduates Wolfman was "Dr. Wolfman," for he had a Ph.D. To Wolfman, we were "Miss"—"Mister."

He was unfailingly courteous with us, if often slightly ironic; he seemed not to trust us to quite understand his jokes. And he seemed to have many jokes. He was quick-witted, funny. With the acuity of one swatting a fly while his attention seemed to be elsewhere, he could rebuke a skeptical student.

Wolfman attended Professor Axel's lectures on Tuesday and Thursday mornings at 10:00 A.M., seated in the front row of the lecture hall with several other assistant professors and departmental graduate students. With few exceptions among the graduate students, they were all male: there were virtually no female faculty members at Wainscotia.

Of Professor Axel's young assistants Ira Wolfman must have been the most trusted, for when the professor was forced to miss a lecture Wolfman took his place. On those mornings there was a quickened sense of anticipation in the lecture hall, for most of the students preferred dashing young Ira Wolfman to the renowned A. J. Axel.

Sometimes following Wolfman's lectures there was a scattering of applause in the lecture hall. Unobtrusively in my seat I clapped with the others, thrilled and breathless.

In our class Wolfman described psychology experiments in which he'd participated: the systematic precipitation of "learned helplessness"—a term for "breakdown"—in pigeons, rats, and small primates (monkeys) after a volley of stimuli followed by random and unpredictable consequences. Initially, the subject tries to make sense of the stimuli, to perceive a pattern amid randomness; the subject tries to "control" the situation, but of course fails, and eventually suffers a breakdown.

After the "learned helplessness," the subject can be reconstituted through a volley of new stimuli, with predictable consequences.

Like animals, human beings require order, coherence, predictable consequences. If they fail to experience these, they suffer breakdown.

It was a fact of behavioral psychology. No one could contest this (cruel) fact.

I thought—*That is our condition. If we are lucky.*

LONELY

Before Wolfman I'd been so lonely: I had done something reckless.

It was foolish, futile—I'd known this beforehand. And it was reckless.

Kneeling in the shadowy corner of my room, beside my bed, and pressing my forehead against the wall, in an effort to summon back memory—memories—of my old, lost life—biting my lower lip to keep from crying—and I could not, *could not remember*—my mother's face, or voice; my father's face, or voice; my own face, in the bureau mirror in my room that I'd had since I had been a little girl, when Daddy had painted my room a pale rose color, and Mommy had painted the frame around my mirror a color she called creamy-white . . . And I felt that my head might explode with the effort; for the microchip in my brain was blocking the very memories that would nurture me, like a kind of mental asthma, suffocating . . .

You can't. You are forbidden. You are in Exile. This is your punishment.

And so I gave up. Several times in succession, I gave up.

And one day when I'd given up trying to remember, when loneliness was choking me, I threw on my fleece-lined jacket and went out again in the late afternoon, at a time when other students were leaving campus, returning to their residences for dinner; and hiked over to the university infirmary on the far, eastern edge of the campus; and there, I searched for the nurse whom I'd seen in my first hour at Wainscotia, in the immediate aftermath of the *teletransportation*—I didn't remember the woman's name but I remembered her face, her darkish blond hair brushed back behind her ears, her eyes that were kindly but wary—*Don't ask me any questions, Mary Ellen Enright. Just go away.*

I asked the receptionist if there was a nurse who fitted that description, in her early thirties maybe; and the receptionist shook her head *no*; she had no idea whom I meant. And I said, "But could I look around for the nurse—there's really something I need to ask her."

The receptionist told me that at the present time no one else was on duty in the infirmary except her. The resident doctor was "on call"—but not on the premises.

I glanced about the waiting room, which was empty. It was a small, cramped space with just three chairs. (Curious to see ashtrays in the infirmary waiting room!) There was a close, stale, congested smell of something medicinal and sickly, as well as cigarette smoke, and in the near distance, in the adjoining ward, a sound of despondent coughing. I remembered now—some sort of sickness was sweeping the campus, "Asian flu"—several girls in Acrady Cottage were stricken.

Since I seemed so indecisive the receptionist said again that she didn't know whom I meant, and she knew all of the nurses in the infirmary.

Hearing this, I simply chose not to absorb it. No.

"Are you sure? Could I—look around? Is there a nurses' station?"

The nurse-receptionist looked at me as if I had to be joking.
"Nurses' station? *Here?*"

"Her name might have been"—I was straining to remember, despite something knotty in my brain that blocked my thoughts— "something like Imogene?—Irma?"

"'Imogene'—'Irma.' No."

Quickly the nurse-receptionist spoke. Too quickly, I thought.

And I thought—*Is this the nurse? This woman, who is speaking to me?*

I saw, now. She was older than I'd remembered—about my mother's age. Her (dark blond) hair was all but hidden behind a nurse's starched white cap. She was wearing a bulky cardigan over her white nylon nurse's uniform for it was cold in the infirmary, and her plastic name tag was hidden beneath the sweater. I saw that she was the woman—the nurse—who'd helped me wake up, in my first, terrified hour in Zone 9.

"I think you know me? 'Mary Ellen Enright'? You were kind to me, when I was brought here . . ."

Now the receptionist spoke sharply, with a mirthless laugh. "Miss, I said *no*. I've never seen you before, and you've never seen me. And now it's time for you to leave."

"But—aren't you 'Imogene'—or—'Irma'? Please—"

"*Now.*"

"—'Irma Kazinski'—'Krazinski'—"

"If you don't leave, miss, I'm going to call—security. You've been warned."

Her eyes on my face, and not friendly. The way in which she enunciated the word *security* allowed me to know that it wasn't *security* she meant but something far more devastating—*Deletion, DDS, vaporization.*

For a moment I stood paralyzed, unmoving. I could not contemplate leaving the infirmary and returning to Acrady Cottage, alone.

"There is no one here who—knows me. I was—I am—Adriane Strohl—I am not 'Mary Ellen Enright'—I was sent here—brought here—from NAS-23—do you know anything about me? Anything at all—you could tell me?"

Nurse Irma's face was shut tight. The irises of her eyes were tiny as peppercorns, as if sightless. Her mouth was an angry scrawl.

"You're raving. You must be feverish. But we can't take you here—we're filled up with flu patients. Don't you know there's flu sweeping the campus? Better for you to get out of here while you can. While you can walk. Before you get sick. Before you get very sick. Miss Enright—is that your name?—be sure to shut the door firmly when you leave. Do you understand?"

"Please—anything you could tell me, just—anything . . . About what is happening there—in NAS-23? Has anything changed? Is there still Homeland Security, and the president? And the Army, and the Wars for Freedom? Do you know anything about my parents—Madeleine and Eric Strohl? We live—lived—in Pennsboro, New Jersey. Don't send me away, I'm so lonely . . ."

Nurse Irma was furious now.

"I said, miss—*Do you understand?*"

I did not understand but yes, I retreated. I obeyed and left the infirmary. I shut the door firmly behind me. And I thought despite my disappointment, and my hurt, that another time, the nurse named Irma with the dark blond hair, the nurse who'd tended to me in my first hour in Zone 9, had been kind to me: she had not reported me to the authorities in lethal violation of The Instructions.

The proof was, I was still alive.

POSSIBLY

Possibly, in my terrible loneliness I had loved Ira Wolfman for weeks. Before our exchange at midterm.

Possibly, forever.

It was a feeling much more powerful than anything I could feel for Nurse Irma Krazinski.

From the first time I'd seen him in our classroom in Greene Hall. Entering the room briskly, dropping his briefcase on the table at the front of the room, casting his eyes like a fine-meshed net out over the rows of desks, our attentive faces . . . *Hello! My name is Ira Wolfman, I am your section instructor for Psychology 101.*

And there was the man striding to the podium in the lecture hall when he substituted for Professor Axel. For a young professor taking the place of a much-esteemed elder, Wolfman had not seemed hesitant, or lacking in confidence. Happily he'd smiled out at his audience like a diver on a high board before he executes a perfect dive. This acknowledging of an audience as if we were companions together, embarking upon an exciting journey, was not the custom of white-haired Professor Axel who more or less read his lecture notes without troubling to glance out at us, to establish that we were there.

Before it was revealed to me that Wolfman was an Exile like myself, I'd seemed to have known—something.

He is the one. He will save me from Zone 9.

As I did not appear to be susceptible to the flu sweeping the campus, so I was not so susceptible to *falling in love* like girls my age. I could take pride in this.

My roommates at Acrady Cottage talked excitedly, tirelessly of boys they were dating, or hoped to date; *fraternity men* with whom they were already in love, or were desperate to hear from, their happiness depending upon the next telephone call . . . But I was not one of these, for I was not that young any longer.

My feeling for Wolfman was different. Mine was the desperation of the drowning who will clutch at one who comes near, who has the power to save her from a terrible death.

DEAN'S LIST

"Mary Ellen?"—the voice was startlingly near. Miss Steadman.

It was the seventh week of classes, just after midterm. Early November. A late, dark, sleet-lashed afternoon. The resident adviser appeared to be waiting for me as I hurried into the residence in my hooded fleece-lined jacket, bound for the stairs, past the wall of mailboxes at which I rarely glanced.

No mail ever came for Mary Ellen Enright of course. But often there were flyers and announcements in my mailbox which cruelly resembled mail.

I wasn't envious of the other girls' letters. I never thought of it, any longer.

My daytime anxiety was concentrated on my course work. There were five courses: five instructors. That one of them was Ira Wolfman was a coincidence.

I tried not to think of anything else. I shrank from the attention of others. Except, I could not be rude to any adult. I could not rush past Miss Steadman in her doorway, smiling at me.

Silently begging the woman—*Leave me alone, will you! All of you! Please.*

Such fierce thoughts raced through my head. It seemed amazing to me, that others could not hear them and recoil from me.

"Mary Ellen? May I speak with you—briefly?"

I could not turn from Miss Steadman with a muttered *No*. Meekly I followed her into her apartment: into her sitting room which opened out onto the smaller of two lounges in Acrady Cottage, where there was a console-model television, a "Philco"— the smallest television I had ever seen, with a miniature screen, a picture in tremulous shades of gray, broadcast over just three channels.

In the evenings, some of the girls of Acrady Cottage watched this ridiculous little television set! And Miss Steadman who seemed to crave company sometimes watched, too.

Ardis Steadman was a tall rangy woman with sand-colored hair, eyebrows, skin. She was plain-faced, earnest. Her eager smile revealed pale-pink gums. Her eyes were large, brown, brimming with sincerity. She'd introduced herself to us at the first meeting of Acrady Cottage as an assistant in the office of the dean of women. She was a Ph.D. candidate in Public School Administration, in her mid-thirties perhaps, or perhaps younger, for Miss Steadman was one of those women who, as a young girl, is already praised as "mature"—"responsible"—"a leader." My roommates spoke of trying to avoid Miss Steadman who was *really nice, but bor-ing*.

Especially, they pitied her as unmarried: *spinster*.

(Such words as *spinster, old maid* were new to me, for in NAS-23 marriage was hardly more common than divorce. But I understood the meanings of these words, and something of the panic that underlay them in Zone 9.)

Miss Steadman smiled happily at me. In her warm eager voice she asked how I was?—how my classes were going?—and I murmured polite replies. I tried to sound cheerful, upbeat: it was a phenomenon of Wainscotia—"upbeat." But I was not very skilled at "upbeat" and I could see that Miss Steadman was not deceived.

She asked me in some detail about my courses. For I was a "scholarship girl"—and so of particular interest to her. (I'd learned that, at Wainscotia, I was a University Scholar—and not a Patriot Scholar. Evidently, in 1959, "Patriot Scholar" did not yet exist.) Miss Steadman had heard of most of my instructors, and was enthusiastic about them; particularly, she was enthusiastic about Professor A. J. Axel who'd collaborated with the great B. F. Skinner at Harvard—"Professor Axel has developed his own experimental project, with the aim of curing antisocial behavior—the 'aberrant,' the 'perverse,' and the 'subversive.' Wainscotia is to establish a center for this new field—'The Wainscotia Center for Social Engineering.' Each October, we expect Professor Axel to receive a Nobel Prize. This will happen soon, I predict!"

I asked Miss Steadman what sort of "aberrant," "perverse," and "subversive" behavior did she mean?

Miss Steadman paused. A faint blush rose into her face. With a frown she said, "Oh—the kind of shameful behavior you can imagine, Mary Ellen. Or, rather—you can't imagine. And I can't, either." She shook her head vigorously. "It's mostly among men, I think. Between men. But Professor Axel will change all that."

How would Professor Axel change such behavior? I asked.

"Oh, I think—electric shock treatment." Miss Steadman spoke vaguely.

I wondered if Wolfman was involved in this Center? I had to suppose he was, as an assistant to A. J. Axel.

But how ironic, an Exiled Individual enlisted to cure *antisocial behavior!*

Miss Steadman had much praise for the chairman of the philosophy department, Professor Myron Coughland, whose theory it was that the history of philosophy and linguistics from the Greeks onward had been a steady progression to present time, mid-twentieth-century Christian United States—"It has something to do with 'practical'—'pragmatic'—ethics, and with 'democracy'— 'the greatest good for the greatest number'—plus Christianity of course. Our American beliefs in the age of Sputnik, which are very different from the beliefs of Soviet Russia! Professor Coughland has a hundred-thousand-dollar grant from the National Science Institute to pursue his research. He's been on the front page of the student newspaper, I'm sure you must have seen the articles."

Vaguely I murmured yes. Maybe I'd seen the articles. Since my exile in Zone 9 I had tried to fill in the considerable gaps in my knowledge of history—there'd been very little in our Patriot Democracy History courses about the Soviets' early success in sending a small, unmanned satellite ("Sputnik") into space, and their development of nuclear weapons, for most of high school Patriot Democracy History focused upon continual threats to American democracy, and the triumphs of American democracy over its countless "terrorist" enemies around the world.

"It is obviously true that our American philosophy is the culmination of thousands of years, and that human beings are 'more civilized' than ever before—don't you think? Who could possibly disagree, who has listened to Professor Coughland? He argues that our American presidency is the 'high point' of political history, and that Dwight Eisenhower is the greatest world leader so far."

I knew little of the presidency of 1959, except that the blandly smiling, golf-playing president had been a general in World War II, and was a popular favorite with the American people, like our president of NAS-23, whose approval ratings in the polls, posted every morning on the Internet, was in the range of 95 to 99 percent.

In 1959, there appeared to be two major political parties, Democrats and Republicans, that struggled with each other for dominance; by NAS-23 there was just the Patriot Party, funded by NAS's wealthiest individuals, which appointed all political leaders as well as the judiciary. CVs—"Citizen Voters," a rank determined by income—could cast ballots for the Patriot Party candidate, represented by a smiling emoji, to which a name was affixed, in an election that was both preliminary and final, for the Patriot Party candidate, with no opposition, was inevitably the president. (Dad had said that within his memory there'd been elections with not just one emoji but two emojis on the ballot. The voter, in the privacy of the voting booth, was "free" to vote his choice.)

(It had been explained to us in Patriot Democracy History that, in past decades, hundreds of millions of dollars had been spent on "campaigning"—a largely useless gesture since the presidency invariably went to the candidate with the largest campaign fund; so election procedure was modified to determine which Patriot Party member could amass the most money, and this individual was presented on the ballot as the party candidate, with no need to actually spend the money.)

BADLY I WISHED that I could trust Miss Steadman, to confide in her something of my lost life in what was, to her, the "future"; but even if I'd dared to violate The Instructions, Miss Steadman would not have believed me, and would have thought that I was *mentally ill.*

Already in 1959, as I'd learned in my psychology lecture class, it had become a technique to discredit "rebellious" individuals by suggesting that they were *mentally ill—emotionally unstable.*

"Borderline personality"—which made me wonder who controlled and defined the border.

In her pleasant but dogged manner Miss Steadman was asking if I found logic "difficult—a brainteaser," as she had as an undergraduate, and I said yes, very difficult. Miss Steadman said: "Logic isn't a course of study for women. Like math, physics, engineering—our brains are not suited for such calculations."

Did I believe this? Such thinking was self-destructive, as it was mistaken. In NAS-23 it was axiomatic that "all sexes are equal"— which is to say, no sex is allowed to be "handicapped"—no individual of any sex merited special consideration. But my objection wasn't very convincing, and Miss Steadman ignored my remarks.

In fact since coming to Zone 9 I'd often felt that something was wrong with my brain. The microchip and the *teletransportation* had injured my ability to think. Laboring on the problems in my logic textbook I felt a visceral misery, as if "logic" were a kind of virus that had infected me, from which I couldn't be purged; after a few hours I was left fatigued, despondent. I had noted that not one professor in the philosophy department was a woman, and that not one philosopher in the Intro to Philosophy anthology was a woman. *It was as if the female did not exist.* I wondered if an immersion in logic might result in a strong wish to commit suicide.

I could not tell Miss Steadman this but I did tell her that, of the several courses I was taking, it was logic I was most worried about failing.

"Oh, Mary Ellen—*you* won't fail! I'm sure."

(It was probably so, I wasn't likely to fail the course. My lowest grade in our several quizzes had been A–. Yet still, I anguished over the possibility of failing, for, in logic, there is much unhappiness that is possible, beyond the range of the merely probable.)

Miss Steadman went on to speak, with the breathy enthusiasm of an administrator, of Wainscotia in the "forefront" of academic research. Philology, mathematics, sociology, physics—"You've heard of Amos Stein? Originally of the Institute for Advanced Study in Princeton? No?" Miss Steadman seemed disappointed in me. "Professor Stein was also featured in the student newspaper recently. He's the director of the 'Hoyle project'—a team of first-rate physicists and mathematicians working on a rebuttal of Einstein's theory of relativity. And there is something called the 'Big Bang' theory—the universe began with an explosion, and is 'rapidly expanding.' But our Wainscotia team believes that the universe is infinite and unchanging—a 'steady state.' No beginning and no end. If you believe in God, only the 'steady state' is sensible. For after all—what could precede God? All the professors need is mathematical proof. It's said that they're working on the problem virtually day and night—and that they have enlisted a computer in the project, in Greene Hall. The computer is so large, it takes up half the ground floor of Greene Hall! Everyone in the intellectual world is eagerly awaiting their findings—we're hoping that Wainscotia will be a site for world-renowned physicists and mathematicians to work together in the future. Einstein also argued that 'all things are relative'—that 'time can bend'— obviously impossible. As if God could 'bend'! Professor Stein says such reasoning is 'Jewish logic'—to confound, and not to illuminate." Miss Steadman spoke so vehemently, her lips were damp with spittle.

Yet, her remarks were exciting to me. I did not understand anything of Einstein's theories—though I had heard of the "steady-state universe"—(I wondered if it had originated in the Midwest?)—I didn't understand it either. But I felt Miss Steadman's hope, that Wainscotia State would be so rewarded.

I recalled S. Platz describing the university to which I would be sent as "excellent"—she had sounded so positive, and hopeful for me! When I returned from Exile, I would be well trained; if I received high grades I might qualify for a good job, and could help out my parents financially.

My life-after-Exile loomed before me like a mirage at the horizon. If my present life was difficult, and lonely, I had only to think of this life—this mirage—to feel more hopeful.

Miss Steadman was speaking of Wainscotia's "preeminent" biologist Carson Lockett II, who'd been trained at Oxford in "life sciences"—a world expert in the work of Alfred Russel Wallace, the Victorian scientist who had anticipated Charles Darwin's evolutionary theory; but had gone beyond Darwin in hypothesizing a "special creation" for the human brain which, Wallace argued, could not have possibly emerged from "natural selection" over millions of years—"Dr. Lockett and his colleagues are engaged in a refutation of Darwinian atheism in the most objective, scientific, 'evolutionary' terms." And Miss Steadman was speaking of Wainscotia's resident poet H. R. Brody—did I know his work? Had I seen his photograph?—"H. R. Brody has white hair like Robert Frost. His poems are rhyming poems, about nature—like Frost's. They are not those strange little lower-case-letter poems by the man 'cummings'—'c. c. cummings'—or is it 'e. c.'?—who doesn't even try to rhyme. I don't understand much of contemporary poetry but H. R. Brody's poetry is *beautiful*, and *wise*."

I tried to remember if I had ever heard of H. R. Brody as a poet, in our English classes in NAS-23. I didn't think that I had. Eighty years later, H. R. Brody had been totally forgotten.

"Are you interested in poetry, Mary Ellen? I think you must be."

Why would Miss Steadman say this? All I could stammer was *no*.

In fact I'd tried to write poetry in high school. In our English class we were given poetry-formulas, with rhymes to fill in—

I'm Nobody! Who are ____?
Are you—Nobody—___?

And

I think that I shall never see
A poem as lovely as a ___ .

Poems are made by fools like ___,
But only God can make a ___.

My own poems were not so smoothly composed. A twenty-line formula-poem, for me, might spill over into thirty lines, or end prematurely at eighteen.

I'd also tried to write what were called "stories"—following the pattern of the Nine Basic Plots we were provided, along with vocabulary lists and recommended titles.

We were not allowed to take books out of the public library marked A—for Adult; we were restricted to YA, Young Adult, which had to be approved by the Youth Entertainment Board, and were really suitable for grade school. My parents had had Adult Books at one time, but I had never seen them.

I did try to compose a graphic novel with animals, not people. My illustrations were clumsy and childish and the project, that had begun with much enthusiasm, gradually dwindled away like ice melting.

I remembered that my brother Roddy had gone through a phase in middle school when he'd built kites out of papier-mâché. These were sort of amazing kites, dragons, eagles, giant butterflies, you'd never expect from Roddy.

For a while I'd helped him make the kites. It had been exciting to do something with Roddy instead of mostly trying to avoid him. Our parents were impressed, too. But eventually he lost interest, or became discouraged for some reason saying with a shrug *Who cares? Making stuff is a waste of time.*

Miss Steadman was smiling at me as if in the expectation that I might share with her something of my "poetic" nature but I knew better than to reply. She then asked how I was getting along with my roommates?—and I told her that they were very nice, and I liked them very much.

"Well, good! They are very different from you, Mary Ellen, I think."

I should hope so! My roommates are zombies.

To hide my surprise I smiled. (In the Happy Place, I was smiling nonstop!) For this had been an unexpected remark of Miss Steadman.

"Your roommates are religious girls," Miss Steadman said, "like most of the girls in Acrady, I believe. But you—some of us have wondered—what is the church of your choice?"

Church of your choice. A strange and coercive phrase.

"I—I don't go to 'church'—actually. But I am a—a—what you'd call a—'religious believer.'"

"Well, I thought so, Mary Ellen." Miss Steadman frowned thoughtfully as if the issue (was Mary Ellen Enright "religious") had been gravely pondered. "And you are—a Christian?"

This was blunt! No one had ever asked me such a question before. In NAS-23 it was assumed that all citizens were "Christian" but that "Christian" did not mean much more than "NAS-citizen." Certainly, no one ever spoke of "Christian values"—doing good, helping the less fortunate, being "selfless."

"Sometimes, Mary Ellen, you seem just slightly"—Miss Steadman searched for the appropriate word—"sad. Your roommates have said that you've been homesick."

"No."

"Well—good! That's good to know."

It was disconcerting that my roommates would speak about me to the resident adviser behind my back, but equally disconcerting that the resident adviser would allow me to know that they had.

"You've come a distance to Wainscotia, I think?"

What did Miss Steadman mean by this? I was becoming very uneasy.

(How much information had Miss Steadman been given about me?)

"We're all homesick in new places. Even those of us who've lived away from home for years."

Miss Steadman spoke gently, warmly. Encouraging me to share my feelings with her. Of course, I would not!

At the front of the residence there were voices, laughter. It was mysterious to me what my roommates and the other girls found so hilarious, so often.

Gathering by the mailboxes, loitering in the foyer. Drifting upstairs to their rooms. Such hilarity! From a little distance I could not decipher a word they said. Some of them brayed loudly, some were convulsed with giggles. It was nearing 6:00 P.M.: soon they would troop off to the dining hall for the evening meal. (Acrady Cottage was too small to provide meals for residents. The deafening-loud dining hall to which we were assigned was two blocks away, and served both male and female students.) Usually I sat alone in the dining hall at a table with solitary individuals like myself, engrossed in a textbook and pretending not to notice girls from Acrady who might have invited me to join them. My invisibility was like a magic coat that kept me warm but also compounded my loneliness.

I was wearing the navy-blue jumper that was too large for me, over a gray turtleneck sweater with a stretched neck; the rubber boots, which were scruffy and stained. I wondered if Miss Steadman could see that the clothes in the box had been secondhand? Not mine, only just addressed to me? And did she know about me? Nurse Irma had known about "Mary Ellen Enright"—to a degree. Certain key administrators at Wainscotia had to know.

But possibly not. Possibly inhabitants of Zone 9 knew nothing about NAS-23 and its procedures. It was likely that they were given minimal information: an Exiled Individual, an individual with no past, no background, no family, no history, appearing among them by government decree.

For all they knew, "Mary Ellen Enright" was an orphan whose parents had died in a natural disaster, or in some act of war. Why would they have questioned this?

As Miss Steadman spoke of homesickness I began to think— *She wants me to confess to her, and then she will report me.*

Or, maybe—*She wants me to confide in her, and then she will comfort me.*

Or—*Miss Steadman is in Exile herself!*

Except I did not believe this. Miss Steadman was hardly the Treason-type.

Oh, I hated this! Trapped in my thoughts.

Exile means you can't free your mind to think of anything except Exile. While others never question the terms of their existence you question the terms of your existence constantly. *Why am I here, when will I be taken from here, who is watching me, who is monitoring me, is it this person who invites me to trust her? What will she report of me? What will be the summary and the judgment?*

Miss Steadman was asking where I was from. A quite natural/friendly question yet I stammered trying to give a plausible reply. Trying to recall The Instructions—what was not forbidden, and was therefore allowed.

"I—I'd rather not say, Miss Steadman."

Miss Steadman stared at me, surprised. "But—why not?" Her manner was guileless, innocent. I had truly surprised her.

"It makes me sad to think about it. I'm sorry."

"Well! I'm sorry that this is upsetting to you. I'd been told that you have a scholarship from out of state, I'd had the idea it was one of the eastern Atlantic states like New York, New Jersey . . ."

"I'm an 'orphan,' Miss Steadman. I'm not from anyplace really. And my parents—my parents who adopted me—are both dead."

These words were shocking to me. I had never spoken them aloud before. Yet, I'd rehearsed them. I thought—*If this is being recorded, I have not said anything wrong.*

Miss Steadman's manner seemed to suggest that my remarks were utterly unexpected. If she had a file on me, it was incomplete. Or, she wanted me to think so.

(I wondered, too: were tape recorders available for ordinary individuals, in 1959? Was it likely that Miss Steadman could be recording this conversation? There were few electronic devices in this primitive culture but agents of the State might be equipped for surveillance.)

More girls had entered the residence and were congregating around the mailboxes like a flurry of squirrels. Ordinarily Miss Steadman would appear at the front of the house around this time in the hope of waylaying a few girls to chat with her, but not this evening.

"Did you say that you have no family, Mary Ellen? Is that what I understood you to have said?"

"Yes. I mean no—I have no family. I was adopted and—my 'adoptive parents' died."

These words came glibly. Almost, I wanted to laugh at the solemn expression on Miss Steadman's face.

"That's so very sad, Mary Ellen! When did this happen?—I mean, when did your parents die?"

When? I had no idea. The Instructions had not prepared me for such an interrogation. Like an actor who has been handed only a skeletal script I would have to improvise.

"A long time ago—I think. I don't really remember."

"Well—that's very sad . . . And very unusual."

Unusual that an orphan's parents have died, or *unusual* that the orphan seems not to remember when they died?

"So, did no one bring you here to Wainscotia? Did you come alone?" Miss Steadman seemed embarrassed by my faltering words yet persevered in her questioning. In her plain earnest face was a look almost of hunger.

"Yes. Alone."

"So young, to be traveling alone! I didn't think I'd seen anyone with you, that first night. You seemed very—independent."

Independent! I smiled inanely.

"I wanted to help you with your steamer trunk, but you told me you didn't want help. Yes, you were very—precociously, I'd thought—independent."

Steamer trunk? What was this woman remembering?

Somehow confusing a badly wrapped cardboard box with something called a *steamer trunk*. Had there been, over the years at Acrady Cottage, more than one "Mary Ellen Enright"?

Miss Steadman was assuring me that she, too, had been homesick when she first went to college—not at Wainscotia but a smaller college, in northern Wisconsin. But she found her friends, eventually—"As you will also, Mary Ellen."

To this cheerful remark I had no reply. Trying to envision "friends" hiding on the Wainscotia campus whom it would be my task to discover and if I did not, it would be a failing of mine.

Somehow, I'd missed the transition, Miss Steadman was speaking of food. Dining hall? Home-cooked?

Often, I forgot to eat. Or, I had no appetite for the food in the dining hall. (Much fried food, greasy fatty meat, vats of lumpy mashed potatoes, "fruit cup" and Jell-O desserts.) My roommates had noticed—at first—and offered to bring back food for me from the dining hall, but I hadn't encouraged them.

"Would you like to have dinner with me, Mary Ellen? I mean—tonight?"

Stammering thanks, but—"I don't think so, I have so much homework . . ."

"I mean in my apartment here. I like to cook. I specialize in simple meals." Miss Steadman smiled shyly. "You're welcome to join me, Mary Ellen, if you wish."

"Thanks, but—I guess I can't."

"Well. Another time perhaps."

With an air of disclosure Miss Steadman went on to say that she'd had an "advance report" from the dean's office regarding my midterm grades.

Grades? Now I was concerned.

In my stressed state the very mention of grades made my heart trip.

"Oh dear, don't look alarmed, Mary Ellen! You've done very well. The dean makes it a point to go over reports from the faculty at midterm to see if anyone is having problems, and to offer help with these problems before it's too late. And I hope you will keep the news to yourself, as it is confidential."

Trying now to feel relief instead of acute anxiety.

"Your instructors have reported that you've earned straight A's, it seems. Well—one A minus, in logic. Wonderful work, Mary Ellen! Congratulations." Miss Steadman smiled at me. If I had not drawn away she would have seized my hands and squeezed them in an outburst of girlish enthusiasm. "You are currently at the very top of Acrady Cottage, and Acrady is usually at the top of the freshman women's residences."

At the top. What did this mean? I was beginning to perspire, for it seemed that once again without thinking I had insufficiently *held myself back.*

But the situation was different. Now it was good to be praised, not a warning. If but by this friendly stranger who had no idea who I was.

I thought *My parents would be proud of me.*

"Oh, Mary Ellen! You aren't crying—are you?"

"N-No."

My face was very warm. By now I was desperate to escape.

At the doorway of her sitting room Miss Steadman asked if I liked chamber music and I said yes, yes I did like chamber music, for it seemed a reasonable reply; though really I knew little about chamber music, and might have thought that this was music played on instruments like organs, pianos, and harpsichords, in special music-chambers. If I was interested, Miss Steadman said, there was a chamber music quartet recital at the music school on Friday evening, playing compositions by Bach, Brahms, and Ravel; she was going, and had an extra ticket if I would like to join her.

"We could have dinner before at a restaurant on Moore Street."

Seeing how alarmed I looked at the prospect of a restaurant dinner, Miss Steadman said quickly that she would pay for both our meals—"If you're interested, Mary Ellen. But I know you spend most of your time on your courses."

I told her yes, I would be interested. For I wanted to learn all that I could, and "chamber music" was new to me.

At the same time I did not want to spend time with Miss Steadman. I did not want to risk intimacy with another person. And I did not trust Miss Steadman, finally.

"Good night, Mary Ellen! It was good to talk to you, at last."

At last. What did that mean?

I ran away upstairs. So grateful to have escaped without saying a wrong thing that might be used against me.

Fell onto my bed, exhausted. Thank God my roommates were at the dining hall.

Thinking—*She wants to be my friend. Why can't I trust her!*

THE SPELL

Truth was: I had no time in my emotional life for anyone except Ira Wolfman. I had not time to *fantasize* about anyone but Ira Wolfman.

Like a small comet in the wake of a larger comet, helpless to resist its gravitational field, I fell under Wolfman's spell.

The microchip in my brain blocked me from summoning my parents. I could not summon my closest, lost friends. And during my sleepless nights I began to think of Wolfman compulsively.

He knows me. He will acknowledge me, soon.

But when? A week passed after midterm. And another week.

In Professor Axel's lecture Wolfman sat in the front row, to the far left as you entered the hall from the rear. I did not want to distract Wolfman by sitting near him, or in his line of vision if he turned his head, so I sat several rows behind him, with a clear view of his profile. In fact I stared mesmerized at Wolfman's profile, as A. J. Axel lectured of the distinction between Pavlovian and Skinnerian "conditioning," and the significance of "operant reinforcement" in terms of a "social utopia." I saw that Wolfman was glancing around, into the audience of mostly undergraduates—as he hadn't done previously, that I could recall.

Was Wolfman looking for me? In my naïveté I wanted to think so.

Here was a purely classical, Pavlovian response: when Wolfman turned to glance back into the audience, my heart gave a painful leap, and a sense of alarm tinged with acute pleasure, or acute pleasure tinged with alarm, ran through my body like an electric current. At once I felt faint, and very happy.

Yet, if Wolfman's eyes moved over me—quickly, with a pained expression—he did not seem to wish to see *me*. This was a clear rebuff. My physiological symptoms abated at once, like air escaping a balloon.

A Pavlovian response, involuntary.

Also, I noted that while Wolfman was courteously friendly to other students in our quiz section, whose names he certainly didn't know, he seemed never to notice me if we happened to encounter in the lecture hall, or in the corridor outside. He seemed stubbornly blind to the physical fact of *Enright, Mary Ellen*.

This had to be voluntary. Deliberate.

In our class Wolfman was not so composed as he'd appeared at the start of the term. Though always prepared for the class, with notes, much-annotated texts, and a seemingly endless supply of anecdotal experimental-psychology material, he often appeared distracted, like one in the presence of—who? What? An enemy? A spy? He was edgy, uneasy. He lit cigarettes and smoked hurriedly. (In Zone 9 it was common for professors to smoke in classrooms; students smoked everywhere, as if by natural right. A dull-blue haze drifted freely. Did no one know of the hazards of smoking? Secondary smoke? Did no one associate lung cancer with all this smoke? After a few days in these barbaric circumstances I'd learned to arrive early in my classrooms to take a seat near a window if possible, or near a door that might be opened a crack. I would never learn to stifle my coughing as clouds of toxic smoke drifted languidly in the air in which for fifty minutes nonsmokers were trapped. It was a symptom of Zone 9 that nonsmokers never dared complain of smoke. If we waved smoke from our faces it had to be with an air of apology, for our silent protests antagonized smokers and smokers were the great majority.)

Wolfman was an energetic instructor, always on his feet. As he lectured he wrote decisively and dramatically on a "blackboard." He paced about the room which was wider than it was deep, with long rows of seats stretching the width of the space; as Wolfman faced the class he had no need to see me for my seat was far to the left, in what would be the periphery of his vision.

Virtually everything Wolfman said was fascinating to me. Yes including the words *the, and, a.* It's as if he were lifting a flap of human skull and peering into the workings of the brain coiled inside. He spoke of animal and plant characteristics that had "several times evolved" through the millennia—"bioluminescence," for instance. ("Obviously, 'bioluminescence' is important," Wolfman said. "Without it, fireflies would have to mate in less exciting ways.") Wolfman spoke of Darwin's initial bewilderment at the extravagant display of the male peacock's tail before Darwin had "got it" that such showy behavior had to be linked to natural selection—"Smart as Darwin was, he hadn't taken a clue from eighteenth-century male attire in England and Europe—wigs, lace, satins and silks, even makeup. He was too Victorian to make the identification." Enchanting to me, who'd had dull, careful, cautious, not-very-intelligent or well-educated teachers in high school, to hear a man of Wolfman's intelligence speak—about anything.

"It's a phenomenon of mental life," Wolfman said, one memorable hour, "that the dreamer is always convinced that he is awake, no matter how surreal a dream may become. There's one test, though. See if you can remember it when you're dreaming, when you believe you're wide awake. Look into the distance, if you can. Out a window, if there's a window. In a dream you won't see the detail you see right now, looking out this window—the foliage, for instance; the intricate network of leaves. I'm sure there are no 'cloud formations' in dreams. And if you're trying to read something in your dream, you will discover that you can't—instead of print there's just a jumble of hieroglyphics, or nothing at all. Go to a mirror, if you can—you'll see, there is no reflection in it. There is no *you* in a dream, just firing neurons. These are ways you can be absolutely certain if you're asleep, or if you're awake—which is the state you are in now: awake." Wolfman snapped his fingers, loudly.

In the lecture hall students laughed uneasily. Was Dr. Wolfman serious, or joking? Somehow, Wolfman's jokes were only funny if you didn't think about them too carefully.

Another time Wolfman told us: "There are no jokes, students. When you laugh, ask what are you laughing at?"

From time to time I began to raise my hand, to volunteer answers to Wolfman's questions. Sometimes Wolfman called upon me—"Yes, Miss Enright?"—and sometimes Wolfman ignored me. Each time I raised my hand, each time I spoke, was terribly exciting to me. Each time, my voice faltered. Sometimes, I choked and could scarcely speak. Yet, I persevered, I plunged forward. I felt like the valedictorian of my class—I *would not be discouraged*.

In the lecture hall, it did not seem quite right that a girl should speak so frequently, or rather so—impressively. Even Wolfman sometimes frowned. *Enough! For now.*

I WONDERED WHEN Wolfman would ask to speak with me after class in his office? I did not want to appear in his doorway without an invitation.

I thought—*He is frightened of me. We are both at risk.*

I thought—*But he knows me! He is the only one.*

Initially, at the start of the term, Wolfman had ignored me. Hadn't seemed to see me.

Not because he recognized me as an Exile like himself. But because he'd had little interest in me, as a "coed."

His manner toward the three girls in our class had been courtly, condescending. He hadn't been malicious, or cruel. Unlike other instructors of mine who made casual sexist remarks in their classes—("sexist" was a concept unknown in Zone 9, I'd quickly come to realize, for "sexism" prevailed everywhere here, like the atmosphere: it was degrees of "sexism" that mattered)—Wolfman wasn't insulting. It was just that the category *female* was freighted with a kind of comic irrelevance in the field of psychology— (except for abnormal psychology, in which female degenerates were given nearly as much attention as male degenerates, though for different reasons). In our psychology textbook examples of representative psychological subjects were exclusively male; the behaviorist model was male. Professor Axel's lectures alluded incidentally to male experimental subjects and only once did the professor discuss a specifically female phenomenon: the failure of "good mothering" that was thought to cause autism in children.

In our quiz section the next morning I raised my hand to ask Wolfman —"Is there any scientific proof for saying such a thing? That 'bad mothers' cause autism in children?"

Wolfman was taken by surprise. Class had scarcely begun. He had no choice but to look at me.

Tried to smile his genial smile. But his manner was stiff.

He said that frankly, he wasn't an expert. Child development wasn't his field. And "Freudian theory" wasn't his field. He'd like to think that there was proof for Professor Axel's casual remark— but he couldn't personally swear to it.

"However, it's a good question, Miss Enright."

Others in the class glanced at me, curious. It must have seemed to them preposterous, that a freshman girl might question the great authority A. J. Axel.

"It's just that I can't imagine an experimental situation in which 'mothering' was observed by these psychologists, over a period of time. Aren't they just inferring it? Wouldn't there be simultaneous 'fathering' too? I think that psychologists must select autistic children, or children they label as 'autistic,' and try to determine if their mothers had been 'good mothers'—but how would they know? Unless they lived in people's houses with them on a daily basis . . ."

Now this was going too far. Saying too much!

There was a kind of ripple in the room, of discomfort, disapproval. You could see that Wolfman was surprised by my remarks, and the length of my remarks, so unlike those of other students when, not very frequently, they spoke in class.

Yet I'd taken care, for this was Zone 9, "1959," to speak in a soft, unassertive, "feminine" tone, to avoid blatantly offending the professor.

Wolfman acknowledged that this was a good point—

"Autism is a poorly understood mental condition that is presumably 'caused' by something—so, an agent is hypothesized. Why not a neurological deficit? Why blame the mother? You are right. The circumstances don't lend themselves to experimentation, like Skinner's work. Which is why Skinner, and not Freud, is the great scientist of the twentieth century."

This was an odd thing for Wolfman to say. Like wiping away an equation on the blackboard with his sleeve, to change the subject.

"But—but—behaviorism doesn't even try to measure 'subjectivity'—does that mean that 'subjectivity' isn't a proper study for psychology? Ever?"

"It means that 'subjectivity' is 'subjective'—it can't be objectively demonstrated. Behaviorism concentrates on what is objectively available, what *works*."

Wolfman spoke curtly. It was the standard Skinner line. He'd had enough of our exchange for now.

He might have said *Come see me after class, Miss Enright. If you are interested in discussing this further.*

Instead Wolfman turned to another subject. For the remainder of the class he didn't call upon me, and I didn't volunteer to speak.

Nevertheless I went away elated—*He respects me! I'm more than just a "coed" to him.*

LIKE A LITTLE BALLOON, elated. Filled with helium but not quite enough helium, the little balloon doesn't soar only just bobs along the ground. Blown a bit by the wind.

Finally stuck somewhere in a bush. Little balloon pierced of helium, reduced to just rubber.

Unidentifiable as a balloon.

Where do elation, happiness, hope *go*?

B. F. Skinner had not a clue.

Dear Professor,

　　I love you.

Sincerely,
　"Mary Ellen Enright"

PS. No need to reply!

Alone I was assailed with thoughts of Wolfman. *Of course* Wolfman wanted me to see him: he was waiting for me to make the first move.

Composing absurd little love-notes to Wolfman. Such pleasure in writing these notes, folding them up into tiny squares and hiding them at the rear of my desk drawers.

The Wolfman-spell was like a narcotic. In such a state you are not unhappy for you are under the spell and to be under the spell is to be incapable of being unhappy.

But my devotion to Wolfman wasn't selfless. No. Always I was thinking, if Wolfman too were in Exile, he could help me—somehow . . . This I wanted to believe.

The wariness in his eyes, when he saw me—this, I tried to ignore.

He doesn't want a special relationship with you.

Can't you see, he doesn't even want to "see" you.

Where in the past I'd never lingered after class to speak with other students, or to hear their remarks, now I yearned to hear comments on Ira Wolfman. Oh, anything! I liked even to hear the name—*Dr. Wolfman*. Or, just *Wolfman*. The name alone mesmerized.

It was astonishing to me, Wolfman was not universally admired. In fact, Wolfman seemed to be resented. His grading, his sarcasm, his "fancy" vocabulary. Wolfman wasn't *down-to-earth* like certain other professors who told jokes, teased their students and didn't expect them to know much.

Of course, in some quarters Wolfman was much admired. "Brilliant guy"—"brainy"—"cool." (It was notable, "cool" had been a term of admiration way back in 1959.) Even the foolish remarks of some girl classmates were prized by me, comments on Wolfman's looks, clothes, style—"He's from New York City, you can tell. Real smart."

"He'd be a dreamboat if he didn't act so snooty . . ."

"Snotty."

(Not funny. Though others laughed I did not laugh.)

"Obviously, Wolfman is Jewish."

"Oh—that's a *Jew*? Him?"

"Bet you. They're supposed to have little horns on their foreheads—just bumps, by now. After so many thousands of years."

Jewish! I had not thought of this.

Wolfman. Ira Wolfman. My love.

Therefore I could not be anything less than the very best student in Wolfman's quiz section. My hope was to be the very best student in Psychology 101—to earn the highest grades among the more than two hundred students enrolled in the course.

Miss Enright! Dr. Axel and I are—well, frankly impressed . . . We would like to encourage you to major in psychology.

How many times I dared to approach Wolfman's office during his office hours. Invariably there was a student inside: the frosted-glass door was closed. Badly I wanted to press my ear against the window, to determine if the visitor was a boy or a girl.

Feeling a kind of disappointed relief, that I was spared making a fool of myself.

The corridor was a busy one, on the first floor of Greene Hall. I was jostled by students hurrying to classes. To give myself some purpose I stared at a bulletin board festooned with posters and announcements of psychology department events: lectures, symposia, applications for summer fellowships, advertisements for graduate programs at other universities. I thought—*I could apply for one of these. Could I?*

I wondered if I really might excel in psychology. If I might go to graduate school, someday. (But when? And where?) If there was any hope for me, in Exile, to do anything worthwhile with my life.

I clung to S. Platz's words: Wainscotia was an "excellent" university. After I graduated, and was returned from Exile, the State would arrange for me to receive additional training, to prepare me for employment in NAS. I wanted to believe this.

Yet, in the corridor outside Wolfman's office, I felt a sense of hopelessness. What had brought me here? What did I think I might acquire from Wolfman? His respect, his friendship? His love? His assistance, in escaping from Exile? (But—where? "Adriane Stohl" had not yet been born.) There was no *his*; I had no true idea of the man, beyond his charismatic classroom personality which (I had to know) was a public performance. I felt dazed, uncertain. Abandoned. If a stranger had tapped me on the shoulder and said *Come with me. Your sentence has been altered, you will now be Deleted* I would have come unresisting.

I wasn't sure if I was waiting to see my quiz section instructor, as others enrolled in Psych 101 were, or just standing there in the corridor hoping not to be jostled. My insomniac nights and protracted workdays left me in a paralysis like a ball that has been thrown hard, has rolled far, but has lost momentum and has come to a stop . . .

A student emerged from Wolfman's office, somber-faced, clutching at a paper, and departed. And there came Wolfman to the doorway yawning, to see if another student awaited him.

His eyes moved onto me, startled. I saw a look of pained recognition. "Miss Enright. Hello."

Wolfman's usually ebullient voice was wary, flat.

Pleading with me—*No. Go away please.*

I nodded hello. I felt very shy, suddenly. Stricken with shyness like a physical disability.

"Did you want to see me, Miss Enright?"

"N-No. Thank you."

"You don't? Didn't?"—Wolfman was vexed but amused.

Quickly I turned away. Fled. The part-deflated little balloon, the rubber ball given a kick by Wolfman—there it goes!

Afterward the encounter left me shaken. I had to wonder if it left Wolfman shaken, too.

ORPHAN

Embarrassing to confess: how I followed Ira Wolfman, at (what I hoped was) a discreet distance. I did not want to upset him—or anger him—but I could not stay away from him.

I was convinced: he'd recognized me. We were two of a kind in Zone 9: Exiles.

In English literature class we read poetry of the Romantics. We learned of the Romantic concept of *soul-mates*. It was clear to me, Ira Wolfman was my *soul-mate*.

Once Ira Wolfman left Greene Hall by the side door near his office, in the company of two other men—young men—had to be psychology colleagues. In our class he often spoke of his "lab"—an experimental laboratory under the direction of A. J. Axel—and these two might have been laboratory coworkers. So long as Wolfman was with them he appeared to be relaxed and jovial; he did most of the talking, and the others laughed; for Wolfman had a dominant personality, in such circumstances. But as soon as he left his companions and walked on alone his expression reverted, like ice melting, to an expression of sobriety, brooding.

Like one who is in Exile. One who is far from home.

In an entranceway I stood just far enough away so that Wolf-

man wasn't likely to see me. Badly I wanted to call to him—
"Dr. Wolfman, hello! Let me walk with you."

Never would I have spoken so recklessly. Of course.

Observing how Wolfman made his way to a bicycle stand behind Greene Hall where he tossed his briefcase into a wire basket,
unlocked the bicycle and straddled it and pedaled away.

He didn't own a car? Was that possible?

Without breaking into a run I followed a block or so. Watching
the cyclist disappear into traffic on University Avenue.

In Zone 9 bicyclists never wore safety helmets. Yet it surprised me that one who knew so much about the human brain,
its fragility as well as its mysteries, should so expose his head to
injury, as Wolfman did.

This was troubling. This made me wonder whether Wolfman
was of this time, after all. Was I imagining everything about him?
Was I losing my mind?

Blindly I turned away. Stepped in front of a vehicle and was
nearly run over amid a skidding of brakes.

Rawly a male voice yelled: "Look where you're going, stupid!"

The shout was a rebuke to my naïve yearning. But it would not
dissuade me from yearning.

TRIED TO LOOK UP *Wolfman, Ira* in the Wainscotia telephone directory but there was no listing under that name. No listing under
the name *Wolfman.*

Which meant: Wolfman didn't have a telephone. Or, Wolfman had a private number.

How strangely isolated people were, in 1959. If you could not
"look up" someone in a telephone directory you could not locate
him at all.

If I wanted to know where Wolfman lived I would have to follow him on foot. How I could manage that, I had not the slightest idea. And if I followed Wolfman home, what then? Could I have knocked at his door? Waited for him in the street?

How long, in the street?

I was becoming desperate, feverish. At the same time with a part of my mind I understood how irrational I was being. How like a rat in a maze, running, darting, always forward, never turning back, driven by appetite that was never filled. Yet the conviction that Wolfman was my *soul-mate* could not be shaken.

Wolfman had to live somewhere fairly close to the university campus, probably in a rented apartment. For he seemed to have no car, or anyway no car that I'd seen.

I wondered if he lived with someone? A chill passed over me at the prospect of his living with a woman. Or worse, his being married.

A husband? A father? Yet—an Exile?

This did not seem likely.

Certainly, not likely. *Procreation* was forbidden.

(But what a cold clinical term—*procreation!* Thinking of my mother who'd had Roddy and me, *had babies*, how much warmer a way of speech, how much more maternal.)

(Thinking of Mom with a sensation of such loss, sadness. Wanting to cry—*Mom! Dad! I am so sorry, I let you down.*)

(Did they even know if I was alive? I wondered. Maybe Roddy knew, and could tell them. Or hint to them.)

In Zone 9 there were limited ways in which one could learn about another person. Ira Wolfman wasn't a public figure, wouldn't have been listed in the university library catalogue probably. Still I decided to search there, in long clumsy wooden drawers, under the heading PSYCHOLOGY, 20TH CENTURY. To my surprise discovering that *Wolfman, Ira* was one of several authors listed for journal articles written under the direction of A. J. Axel. But there was no personal information about Wolfman.

How strange it was to me, this enormous building filled with *books*! In my former life everyone used eBooks of one kind or another, rarely "paper"-books; in NAS-23 the State controlled all electronic communications and transmissions and so it was not possible to access any title of any book that had not been approved by HSIB—Homeland Security Information Bureau. What would have been "magazines" and "newspapers" that had passed through the censor-filter were exclusively online in NAS-23—unlike the thousands of publications here in Zone 9, in 1959.

How strange, and how wonderful: the Government could not possibly control all these publications! Yet it seemed mysterious to me, that there was, in Zone 9, so much *freedom*—that, in Zone 9, did not feel like *freedom*.

As a university student I was often in the library—a building that stored books. There were no longer "libraries" in NAS-23— these old buildings had been put to more practical uses.

The university library was an old pinkish-red sandstone building with a rotunda, stately columns, and wide stone steps descending to a broad flagstone path; from a distance the lighted rotunda could be seen for miles. The library faced the chapel, the administration building, and the Hendricks Humanities Building across the campus green: this was the so-called historic campus, originally established by a land grant from the federal government in 1831. Nearby was the McCabe Science Building with its several new wings for physics, chemistry, math, psychology—in the post-Sputnik era, the fastest growing departments at the university.

Everywhere were trees for which I was learning names: oak, elm, pine, birch. A particularly beautiful tree, with a beautiful

name: juniper pine. Paths through the trees were sometimes concrete, sometimes earthen. At the edge of the campus was the arboretum—a hilly place of many acres. (My roommates never visited
the arboretum—why would they? Too far away, and nothing to
do but walk. They got enough walking up steep hills on campus!
And I was afraid to walk a mile or so to the arboretum, out of an
irrational concern that I would stray too far from my "epicenter"
and draw unwanted attention from my invisible monitors.)

More than nine thousand students were enrolled at Wainscotia State yet the atmosphere of the campus was rural at its edges.
In the early morning deer were seen grazing on lawns. There
were sightings of wild turkeys, pheasants, foxes, raccoons. There
were rumors of black bears.

Creatures of which I knew, but which I'd never seen except in
facsimile, or as "virtual" images.

When I'd first come to Wainscotia this setting had seemed
like a dream to me—unreal, disorienting. I thought *This is a
virtual place, this is not real.* In Media Dissemination, where my
brother worked, there were skilled hackers who'd been co-opted
by the Government to work for them, to create such utterly realistic VVS—(Visceral Virtual Settings)—that you could not believe they were not real. An entire subfield had emerged in VVS
research, to establish a way in which "virtual" sounds could be
made to seem as if they were coming from outside the human
skull, and not inside; I had to suppose that the same was true for
"virtual" sights. (Since the devastation of so much of the landscapes of North America, there had been perceived a practical
need for virtual landscapes; even higher castes invested in these.)
At Wainscotia I could not shake off the feeling that when I turned
away from the "historic" campus it dissolved like ink fading in
water.

Wondering if, in time, this unreal place would begin to seem "real" to me. And whether it might ever give comfort.

At first, books had seemed very strange to me.

Objects made of paper and cardboard designed to be *read*. How wasteful and clumsy it seemed—an *object* you had to hold in your hand (for reading was dependent upon turning the pages of a book). You could carry only five or six books easily, while on any eBook, you could access thousands of books. But then I realized that if there was a power outage you could continue to "read"—as the book you held in your hand would not vanish but continue to exist. The curiosity of the book was that, as you held it, and "read" it, you felt an intimate connection with it as with a living thing, which you did not feel with an eBook; as soon as you were finished with the eBook, you stored it, or deleted it; you felt no sentiment or particular ownership. You could not *see* it on a shelf or a table, you could not admire its design. In effect, it had been Deleted.

Badly I wanted to ask Ira Wolfman what he thought about these matters. Was there ever to be *comfort, happiness* in this place?

"OH, DR. WOLFMAN! Yes. He is"—lowering her voice so that no one could hear except me—"a *special case.*"

Under the pretext of inquiring about part-time employment in the psychology department I filled out an application form for one of the department secretaries, a friendly young married woman named Bethany. I marveled to Bethany that it must be wonderful to work in a department with such distinguished professors as A. J. Axel and his assistant Wolfman; I asked her about Dr. Wolfman, who was my quiz instructor, and Bethany told me that Dr. Wolfman was generally considered to be the most promising of the younger faculty, who'd published papers with Dr. Axel and gave talks at conferences.

In a confiding tone, as if she didn't think it was curious that I should be asking about Wolfman so much, and Axel so little, she told me that Wolfman had been at Wainscotia, in the psychology department, for five or six years; that he lived alone, and was said to "work all the time"; he was invited to staff parties but rarely came. "I know for a fact that Dr. Axel invites him to dinner at his home sometimes, and Dr. Wolfman accepts—he wouldn't dare turn Dr. Axel down!" Bethany paused, lowering her voice. "He isn't engaged. He's never with anyone except other people in the department. He's always *alone*."

It might have been clear to Bethany that I was in love with my psychology instructor. Such emotions are hard to disguise in a seventeen-year-old. I'd seen the young woman glance at my (ringless) left hand with a look of sisterly solicitude.

And so I dared ask about Ira Wolfman's background: where did his family live? Had she—or anyone—ever seen them? And Bethany said solemnly, "That's just it! Ira Wolfman is so *sad*. He was an orphan, I've been told—he never knew his mother—and the saddest thing, the wonderful people who adopted him were both killed in a car crash, also. He's from somewhere far away in the East—maybe New York City. But he doesn't have any home, he says. He's just—here."

SUDDENLY

And that night, a powerful dream-memory.

After I'd learned that Ira Wolfman was an orphan-in-Exile like me.

For there sometimes came, in Zone 9, at unpredictable times, sudden memory-surges like heat lightning that illuminates the night sky for a split second—then vanishes, in silence.

And that night there came a sudden memory of a long-ago time when I'd been a child of three or four and my parents were walking with me outdoors in lightly falling snow. And Mommy was gripping one of my mittened hands, and Daddy was gripping the other hand, and there came snowflakes to tickle my face, and make me laugh. And there was a gust of wind, and Daddy cried *Whoa!* and stooped to shelter me. And Mommy adjusted a wool cap on my head. *Hold still, honey! Let Mommy fix this strap.*

And I stood very still, and lifted my head, and—

(WHAT CAME NEXT? Nothing!)

(The memory ended abruptly, like a switch turned off.)

(And there was no way to return. No way back. Waking in my

bed on the third floor of Acrady Cottage, in a darkened room with three other girls sleeping nearby, snoring faintly or murmuring in their sleep. And I bit my lower lip to keep from crying. Telling myself that I had lost my parents but only temporarily. And in the meantime I have found Ira Wolfman my *soul-mate*.)

THE DENIAL

"My name isn't 'Mary Ellen Enright.' My name is—"

"No. Don't tell me."

Wolfman pressed his fingers against his ears. This was a gesture that reminded me of my father—a gesture that was both playful and serious simultaneously.

"You don't want me to tell you?"

"No. I don't want you to tell me."

We were in Wolfman's office. At last, after an exchange in class, he'd asked me to see him.

(This was not unusual. Wolfman often invited students to speak with him following class, when they had questions. But he'd avoided inviting *me*.)

I thought—*He will acknowledge me now.*

I was very excited, and apprehensive. For The Instructions were clear: the Exiled Individual is *forbidden to identify herself.*

Violations will insure that the EI will be immediately Deleted.

After Professor Axel's lectures this week Wolfman had taken up the subject of "schedules of reinforcement" in behavioral psychology— how acts (of human beings as of monkeys, rats, and pigeons) can be formulated along lines of behavioral responses to stimuli. On the blackboard Wolfman had drawn graphs. He'd scrawled equations.

For it was not enough to demonstrate that the living being was a kind of automaton, when it came to "operant conditioning," but the terms of conditioning could be reduced to formulae.

Why does the (hungry) rat press the lever that will bring him food pellets? Not because the rat is "hungry"—("hunger" is an internal state, thus not measurable)—but because its response to stimuli has been sufficiently "reinforced."

Why does the (addicted) gambler continue to press the slot machine levers that sometimes, though not frequently, bring him rewards in the form of coins? Not because the gambler is "happy" when he wins—("happiness" is an internal state, thus not measurable)—but because his response to stimuli has been sufficiently "reinforced."

Why does any individual—animal, human—behave as he does, over a period of time? Not because he has chosen to behave in these particular ways, but because his response to stimuli has been sufficiently "reinforced."

There was something wrong with this logic. My heart beat hard in opposition to it, though I could not have articulated my objection.

In our quiz section were undergraduates who had to be, like most of the residents of Acrady Cottage, Christians. Yet not one of them raised his or her hand to object to this mechanical, soulless view of consciousness. Instead, they took notes.

At last I said, trembling, whether with nerves or with indignation, "Dr. Wolfman? Do *you* think that human beings are machines, without 'free will'?"

Wolfman turned to me, politely. His slate-colored eyes were narrowed and wary. "It's a behavioral postulate that mental states can't be examined, so the scientific approach is to examine what is 'objective'—behavior. B. F. Skinner doesn't believe that there is no 'mind'—as his detractors claim—but that, given the principles of behaviorism, what is 'mind' is irrelevant to our understanding of behavior."

This seemed to me an oblique answer to a direct question.

"But—what about 'free will'? Don't human beings have 'freedom' in their choices for themselves?"

Wolfman said, with a shrug, "'Free will' is just a term. It's just speech. A habit of speech. It has no specific referent, thus no specific meaning. And it is not amenable to scientific proof or disproof."

I wanted to cry *My parents taught me there is free will. There is a soul, within.*

Wolfman continued, dryly: "'Free will' is a delusion, for most of us. It's a pleasant delusion, like the anticipation of Heaven which, though it never arrives, is comforting to contemplate. More realistically, 'free will' is like suggesting to a paraplegic that he has the choice of getting up and running in an Olympic competition— what's stopping him?"

This was not right. This was not the same thing at all. But I could not explain. I felt a thrill of sheer dislike of Wolfman, in that instant—the man was not my friend in Exile, but my enemy.

I could hear my parents' voices, objecting—but I could not hear their words clearly.

Trust the inner, not the outer. Trust the soul, not the State.

I was trembling now with emotion. I felt my eyes sting with tears. Ira Wolfman was my only hope—yet, with these harsh words, he was pushing me from him.

Around me, my classmates were yawning, and taking notes.

Seeing the look in my face, Wolfman relented.

"All right, Miss Enright. Come see me after class."

AND SO: I told him.

As I'd so long rehearsed. As I'd excited and frightened myself, imagining.

I told him that *Enright, Mary Ellen* wasn't my name.

And he stopped me from telling him my true name.

"But—why?"

"'Why'? You know perfectly 'why,' Miss Enright."

His pained expression. His narrowed eyes, that would not confront me.

"I—I think—I want to tell you, Dr. Wolfman. Please."

"To what purpose?"

Wolfman's face flushed with heat. He kept glancing behind me, at the opened door to the corridor.

His question was cruel, unanswerable. *To what purpose?—so that I would be less lonely, and less desperate.*

And because I love you.

If Wolfman was in Exile, like me, he'd made a substantial place for himself in Zone 9. Maybe he'd been exiled as a young man, a student—in the intervening years, he'd accommodated himself to his new life, and had acquired a degree of conspicuous success in a competitive academic world. He had an appointment at Wainscotia State University. He was a trusted assistant of the revered A. J. Axel, formerly of Harvard. His office was small and had to be shared with another assistant professor, but it was an office in distinguished old Greene Hall with its high ceilings, elaborate moldings, interior carved staircases, hardwood floors.

Wolfman's desk was near the room's single window. In a bookcase beside the desk were many books, both hardcover and paperback; predominant were the names *Darwin, Pavlov, Watson, Thorndike, Skinner.* On the other side of the desk on the wall was a mesmerizing art poster, at which I'd been glancing, which I would one day recognize as a reproduction of Vincent van Gogh's *The Starry Night.*

(A strange art-choice for a behavioral psychologist, such beauty and such mystery!)

Half-pleading now Wolfman said, "Look. I don't have any idea what you're talking about, Miss Enright. And for your own good, I think you might just leave my office now, and forget about this conversation. Do you understand? For your own good."

The pleading was in Wolfman's eyes. The words were terse, unfriendly.

Wolfman was a young man, not yet thirty. He was much younger than my father. Yet I saw something of my father in his face—the faint, furrowed lines in his forehead, the anxious eyes that tried to deceive with levity, sardonic humor. And the smile that came and went, fleeting.

"'For my own good'—that's why I'm here, Dr. Wolfman. I need to—know—from you . . . Who you are, and if—if you are—anyone like myself."

"No. I am not 'anyone like yourself.'"

"But—"

"No."

And I wondered then: Was Wolfman my delusion? Had I imagined—everything? Out of my loneliness and sorrow, had I imagined a savior in Ira Wolfman?

Yet, I was convinced that he knew me, deeply. He recognized me. Though he would not acknowledge it, as perhaps, in similar circumstances, to protect me, my father would not have wished to acknowledge the bond between us.

Wolfman wiped at his warm face with a tissue. His hands were strong and protective-looking, his fingertips stubby. His nails were cut short and rounded and reasonably clean.

No ring on the third finger of his left hand. But I knew that already.

Wolfman took up a ballpoint pen and scribbled something on the palm of his left hand. He held up the palm to me, as he touched the forefinger of his right hand to his lips:

GO AWAY PLEASE

Then, he shut his hand into a fist.

I understood. He did not want to speak to me, even to order me away. He had reason to believe that we were being overheard—we were *under surveillance*.

In a level voice, the voice of a reasonable university instructor confronted with a problematic student, Wolfman was saying, "Thank you for dropping by, Miss Enright. Be assured that your work in Psychology 101 is exemplary and that your ideas, though irreconcilable with Skinnerian behaviorism, will not endanger your grade—provided you continue to excel on our exams and know all the right answers to our questions. But we won't need any more personal conferences this semester, I think. And don't speak to anyone about this conversation—of course."

My legs were leaden. My head was heavy, and my heart.

I was shaky on my feet, but would leave Wolfman now.

"Who would I tell, Dr. Wolfman? You're the only person in Zone Nine to whom I could speak."

At the mention of *Zone 9*, Wolfman's face went rigid.

At the door I said, in a lowered voice, defiant, reckless—"And my name, which is not 'Mary Ellen Enright,' is, in fact, 'Adriane Strohl.'"

NOW, HE KNOWS. *He knows with certainty. And I know him, with certainty.*

And what will happen to me, to us, now?

THE WALL

There followed then a feverish time. The Asian flu that had been sweeping Wainscotia struck me at last. Days passed into nights, nights into days. I missed Wolfman's Friday morning class. I began to miss other classes. For I was too ill to get out of bed. I was often despondent. Only waiting until my roommates were gone from the room and then kneeling in a corner pressing my forehead against the wall was a solace of a kind.

Push push push—but the wall was unyielding.

My roommates whispered to themselves. They believed that their *strange, sad roommate* from out-of-state was praying.

They were pitying. They were concerned. They were beginning to be impatient.

For they believed that I was still homesick, and that I was kneeling in prayer to their Christian God, after so many weeks.

Wolfman would regret that he'd sent me from him! I wanted to believe this.

Miss Steadman tried to encounter me, in the front of the house. But I went invisible, and slipped past her.

On my bare knees on the floorboards. Pushing against the wall.

If I tried hard enough, the censor-barrier could be penetrated— flash-memories came through like the blinks of an eye.

Pushing a door open, hesitantly. My old, lost home!

It was our shingle board house on Mercer Street, Pennsboro.

Seeing my parents in the kitchen, at the table. Mom's bench containing geraniums by a window—stunted plants that didn't bloom much through the winter but, when they did bloom, yielded beautiful bright-red flowers.

It had been one of my duties as a child to keep Mom's plants watered. And to pinch off brown, desiccated leaves.

As Mom said sadly A *leaf that has turned brown will never be green again.*

So vivid was this memory, it seemed to me to be happening *now.*

As I knelt pressing my forehead against the wall to ease the tiny snarl in my brain, *now.*

Daddy was whistling. Brightly Daddy was whistling an old favored tune—"Battle Hymn of the Republic." (Daddy had explained to me: this was a famous abolitionist/antislavery hymn of the 1850s, of the old U.S.A.) So many times I'd heard Daddy whistling this tune, it came back to me now, as if I'd heard it only yesterday.

Yet, there was something strange about the tune. It was recognizable immediately—and yet, not quite right.

Daddy was whistling brightly, loudly. In a way to annoy (?) my mother as, preparing coffee, he annoyed my mother by brushing too close to her at the stove.

Mom murmured something to Daddy that I couldn't hear. Daddy laughed, mirthlessly.

(What was wrong with Daddy? His face was hidden from me, like half a moon cloaked in shadow. He was dressed in his hospital-attendant uniform, his slave-uniform as he'd called it, an orderly's white pullover and coat, white work-trousers and white rubber-soled sneakers.)

Mom was setting bowls down on the table—cereal bowls.

Dad's bowl was at one end of the little rectangular Formica-topped table, Roddy's and my bowls to either side of his, and Mom's bowl nearest the stove.

Oh!—My mouth watered: I could smell the oatmeal Mom was preparing which was our favorite breakfast.

Steel-cut oatmeal cooked with raisins, brown sugar, and milk.

Mom had made this breakfast since I'd been a little girl. I hadn't realized how I had missed it.

There was Roddy in the doorway. Short-cut hair, nondescript clothes of a low-level "intern." His face lean and sullen and the eye sockets shadowy like a death's-head.

It was my childish wish that my brother might not be there, in this memory. If this was *my memory*.

There was something cruel that Roddy had done to me—I couldn't remember what it was. When news had come of my Patriot Democracy Scholarship he'd stared at me for a moment without any reaction—then, a clumsy forced smile.

Congratulations, Addie!

A smirk of a smile.

Badly I wished that just Mom and Dad were there, in this memory that was so precious to me.

But I could not dictate the memory—could I? If I tried, I was in danger of losing the memory altogether.

Softly I said *Hello! It's me—Adriane* . . .

Daddy didn't hear me. Mom didn't hear me. If Roddy heard me, he gave no sign.

They were talking about—I couldn't hear what they were talking about. It was a familiar problem—it was not-new—boring, humiliating. Probably, it was a financial problem. Or some problem of Roddy's? Some complaint of Roddy's? Or—Mom was having difficulty with her supervisor at work. Or—(maybe this was more likely)—Dad was having difficulty with his supervisor at the medical center. (It was an old humiliation of Eric Strohl's, that since he'd been demoted from staff resident M.D. to hospital attendant, his income had been decimated; yet, to keep this miserable job, he had to advise ever-younger and more career-minded physicians, and often he had to do minor surgical procedures for them, like the installing or removing of "ports" [portable, catheter-like artificial veins] in chemotherapy patients. He was also on hand to draw blood and assist in radiology. Yet, it wasn't like Dad to complain even jokingly—so this was strange.)

(And strange too: where was I, in this memory? Why wasn't I sitting at the kitchen table with my family? Judging by Roddy's age, I must have been about sixteen at this time. *But where was I?*)

Pressing my forehead against the wall of the third-floor room of Acrady Cottage, Wainscotia, Wisconsin—Zone 9.

Saying, begging—*Mom? Dad? Can't you see me? It's—Adriane. Please—look at me.*

But they weren't looking at me. They were oblivious of me.

Then I saw something that chilled me: my father's face was coarser than I'd ever seen it. Not just that Daddy hadn't shaved, but Daddy was looking disheveled like a man you'd see on the street, homeless, feral: his gray thinning hair uncombed, furrows in his cheeks, a self-pitying downturn to his mouth. And small, bloodshot eyes . . .

And my mother: What had happened to *her*?

Mom who'd always been so slender had definitely gained weight. Her face was fleshy, peevish. A sardonic smile on her lips. And her face heavily made-up, with exaggerated arched eyebrows.

Mom's gaze was sulky, dissatisfied. An expression of barely suppressed rage in Mom's face I had never seen before.

Mom, Dad—don't you love me? Don't you miss me?

It's your daughter Adriane—don't you remember me?

Carelessly, as if she resented waiting on my father and brother, Mom brought the pan from the stove, to spoon out oatmeal into bowls. And now I saw that there were only three bowls on the table. And I saw that their breakfast wasn't oatmeal after all— greedily they were eating some sort of gelatinous substance that was sticky, quasi-transparent, of a sickening fleshy-pink color, quivering in their bowls. This was no breakfast food that I could identify—horribly, it seemed *alive*.

Roddy glanced in my direction, as if he could see me!

Roddy said with a mean little smirk *Where she's gone to now, it serves her right.*

And Mom said *She thought she was too good for us.*

And Dad said *Good riddance!*

They laughed. It was a crude, hellish laughter. There was a swirl of light in the kitchen, as if the Plexiglas barrier were reflecting something shiny, interrupting the scene. I saw with horror that something was very wrong—these people were strangers to me.

The crudely imagined figure meant to be my father was not my father. The crudely imagined figure meant to be my mother was not my mother. And my brother Roddy . . .

It isn't Roddy, either. They have taken away Roddy, and substituted this person for him.

I wondered if this was so: if it hadn't been my brother who'd informed upon me but the other in his place, who'd sent me into Exile.

Suddenly the smell of the quivering jelly-like "oatmeal" was nauseating. Suddenly, I was gagging. The Plexiglas went opaque. The memory was shuttered, gone. I was left alone in a corner of the room, on my knees, pressing my forehead so hard against the wall that the impress of the wood would leave welts in my skin.

"MARY ELLEN!"—someone was calling me.

That ridiculous name, that name I hated—someone was calling it, and pulling at my shoulder.

I woke, astonished and frightened. Where was this? What time was this? I must have fallen asleep, or lost consciousness, on the floor beside my bed, on my knees, and one of my roommates had found me slumped against the wall.

"Mary Ellen, what's wrong? You've been crying—you're upset. It must be the flu. Let me help you up."

My roommate whose name was—(Betsy?)—helped me up, onto my bed. She sat beside me, holding my hands that were like ice. She spoke comfortingly to me, soothingly. I was confused but knew not to say anything incriminating about where I had been. Where my remembering had taken me.

Another roommate came into the room, and joined her. What was wrong with Mary Ellen? What had so upset Mary Ellen? A nightmare? A bad memory?

The second girl was—Hilda? I remembered her last name, which was a charming name: McIntosh.

Betsy and Hilda conferred about "Mary Ellen"—what was wrong with "Mary Ellen," and what might be done with "Mary Ellen."

I was thinking: the ugly memory had been a false memory. The microchip was programmed to interfere with my memory. To provide me with a cruel, false, ugly memory. *To punish.*

Yet, it was hard not to think that the father in my memory, the mother in my memory, and the brother in my memory were "real."

It is terrifying, to lose memory. To lose trust in memory.

What is a human being except the sum of her memories? *Look inward, not outward. The soul is inward.*

I believed this. And yet, if my memories were taken from me, what would happen to me? What would happen to my soul?

My roommates argued with one another: Should I be put to bed, or should I be taken to the dining hall, to eat a proper dinner with them? Betsy and Hilda believed that I was seriously undernourished, and didn't sleep enough; Carly thought that I had "some kind of flu-virus." My skin was clammy, not feverish. But my face was oddly flushed, and my eyes were bloodshot.

My third roommate had come into the room, and joined us.

I told them that I wasn't sick—I didn't want to be put to bed. I had homework to do that evening, I was behind in my work. So Betsy, Hilda, and Carly walked me into the bathroom, ran cold water and washed my face that was streaked with tears. They brushed my hair—("Oh gosh, look—it's coming out in the brush. She needs to drink more milk!"). They insisted that I apply makeup: foundation, powder. And lipstick, which Carly lent me. They were enthusiastic about my looks—"If you'd just smile more, Mary Ellen. If you weren't so tired, and sad-looking."

"We're homesick, too—or anyway we *were*. But Wainscotia is so wonderful, it's time you were over missing home."

Carly lent me a sweater to wear to the dining hall: a soft-woolen heather-colored cardigan, far more beautiful than any of my secondhand clothes. Hilda lent me an actual coat—not a jacket—to wear to the dining hall. And leather boots, instead of my ugly rubber boots.

They sat with me, and watched me eat. They took my plate back for seconds. We talked together, and laughed. Other girls from Acrady Cottage joined us. After a while I was feeling better. That is, Mary Ellen was feeling better.

We returned to Acrady Cottage. There was a light snowfall, an icy crust of snow crackled underfoot. In the near distance, the rotunda of the library glowed with a bluish interior light. I thought—*I have no one now. But I will survive. And I will return home, one day.*

I gave back the leather boots, and the deep-forest-green wool coat, but when I tried to give back the heather cardigan Carly said, with a sweet, pained smile, "Oh no, Mary Ellen! It looks nicer on you than it does on me. It's *yours.*"

(BUT I COULD NOT be their friend. Because I was not "Mary Ellen"—I was someone else whom they had never met.)

GO AWAY PLEASE

I had only to close my eyes and the red-inked words hovered before me.

In melancholy moods, I didn't have to close my eyes. In the very air before me I seemed to see Wolfman's uplifted hand, the palm of his hand, the admonition that had sent me from him—

GO AWAY PLEASE

The gesture was playful—but desperate-playful. Wolfman had been sincere in wanting me gone.

He'd acknowledged me as one like himself, in Exile. And in the same gesture Wolfman had repudiated me.

Of course I understood: he was under a sentence of potential Deletion, like me. He had to follow the Instructions, as I did. He could not be reckless as I was.

Ira Wolfman had adjusted to his Exile-life, evidently. Or he'd learned to give that impression. He was older than I was, much more intelligent, and wiser as well—he knew that there was no way back home for us.

We could be *brought back home*. But we could not *make our way home*.

This was the curse of Exile: you are powerless to change your life, except for the worst. Others can change your life for you, unpredictably.

When I'd been sick, I'd missed some classes. But just one of Wolfman's classes, which met on Friday morning. Since the flu was said to have affected so many students at Wainscotia, instructors were understanding, and hadn't penalized us for absences or late papers. I was determined to make up for my lost classes for I was determined to be an excellent student at Wainscotia State University.

In all my courses but particularly in Psychology 101.

I WENT TO MEALS with my roommates more frequently. I was friendly with other residents of Acrady Cottage. I did not avoid Miss Steadman. (Though I'd disappointed our resident adviser at Thanksgiving by hiding away in the university library, and then upstairs in my room in the near-empty residence, with the excuse that I had too much work to do and couldn't come to dinner with her and other lonely left-behind Wainscotia girls.) I wore the heather-colored cardigan sweater. I wore the Black Watch plaid pleated skirt. I even wore a strand of pale pink "pearls"—a dimestore necklace I'd found on a sidewalk on campus. In Wolfman's quiz section I was an attentive student but I did not raise my hand for Wolfman to call upon nor did Wolfman, in his coolly courteous way, call upon "Miss Enright." In a lapse of good judgment I even agreed to "double-date" (a term new to me) with Betsy and her Sigma Nu boyfriend; my date was a fraternity brother of his whose friends called him "Hedge," from a small town in northern Wisconsin. Hedge was taciturn and blushed easily. It was difficult to know if Hedge was chronically embarrassed or chronically irritated. He seemed not to know how to talk to me, and I certainly didn't know how to talk to him. He was a mechanical engineering major whose grades were mostly C's. He'd been drinking beer before he'd come to Acrady Cottage to pick me up, and, at the Sigma Nu "keg party," in the basement of the sprawling red-brick fraternity house with its carpet stained from years of rough usage, he continued to drink with his friends, like one determined to get drunk as quickly as possible. The evening passed in a blur amid deafening music, voices and laughter, a smell of pizza and spilled beer. It wasn't surprising but it was dismaying to see Betsy drinking so much, "necking" with her date, and dancing drunkenly with him; after an hour or so, Betsy disappeared into the crowd, and I never saw her again that night. Hedge, flush-faced, seemed to want me to dance with him, too—or at least, he wanted to show his frat brothers that he was dancing with me, or clumsily pressing himself against me. He was very drunk: his belches stank of beer like popped balloons. At one point I hid away in a lavatory marked GIRLS ONLIE. I comforted freshman girls who were very drunk, and vomiting. I thought—*If Wolfman would love me. If Wolfman would just acknowledge me.* I wondered if, if I behaved recklessly, in violation of The Instructions, I would be "vaporized" by a Domestic Drone Strike here in Zone 9, and disappear before I felt pain or even fear. When I had the opportunity, and Hedge was nowhere in sight, I slipped away from the fraternity house through the wide-opened front door, and ran stumbling back to Acrady Cottage.

How wonderful it was, to have escaped my first—(and last)—

fraternity keg party! How wonderful to be alone, but not now lonely, running through a lightly falling snow back to my residence, my breath steaming! My roommates were all out on "dates"—I could have wept with relief in the room.

Relief is happiness for those who, otherwise, would have no happiness. But relief can be an exquisite happiness, even in Exile.

BETSY NEVER SPOKE to me about the keg party at Sigma Nu.

Betsy did not speak to me very frequently, or very warmly, about anything at all, for the remainder of the time we roomed together.

(Though I'd overheard Betsy complaining bitterly about me, for I'd embarrassed her with my "selfish" behavior, in front of her Sigma Nu friends; and I would learn sometime later that the night of the keg party had not been a lucky night for Betsy, who would withdraw from Wainscotia during winter break, and would not return.)

When, all too often, I saw Hedge on campus, and Hedge saw me, we quickly looked away from each other.

The sudden, cruel thought came to me—*What does it matter if they hate me? They are both in their seventies by now, if they are still alive.*

THIS WAS THE TERRIFYING SECRET of Zone 9, of which its inhabitants were blissfully unaware: in the time that, to me, was present-tense, in the twenty-third year of the Reconstituted North American States, their lives were nearly over. If they were alive at all.

THE MUSEUM OF NATURAL HISTORY

In December, I began part-time work at the Van Buren Museum of Natural History which was a sepulchral stone building adjacent to the Greene science building.

The Museum of Natural History was a shadowy hushed place which few people visited. In its dark interior, Time seemed to have stopped decades ago; in the rooms where fossils and ancient bones were displayed, thousands of years ago.

My work was carding and shelving books, for the museum included a special collection of rare natural-history books. My work was typing letters, documents, labels identifying items in glass display cases.

In Zone 9, much of work-life was reiteration. Office work was mechanical, robot-like. A typist was a kind of robot. You typed words directly onto a sheet of paper—the words were formed by (black) ink on a "ribbon" that wound through the typewriter on twin spools. Often you used a "carbon copy" to make a second copy, for there were no photocopying machines—there were no personal computers and printers. Everything was done *by hand*.

Everything was one-time-only, and yet would have to be repeated, usually. It was a kind of madness that could not have been explained in NAS-23. Office work inched slowly along deep-rutted grooves. It was commonplace to be required to type a second "ribbon" copy. There was terror in such repetition to no purpose save a replication of the original which might have been achieved by merely mechanical means—except, in 1959, these means did not yet exist.

Much of my part-time work involved (re)typing labels for exhibits. In NAS-23, such labels would be printed out within minutes, or seconds. Yet the task occupied hours of my time. My salary was one dollar an hour—before taxes.

In NAS-23, currency had been "reconstituted" in order to combat inflation. Still, everything was much more expensive than it had been in my parents' memory, they'd said—while their salaries, which were modest salaries, had been frozen for years. It was a profound surprise to me to receive so small a salary from the university—(equivalent to approximately one-half of a cent in NAS-23 terms)—and yet more of a surprise when I discovered that, after taxes, I was earning less than sixty cents an hour!

When I realized this, I burst into tears. My supervisor Miss Hurly said curtly, "Every one of us is taxed, Mary Ellen." She meant to encourage me by saying that if I continued to work well I might receive a raise of as much as twenty cents in the next semester.

Twenty cents! I laughed.

This was training, however. This was "experience."

Though the term did not exist in Zone 9 as it existed in the twenty-first century I was an "intern"—of sorts—I was accumulating skills, and a résumé, and, if I needed them, references for future employment. On Hilda's borrowed typewriter I had learned to type competently; in fact, it was not a very different sort of typing from what I'd been doing since the age of two at a computer.

And then, I'd learned to operate the somewhat fantastical "office-model" Remington in the museum, which must have weighed twenty-five pounds. This was an enormous black machine with steel keys that could be made to fly through the air in arcs of about three inches, striking white paper and "printing" black-ink letters. Miss Hurly had taught me to remove old, used-up ribbons and replace them with new ribbons; these ribbons, amazingly, came equipped with both black ink (top) and red ink (bottom); it was not possible to remove a typewriter ribbon without getting my fingers smudged with ink, but I was proud of myself for having learned at all. Miss Hurly taught me also to clean the keys of gummed-up ink until they shone as if new.

Of my lost world of computers, cell phones, electronic pads and "readers"—I could have explained very little to any inhabitant of Zone 9. Even my memory of what these had meant to me, how I'd been habituated and addicted to them, seemed to be fading, like memories of my family and friends.

(And I wondered: Can you still love someone whose face you are forgetting? Whose voice you can no longer hear?)

(And I had to concede: If I had my cell phone here in Wainscotia, whom would I text or call? There was no one.)

Unexpectedly, I was beginning—almost—to like the typewriter. I could see why Hilda was so proud of her portable model, set beside which the massive office-Remington at the museum was high-tech. The most extraordinary fact regarding both was that they did not have to be *plugged in*—here were machines so primitive they didn't require power. I had learned to "set margins"—to "backspace"—to anticipate, like a subject in a behavioral psychology experiment, a tiny bell ringing near the end of a line. Most importantly, my fingers that had been accustomed to the light touch of a computer had learned with surprising alacrity to strike the keys down, hard. In the more popular keys, *a, o, s, t,* you could see the faint indentations of typists' nails.

Ghost-typists had preceded me in the shadowy interior of the Museum of Natural History.

My supervisor was a white-haired woman in her mid-fifties named Ethel Hurly who spoke in a hushed but stern voice as one might speak in a mausoleum. She had a large soft descending bosom and wore polka-dot blouses with bows at the throat. Her superior was the museum director, Professor Morris Harrick, whose Ph.D. was from Princeton in the field of "classical science" and whom I rarely saw. (Like so many of the higher-regarded Wainscotia faculty, Professor Harrick was from what had used to be, until the controversial Reconstitution of Higher Education in NAS-20, an Ivy League university.) I perceived that Miss Hurly was in love with Professor Harrick, who was himself white-haired, and in his fifties; with polished glasses, a distracted look, and a habit of blowing his nose loudly into white cotton "handkerchiefs." (These were actual clothes, small squares of cloth, invariably white, which men of a certain caliber or class used conspicuously, to signal, it seemed to me, that they had in their power or in their hire, in private, a female so trained that she would not bridle at laundering and "ironing" them, for his one-time use. Fortunately, for the rest of us, paper tissues had been invented by 1959.) Professor Harrick seemed to be married, for he wore a wedding band on the third finger of his left hand, and on the desk in his office there were small framed photographs of family members including young children. So Professor Harrick was a father, too! And very likely a grandfather.

It was touching to me that Miss Hurly adored Professor Harrick with no hope of his returning her feeling, or even becoming aware of it. I felt sorry for her even when she was impatient with me, surprised at my lack of skills and my general naïveté. (Once she'd said to me, "Mary Ellen, were you born in the United States? Sometimes you don't seem like an *American*.")

Morris Harrick was a dignified gentleman who not only carried a freshly ironed white cotton handkerchief in his vest pocket, but also wore a vest, a tweed jacket with leather elbow patches, a white cotton shirt and proper necktie, each day he appeared in the museum, or taught A History of Western Science in Greene Hall, immediately following Professor A. J. Axel's psychology lecture. Professor Harrick rarely spoke to me directly and scarcely "saw" me at all—he'd several times called me "Dolores," which was the name of another undergraduate girl–employee in the museum, whose work-hours didn't overlap with mine.

As I loved Wolfman at a distance, or consoled myself in my loneliness with the possibility of loving Wolfman, so Miss Hurly loved distracted Professor Harrick, for whom she was always typing letters, documents, and manuscripts to be sent to learned journals in the professor's field. Miss Hurly showed me Professor Harrick's publications in these journals, which were impressive, and all but unreadable, as well as the professor's several books with such titles as A *History of Natural Philosophy from the Pre-Socratics to the Enlightenment*, published by the University of Wisconsin Press. Miss Hurly echoed Miss Steadman in her enthusiasm for the male academic intellect: "Professor Harrick has devoted his career to examining how 'false' theories of science have been replaced by 'true' theories through the centuries, to our present-time, mid-twentieth-century America. I don't know the details of course—but Professor Harrick's argument is very persuasive. He is another candidate from Wainscotia for the Nobel Prize."

I asked Miss Hurly if she'd heard of A. J. Axel and she said yes, of course—"One of Wainscotia's great minds!" I dared not ask her if she'd heard of Ira Wolfman, for a quaver in my voice might have given me away.

It made me swallow hard, struck with sadness to think of Miss Hurly and Morris Harrick and to realize that, in NAS-23, each had been long dead; and whether Professor Harrick had won a Nobel Prize, I did not know.

IN THE MUSEUM, often I worked late. I did not like to return to Acrady Cottage where I was obliged to impersonate "Mary Ellen" to my roommates and sister-residents.

Work was a narcotic! Work was my way of sanity, that prevented me from brooding upon my parents, my lost friends, Ira Wolfman—the colorful, tattered kites that Roddy and I had once made together, that were now long destroyed.

Daddy had once advised *One day, one hour at a time, honey. One breath. We can do it.*

(Not to me, Daddy had said these gently urgent words, but to my mother, I think. She'd been crying at the time, in their bedroom with the door shut.)

Resolutely I typed—retyped—labels for the display cases: the Latin names of flowers, fungi, birds and mammals; beautiful names, they seemed to me, because exotic, in a long-dead language. (In NAS-23, Latin was not taught in even the most esteemed schools.) And when I'd had enough of typing on the massive Remington machine, when my fingernails ached with the pressure of striking the keys, I turned to my textbooks, and my notebook, and worked on my assignments. It had not been difficult for me, despite my anxieties, to receive uniformly high grades in all my subjects, including now Intro to Logic; in the narrowness of my Exile-life, which yielded a private, secret depth like that of a chasm in the earth whose depth is not visible from the surface, it was not difficult for me to excel—most of my classmates seemed only moderately engaged with their course work, even scholarship girls at Acrady Cottage. The great, public life of the university was of the surface: football and other sports, "Greek life" (fraternities, sororities whose mansion-sized houses were predominantly ranged along the hilly end of University Avenue), "pledging," "dating." If I worked in the museum on Saturday afternoon I could sometimes hear, in the distance, a roaring of cheers like a frenzied and frothing waterfall from the football stadium at the far side of the campus, which was said to contain more than twenty thousand spectators! In the museum, empty except for me, it was scarcely possible to imagine twenty people, let alone twenty thousand.

Among Wainscotia undergraduates it was considered "square" to study hard, and to seem "serious"; to spend much time on studies was considered a kind of treason, if you were a "Greek." My grades of A and A+ were best kept secret as they were disfiguring to me as acne—(I knew: my roommates had frankly told me). Since the Sigma Nu beer party which I'd fled, and since Hedge, and my roommate Betsy's disappointment in me, I had had no "social life" at all—to my relief. For nowhere are you so lonely as in the midst of a party.

And nowhere do you feel so unloved as when couples are drunkenly pawing at each other, in a crude parody of love.

And when I'd had enough of my schoolwork, and my head and eyes had begun to ache with the effort, and something of the futility of this effort, I prowled the museum, switching on lights in the shadowy recesses—room after room, exhibits on platforms and on the walls, in display cases, what seemed like acres of fluorescent-lit gloom. In winter, when the sun disappeared in a bank of discolored clouds at about 5:00 P.M., it was pitch-dark by 6:00 P.M. as if it were midnight.

Even during the day there were few visitors to the Museum of Natural History. Some of these were professional guests whom Professor Harrick and a colleague or two might be showing through the museum, concentrating upon special exhibits. Some might be alumni, or the parents of students. These visitors rarely stayed long. Their voices too were hushed in the sepulchral regions of the museum. Between rows of display cases they moved like wraiths, staring briefly, then moving on. If they glanced in my direction—shelving books, or perched behind the giant Remington at a desk adjacent to Miss Hurly's desk—they appeared startled, as if one of the stuffed-creature exhibits had stirred to life.

Wandering the museum after hours! At times I felt an unexpected sort of happiness, at other times a more profound sense of futility.

No matter how much you learn—how high your "grade-point average"—you are alone here. And no one cares whether Adriane Strohl lives or dies.

One of the famed exhibits in the museum was the Van Buren Glass Flowers, which exerted a kind of spell upon me. I thought of my mother—I tried to think of my mother, Madeleine—(yet how strange this name sounded to me now: "Madeleine")—who would have admired the extraordinary flowers, ingeniously wrought in glass of the most subtle hues: the most beautiful were orchids, lilies, exotic tropical blooms of the size of a human head. Yet the flowers were curiously scentless of course. A faint patina of grime lay upon even the most polished specimens, if you looked closely.

The most intriguing of the exotic glass flowers was a "carnivorous" plant from the Amazon rain forest—its elongated flesh-colored petals resembled a crocodile's (opened) jaws, and, in life, were said to be sticky-sweet to attract insects and small mammals. Out of childlike curiosity I lay my fingers on the (opened) petal, to see if the carnivore-plant would shut its jaws on me; but it did not, for it was made of glass, and never moved.

I tried to remember my mother's indoor plants: stunted specimens, which failed to thrive in weak winter sunshine, but which flourished, to a degree, in warm weather, when we set them onto our back steps. But what was their name? A very common name . . . And the blossoms were small, and bright-red. And had no scent.

I missed Mom so much! And Daddy.

And Paige, and Melanie, and . . . what was her name, whose father had been arrested? Carla . . . ?

The punishment of Exile is loneliness. There is no state more terrifying than loneliness though you would not think so, when you are not lonely; when you are secure in "your" life.

In these after-hours in the Museum of Natural History it seemed to me that there might be someone who'd stayed behind, until the museum had officially closed; or, more unnerving, a being in the museum whose presence predated my own. My heart began to beat quickly as I moved from room to room—switching on overhead lights, seeing shadows leap in the corner of my eyes. On the high-ceilinged walls were skulls, bones, near-complete skeletons of ancient birds and beasts; in display cases, rocks containing fossils, and more skulls, bones, and small skeletons. And there were stuffed and mounted creatures—such birds as hawks, owls, falcons, shorebirds and songbirds, a glassy-eyed bald eagle; such small mammals as red foxes, raccoons, squirrels, lynxes and bobcats, and on a wall an enormous elk's head with twelve-point antlers and shining glass eyes.

And a beautiful wolf with silver-tipped fur, a wise sharp dog-face, intelligent-seeming glassy eyes. *Canine x, indigenous to Wisconsin.*

Intensely the (dead) creatures seemed to be observing me. Theirs was a profound sorrow, all the more piteous for being mute. I saw that the labels attached to many of the displays were yellowing and supposed that, soon, I would be asked by Miss Hurly to retype them. I smiled at the futility of the museum—this place that felt underground, though it was on the first floor of the building; this place that Time had forgotten. For in the Museum of Natural History it was not even 1959.

A soft film of dust lay over all things. I expected to see, when I walked along the aisles, my footprints on the floor behind me.

I was smiling, in order not to cry. In one of the glass display cases I saw my own wan reflection floating, superimposed upon an exhibit of aged turtle shells, from which the living turtle-bodies had departed. Yet, I thought I saw one of the turtle shells quiver. Something was moving, reflected in the glass . . .

To my shock I saw then that a man was approaching me in the shadowy light, his hand extended. Wolfman! Smiling at me, a forefinger to his lips and the palm of his hand held out to me, upon which he'd written in bright red ink:

FOLLOW ME PLEASE

SHELTER

In silence Wolfman led me deeper into the interior of the museum.

Quickly I followed. I would have followed Ira Wolfman anywhere.

Like a sleepwalker who is but dimly aware of her surroundings but determined to behave as if she were in control of her movements, I followed.

Wolfman had come to *me*! I'd seen in his face a frowning smile, or half-smile—a pained sort of tenderness. *He is risking his life for me*, I thought.

I vowed then, I would love Wolfman with all my strength. I would die for Wolfman.

He knew the Van Buren Museum of Natural History, it seemed. He was not a stranger to its vast subterranean-seeming reaches. And he knew that I worked here—he must have made inquiries about me, to have such specific knowledge of my work hours.

Shyly I smiled at Wolfman. My heart was beating rapidly in my chest like a trapped bird!

Still without a word Wolfman took my hand, and pulled me gently—firmly—forward. His fingers clasped mine with startling familiarity. As in a dream of surpassing wonder and beauty we passed through rooms I had never seen before: entire walls festooned with songbirds perched on replicas of tree limbs, and a great flock of brightly colored warblers; a hall of "wetlands" creatures—rubbery stuffed frogs and toads, turtles of all sizes, snowy egrets perched on a single leg, mute swans, Canada geese, mallard ducks frozen in dark gelid ponds. Clinging to the underside of a tree trunk, a weak-eyed opossum. All these creatures seemed to be watching us, alerted by the life in our bodies.

Wolfman mouthed the words—*Don't speak! Not yet.*

We passed through a display of large mammals of the American West—antelope, deer, buffalo, bison, mountain lions, bears (black, brown); a display that was an enormous whale skeleton, through which we walked furtively, eagerly. And there was an enlarged photograph of the gigantic creature, on a wall facing us. Where was Wolfman leading me? I felt dazed with happiness, or with dread. In my fantasies of Ira Wolfman I had not made out what we would say to each other, if we were truly alone together. How we would declare ourselves to each other, if we dared. I had not allowed myself to imagine Wolfman actually taking my hand.

I had not allowed myself to imagine Wolfman holding me, kissing me. My dreams of Wolfman had always faded at that point.

As if to obscure our trail Wolfman switched on and off lights as he led me through the museum.

At last we were in a remote corner of the museum, far from the entrance. Here was a dimly lighted room crowded with a miscellany of exhibits as if the museum curator had run out of space and despaired of his mission, like a distracted Creator, and had stored here crudely stuffed animals (small deer, lynx and bobcat, rodents) whose fur was matted and whose glass eyes were sunken in their sockets; trays of fossil rocks whose labels had peeled away; unidentified skulls and bones; and rattlesnakes coiled together on sheets of shale, looking so uncannily alive with their small beady glass eyes that I shuddered at the sight of them, and tried to pull my hand from Wolfman's as he laughed at me—"They can't hurt you, 'Mary Ellen.' They're under a spell, too."

At the rear of the room, partly hidden behind a large display case, was a stairway, leading down; at the foot of the stairway was a squat-looking door cut into the wall. Set into the door was a combination of the kind affixed to safes and lockers and this combination Wolfman turned, and turned back, and again turned, with careful fingers, until the lock sprang open, and the door swung just slightly inward. And Wolfman took my hand more firmly now, and pulled me inside.

Wolfman shut the door. "Now, we're safe! No surveillance."

He'd switched on a light. Fluttering fluorescent lights, overhead. There was a smell here of the earth damp, rich, rotted. We were on a cement landing. More steps led down, to a darkened lower level.

Wolfman tugged at my arm. I felt a moment's panic, that I must follow him—where?

Wolfman preceded me, down the steep cement steps. So that, presumably, if I grew faint, or lost my balance, he could turn and steady me.

We came to another, much lower level. The air here smelled ranker and was fiercely cold.

I rubbed at my eyes. My vision seemed blurred. I saw that we were in the outer room of a basement of some sort, with cement walls, cement floor and (cement?) ceiling; on the floor was a metallic-gray carpet, and on the walls signs containing elaborate instructions NUCLEAR ATTACK EMERGENCY PRECAUTIONS. Close by the entrance was a life-sized dummy equipped with gas mask, thick gray clothing and gloves and heavy boots.

It was impossible not to think that the uniformed dummy was somehow observing us, through his goggle-glasses.

"This is a nuclear bomb shelter, 'Mary Ellen.' You've probably never seen one before. There's another, larger shelter, better-equipped, for more elite members of the university community, in the basement of the administration building, which I know about but have never seen."

In wonderment and in dread I looked around. I could not help but think that, despite our distance from the surface of the earth, Homeland Security was recording us, and that we would be terribly punished.

Wolfman said that there were still bomb shelters in our time—in NAS—but only for Government officials. The average citizen knew nothing about them. But in the 1950s bomb shelters were much more common, featured in mass-market magazines like *Life* and *Time*, and on television; many private citizens built their own shelters in their basements—"Like an extension of the 'family room.' A place in which to be cozy though you would want also to be armed, to keep out neighbors."

Wolfman remarked that the nuclear bombs and missiles had been expected to come from the USSR—Soviet Russia, a "political behemoth" that no longer exists in NAS-23. He asked me whether in present-day NAS Russia was still one of the "terrorist enemies of democracy" and I told him that I thought so, yes, for there were a number of "terrorist enemies"; but Russia was an enemy with whom the North American States made treaties, in opposition to China which was NAS's great and abiding enemy.

Wolfman asked me who was president of NAS-23 but when I pronounced the name, Wolfman didn't recognize it.

Presidents of the Reconstituted North American States were heads of the Patriot Party. The general population knew little about them though they were believed to be multi-billionaires, or the associates of multi-billionaires. Their names were often invented names, fictitious names, attached to individuals or animated human figures replicated endlessly online and on TV; you were conditioned to "like" them by their friendly, smiling facial expressions and by ingeniously addictive musical jingles that accompanied them, as you were urged to "dislike" other figures. To attempt to learn facts about them was in violation of Homeland Security Information and could be considered treasonous.

Wolfman was saying that, for a long time after World War II, Americans had lived in terror of a *nuclear holocaust.* Schoolchildren as young as five or six were drilled in what to do if there was a sudden "nuclear flash"—they were to scramble beneath their desks, bow their heads and cover the backs of their necks with their clasped hands.

"For those fortunate enough there were shelters, stocked with provisions like this one. 'Shelter' was a booming business, for a while."

"But there was never any 'nuclear holocaust'?"

"As it turned out, no."

"But now—here in Zone Nine—they still believe it might happen? Russia might drop a nuclear bomb on the United States?"

"No, Mary Ellen. You don't say 'they still believe'—it's 1959, and it's natural for the citizens of the United States, in 1959, to believe in the possibility of a nuclear holocaust, not an aberration."

I felt a twinge of pain behind my eyes. I'd learned to deal with the perplexity of living in this bygone era *as if it were the present-tense* and not past; but I had not articulated the logic of such a life, and of course I'd never spoken to anyone about it.

Amused by my naïveté, or slightly exasperated by it, Wolfman said: "Yes, the citizens of the U.S., at this time, have a prevailing expectation of nuclear war. They can't comprehend it, really—any more than we can comprehend our own deaths, and the extinction of our identity. We are not capable of imagining ourselves, or our loved ones, *non-existing*. We are not capable of imagining tens of millions of human beings murdered in the twentieth century, in the great 'world wars,' in Soviet Russia and in Communist China. But the populace has been brainwashed to believe in the Communist threat, and to buy into bomb shelters and enormous stockpiles of weapons. Though you haven't had much history education in NAS since the Cultural Revolution of the Public Schools you must know about the Russian satellites of the 1950s—the two 'Sputniks'—and nuclear testing in Russia as in our American Southwest. It's a time of nuclear fetishizing. The Americans of 1959 don't have access to the 'future'—as we do—to know that there never was a nuclear holocaust and no bomb shelters were ever used by anyone. Nor was there a 'Communist takeover' of the U.S. government—or anything remotely approaching it."

"But—that's good news, isn't it? I think it must be."

Wolfman laughed. "You're a sensible girl, 'Mary Ellen.' Of course, you're right. Our parents would not have been born, consequently we wouldn't have been born, in the 'future,' if there'd been a 'holocaust' in this past. So yes, you are absolutely correct."

Strange that what Wolfman meant as "future" was in fact "past" to me—the past from which I'd been expelled. And for Wolfman, who'd been Exiled much longer than I had, this "future" was yet more past.

Seeing my look of confusion Wolfman spoke of post-war American politics: the so-called Cold War, the sinister power of

Senator Joseph McCarthy, the army-navy congressional hearings, the "witch-hunting" of Communists by such "patriotic" federal justices as Harold Medina in the late 1940s and the defeat of the intellectual Adlai Stevenson by the popular ex-general Dwight D. Eisenhower. "The history of the United States was always a struggle between 'them' and 'us'—capitalists, and their wealth— and the rest of us. Not surprising, 'us' never had a chance."

Wolfman laughed, and shrugged. In this melancholy underground place, what did anything so abstract as *history* matter? You were aware of breathing, and of the need for oxygen; if something went wrong with the ventilators, you would soon cease to exist.

In such a situation, it was natural for me (I see that now) to gaze upon Wolfman with eyes that welled with both tears and adoration.

Here is my friend! My only friend.

This, like any responsible adult man confronted with the infatuation of a seventeen-year-old, Wolfman hoped to ignore.

"Just be calm, 'Mary Ellen.' This is ordinary life now. *Don't* become emotional—there is too much for us to lose."

With the air of an earnest TV advertiser Wolfman took me on a brief tour of the bomb shelter. He opened storage closet doors to show me the goods inside: floor-to-ceiling shelves with cans of tuna fish and salmon, peas, corn, spaghetti, and "fruit cocktail"; Campbell's soup—tomato, chicken noodle, cream of mushroom; boxes of Cheerios, Wheaties, and Rice Krispies; powdered eggs, powdered milk, boxes of sugar, and salt. A gigantic refrigerator (not turned on, and lacking a light) stocked with Pepsi and mineral water. In another storage room, more gallons of mineral water, gallon-sized containers of liquid soap, bleach. In another storage room oxygen tanks, bandages, stretchers, bedpans, canes, crutches, walkers, several collapsible wheelchairs. Rows of lockers. Lavatories, bathrooms. An unnerving sight of dozens of gray uniforms hanging together on a heavy bar like the husks of great insects. In the event of the nuclear holocaust, survivors were obliged to wear this gear? Including the thick-goggled masks? The prospect was so depressing, I could not bear to think of it.

Wolfman said, "They are like our ancestors, aren't they? They seem so innocent. But fortunately, nothing turned out as they'd feared."

The bomb shelter did not seem to be dust-free. You had to wonder what sort of long-dormant germs, bacteria, dwelt here biding their time. I had an impulse to glance behind me, to see if I'd left footprints on the floor. The air was barely circulating through vents and smelled stale, like soiled clothing.

In an assembly room a single console-model dwarf TV sat against a wall, gray-blank-screened, facing rows of seats. I counted the rows: fifteen. Seats in each row: twelve. More signs on the walls, with numbered instructions and cartoon figures. Here was an air of suspended drama as if a vital scene had been interrupted. On the metallic-gray carpeted floor, as if it had just recently been tossed down, was the crumpled cellophane wrapper of a Milky Way candy bar. Adjoining were two dormitory rooms containing as many as fifty beds each—"One dorm for each sex," Wolfman said. A grim sort of merriment shone in his eyes.

I asked if he had a place in the shelter, in case of an emergency.

"Of course not. I'm an assistant professor, I'm of no consequence at Wainscotia. I just happen to know about the 'nuclear bomb shelter,' as I make it my business to know about many things, and so I tracked it down, and learned the combination. In my former, 'subversive' life in NAS"—Wolfman lowered his voice, with a tilt of his head—"I was an 'outlaw-hacker.' I was pretty damned brilliant for my age if I say so myself—I'd hacked the computers at Homeland Security, Youth Disciplinary, Media Dissemination, Patriot Insurers, my own high school in Manhattan, and a few others. When they finally caught up with me I was twenty years old—but I'd been a star hacker for six years, and would've never gotten caught except a 'friend' informed on me."

Wolfman spoke so openly about his case! After his initial stiffness with me, it seemed scarcely believable.

Wolfman asked what I'd done to become Exiled—"It must be a paranoid time in NAS for them to Exile someone as young as you. Usually it's just incarceration at a youth facility and 'rehab'— 'co-opting.'"

Apologetically I told Wolfman that I hadn't done anything nearly so "subversive" as he'd done—"At least not intentionally. I was valedictorian of my high school class and I'd written a speech—a series of questions—that our principal was afraid would get him into trouble with Homeland Security, I guess. No one gave me any warning or asked me to modify the speech— they just arrested me at commencement rehearsal, and took me away." My voice quavered. "I never saw my parents again."

Wolfman regarded me with concerned eyes. Though he meant to keep the mood between us light, matter-of-fact.

"What's your sentence?"

"Four years."

"Four years! That's nothing. Just enough to get a useless college education, and be *teletransported* back."

I didn't want to register *useless college degree*.

"How long is your sentence, Dr. Wolfman?"

"Call me 'Ira,' please. As we're kindred in Exile, so we're on a first-name basis. My sentence is eleven years of which I have just two to go before my case can be adjudicated—God knows what will come of that. I've heard of adjudicators who double the sentence, or worse. It's totally up to a five-man panel, you know."

"It is? I didn't know . . ."

I was feeling sick to hear this. But Wolfman tried to soften the blow by smiling at me.

"All decrees of Homeland Security are provisional. But they can be vacated, too—if your people can pay the requisite 'fine.' Didn't you know that?"

"N-No . . ."

I wondered: Did my parents know? Did my father know, that his MI status might have been revoked?

"But you'll do well, 'Mary Ellen'—you have plenty of time to prepare your case. As for me, I'm not eager to return home. I had enemies there. I was betrayed by people I'd thought were my colleagues. I've tried to make my peace with Zone Nine."

"Don't you miss your family? Your friends?"

"Of course, I did. For a long time I did. Most EIs suffer from severe depression for the first year of Exile, and are at risk for suicide. But I've been an 'orphan' so long, and my parents 'deceased,' I've come to believe my Exile-identity, almost. It's hard for me to think seriously of being 'reconstituted'—I'd be at risk of falling into my previous pattern—'L and S'—("Liberalism and Subversion")—and getting arrested again, and 'vaporized.' No one goes into Exile twice."

Rapidly I'd calculated: if Wolfman had just two more years to his sentence, I would be left behind by him. Already I was feeling a mild panic at this loss.

"I know that you've been taking your transition hard, 'Mary Ellen'—I've felt sorry for you but there wasn't much that I could do. Encouraging you to meet with me in my office, and to cry over your fate, would not be helpful. I've tried to keep our relations professional, and I will continue to do so. Tell yourself: Zone Nine isn't so terrible a place, set beside NAS. It may take you about eighteen months to become adjusted, as it did me. During that time I felt so totally estranged from everyone, it was like they were dead, and didn't realize it; or I was dead, and didn't realize it. I can't shake a feeling of pity for them—though I've come to like some of them. I even admire Axel—the die-hard behaviorist. I earned my B.A. from Wainscotia, for whatever that's worth, and my Ph.D. in experimental psychology; the department was so impressed with me that they hired me immediately as an assistant professor, due to the influence of A. J. Axel. In Zone Nine I have a kind of career I couldn't have in NAS since my personality is naturally 'subversive'—that's to say, skeptical—and 'questioning of authority.' It was astonishing to me that I hadn't been Deleted instead of just Exiled—I think they must have felt that my hacker skills were too valuable to destroy."

"Are there others like us here?"

"Scattered through Zone Nine, yes. Since I've come here I've encountered individuals, mostly male, who'd seemed to me fellow EIs—obviously. My first year, I looked for them everywhere. But I was wary of approaching anyone, or identifying myself. And the others were terrified, and avoided me. You're different of course—distinctive! You have the courage of youth, as few of us do."

Courage of youth. That did not seem flattering.

"Is there anyone you know like us—Exiles?"

"Well, possibly. I have suspicions. But I've kept my distance, as I've said. For of course you can't tell if someone is a spy, even in Zone Nine. I assume there are spies here, as well as agents of our Government. How they communicate with one another, how they travel from one zone to another—if they do—I don't know. Recall, cyberspace is 'eternal'—'timeless'—if you know how you can traverse it in any direction. My parents were scientists who'd been drafted to work for the Government—so I know some things about NAS cybertechnology—though my information has got to be dated. I do know that Wainscotia State University is the default university for people like me, and you, who have intellectual pretensions and have been 'subversive.' Wainscotia is the teeming petri dish of mediocrity."

Wolfman went on to speak with contempt of Wainscotia's "preeminent" men: Amos Stein and his team of physicists and mathematicians who were working on a "proof" of the steady-state universe, in order to rebuff those astrophysicists who believed that the universe was infinite, and infinitely expanding; Myron Coughland, an intellectual chauvinist who argued that the history of philosophy from the pre-Socratics to the present day culminated in the trivial babblings of American "positive-thinkers"; Morris Harrick, with his comically chauvinistic history of the "progress" of science, culminating in the present "Christian, Caucasian" era; another historian, C. G. Emmet, who believed also in the "progress" of humankind, culminating in twentieth-century northern European and American civilization, without any acknowledgment—at all—of the Holocaust—"As if it had never been." Wolfman spoke disgustedly.

And A. J. Axel himself, Wolfman's mentor, so steeped in Skinnerian behaviorism he'd stopped thinking as an experimental scientist a decade ago, and had no clue that a "cognitive psychology" revolution was close at hand—"Within a few years, B. F. Skinner will be finished. His 'great achievement' will be history, a fossil. I hope to leap clear of the wreckage if I can."

Wolfman spoke recklessly, defiantly. I was astounded by his words. For months I'd believed all that Miss Steadman, Miss Hurly, and others in the University community reiterated with

such enthusiasm, that Wainscotia was a center of excellence—
that we were all so very fortunate to be here. Now Wolfman was
laughing.

"Why are you looking so surprised? Did you think that Professor Axel *was* a genius? Every place and every time has its resident
'geniuses.' A naïve fool who'd fallen for Walter Freeman's gospel
of the lobotomy—at least Axel is extricating himself from that
debacle after he'd witnessed a few deaths. His zeal now is 'social
engineering'—shocking men and boys who are attracted to other
men and boys until they're reduced to quivering masses of nerves,
unable to be attracted to anyone, or anything, and likely to commit suicide. Which Axel won't include in his data since it falls
outside the perimeter of his experimentation."

Wolfman laughed, seeing the expression on my face.

"But—but isn't Wainscotia—"

"No. Wainscotia *is not*. To punish 'free thinkers' for subversion they sentence us to 'the Good Place'—Wainscotia. One of
those idyllic American campuses in the Heartland where no research or creative work comes to anything. No matter how much
effort is poured into it, how much 'talent' and 'perseverance.'
Perfectly intelligent scientists originally from decent East Coast
universities here take disastrous turns, wind up in dead-ends—
and won't realize it until they're embalmed and can't leave. No
one is 'original' here—no one is 'significant.' A promising young
astrophysicist from Cal Tech gave up his Ph.D. project in 'string
theory' to pursue 'extra-terrestrial life'—that's it for him, until he
retires. Scientists, mathematicians, scholars, artists, writers and
poets—even chemists—nothing they discover in Wainscotia will
outlive them. Nothing they accomplish will have the slightest
value to anyone. Their heirs will hide away their self-published
autobiographies and melt down their gilt 'lifetime achievement

awards.' Their ideas are derivative, or redundant, or just plain mistaken, silly. In the meantime they live exalted lives at Wainscotia, as inside a bell jar, like pampered bacteria. They win awards and government grants administered by their friends. They're featured on the front page of the student newspaper and the local newspaper. They may even make it into *Time*, once. They're invited to give Sunday sermons. Some are *worshipped* by their post-docs, as by the local ladies."

This was stunning. Shocking. I listened in silence to Wolfman's words like buoyant flames. It was clear that Wolfman meant to be funny, and yet—Wolfman was angry too, and sad.

It was true, I'd thought that Professor Axel's behaviorist psychology was limited in its scope and technique, but I'd supposed that this was a limitation of my own; what I'd learned of Freudian psychology hadn't seemed more convincing, and had the added disadvantage of being undemonstrable in a laboratory.

But—poor Morris Harrick of the Museum! I felt particularly sorry for this elderly gentleman who seemed, like a species of underground mole, to have spent his entire professional life in a burrow, to no purpose.

Zone 9 would be my world for the next three and a half years. Its air of collective mediocrity would be the air I had to breathe, to survive.

I felt as if the ground were tilting beneath my feet.

Wolfman's cruel laughter turned into a fit of coughing. His skin looked clammy. I wondered if he was well.

He sat heavily in one of the chairs facing the blank-screened TV. The merriment had faded from his eyes. He regarded me as one might regard a child bright for her age yet handicapped in some way.

I would wonder why Wolfman had come to me at this time.

Near the end of the first semester—when soon, within weeks, I would no longer be his student.

I would wonder if he'd followed me to the museum. If he'd made inquiries about "Mary Ellen Enright" of the Class of 1963, College of Liberal Arts.

I loved Wolfman so much, I needed to believe that Wolfman might love me.

After his initial ebullience in the bomb shelter Wolfman was quieter now, and looking rueful. I saw his fingers fumble for a pack of cigarettes in his shirt pocket—but fortunately he thought better of lighting up a cigarette in this airless space.

I thought how strange it was, Ira Wolfman *smoked*. As if he'd been born in Zone 9, and not *teletransported* to it.

The shelter filled me with a sense of dread. I could not believe that we were so safe in it as Wolfman seemed to believe. For why wouldn't Surveillance have followed us here?—I could not think other than that our Government of NAS-23 was capable of penetrating any barrier.

The terrible thought came to me—*Can you trust him? Wolfman?*

Wolfman said quietly, "Don't be afraid, 'Mary Ellen.' You can trust me, I am your friend in Zone Nine."

I told Wolfman yes, I trusted him.

And I love you.

But you know that.

Wolfman asked me my name, and I told him.

Wolfman asked me where I was from, and I told him.

Wolfman asked me to tell him what was in my heart.

Wolfman opened his arms to me, and I came to him.

And all that passed between us that night went unrecorded in the shelter beneath the Van Buren Museum of Natural Science.

THE SACRIFICE

It was a famous behaviorist experiment of 1920 conducted by John Watson.

The eleven-month infant, Little Albert, had not been frightened of any animals until a gentle white rat was placed in his lap and a sudden loud noise of two steel bars struck together behind his head, several times in succession. Soon then, Little Albert began to cry at the very sight of the rat, as of a dog, or even a fur coat, and to exhibit symptoms of terror preceding the clanging of the steel bars.

We were shown a film in the lecture hall. Old, grainy, jumpy but unmistakable—the infant was convulsed with terror as the steel bars were struck behind his head, and soon he learned to hate and fear the gentle white rat he'd previously seemed to like.

One day I would ask Wolfman—Why hadn't the experimenter de-conditioned the infant, after the experiments? Hadn't anyone thought of this?

Wolfman said he didn't think so. Didn't think that Watson or anyone else had thought of "de-conditioning" at the time.

I asked Wolfman if Little Albert had grown up to be frightened of animals and fur coats and Wolfman said, No. The poor kid hadn't grown up at all, he'd died at age six.

ADORATION

We don't have to see each other to be near each other.
Remember, I am your friend.

Waking in the dark, in the Wainscotia winter. A half-mile away
the chapel bell thinly tolled the hour of 6:00 A.M.

My final morning of exams! It was January 1960.

Hurriedly I dressed in the dark as my roommates slept. It had
been my exam-week schedule to wake early and work downstairs
in the study room, then run through the snow to the dining hall
for breakfast, and then to my first exam which might be sched-
uled as early as 8:00 A.M. Today was my Psychology 101 exam, at
9:00 A.M. in Greene Hall.

I was nervous and excited. So badly I wanted to *excel*.

I wanted Ira Wolfman to be impressed by me. I wanted him to
be proud of me, even if in secret.

Many times I'd studied the semester's material: notes assidu-
ously taken at Professor Axel's lectures, and the textbook from
which the lecture topics were taken, which A. J. Axel had helped
edit. In my agitated sleep, I skimmed columns of (unreadable)

print. I was underlining, taking notes. I woke with a headache, eager to be examined. I wondered if the microchip in my brain would affect my memory generally, or only just "censored" memories.

Wolfman had told me that there was no microchip inserted in my brain! He was certain.

They want EIs to think this. The effect is that we "censor" ourselves, through suggestion.

(I had no idea what to believe, now. Though I wanted to believe that Ira Wolfman was correct.)

My grades, as I went into the final exam for Psych 101, were A's and A+'s. I had virtually memorized my lecture and textbook notes. Yet I feared a sudden reversal of fortune, that would disappoint and disillusion Ira Wolfman and he would no longer want to be my friend.

Wolfman had informed his students that most of the questions on the final exam would be multiple choice, to be graded mechanically. There would be a few brief written answers, and a single essay of approximately 750 words. Wolfman had told us dryly that no new, original thoughts were expected on the exam, or welcomed:—"Each question has an obvious answer, and that's the answer that's correct. The other answers will be marked 'wrong.'"

Dr. Wolfman's attitude toward the course he was helping to teach had become ever more ambiguous during the semester. Often he seemed to be speaking ironically, as if reciting words in which he didn't believe. He seemed to be losing faith in behaviorism, which was the foundation of the Wainscotia psychology department and of the institute which A. J. Axel would be heading, to "cure" aberrant/perverse behavior. I wondered—did his colleagues notice? Did the other students notice? Or was I the only one?

Or was I imagining, in my obsession with Ira Wolfman?

From our State-monitored public high school I knew how to answer multiple-choice questions. At least 80 percent of our education was tested in this way; our teachers had taught us to take exams, essentially. Originality, subtlety, and skepticism were not valued.

The more you knew of the material, the more complexity you saw in it, and so the more difficult it often was to provide a simple, crude answer. Yet, if you were a canny exam-taker, you understood that only one answer could be "correct"—this was the answer you'd been drilled in over the semester. It was understood that Professor Axel's assistants drew up exams that followed his lectures and his textbook scrupulously; often, questions were really just restatements from either, which might be memorized. Only in essays could you hope to be original—but in essays you could also sabotage your own chances.

Wolfman would be grading my exam, since I was in his quiz section. He'd coolly informed me that he graded all student work "blind"—he wouldn't know whose work he was grading until he'd completed all the exams. And he never changed an exam grade—"I get it right the first time."

Since the museum I'd seen Ira Wolfman only a few, fleeting times outside class. He would not telephone me—of course. He'd warned me not to try to call him; or to write to him, or leave notes for him.

He'd said *When the semester is over. When you're not my student any longer.*

That night in the museum we'd remained together until a quarter to twelve, which was my curfew at Acrady.

Curfew! Only undergraduate girls at Wainscotia had curfews, not boys. It was an unexamined principle of male privilege, that no one seemed to have noticed; the precise hours of curfew, which varied from 11:00 P.M. weekdays to midnight on Fridays, 1:00 A.M. on Saturdays, and 10:00 P.M. on Sundays, had been devised by the dean of women.

That night, Wolfman had held me, and comforted me, and encouraged me to talk—to tell him everything, anything.

So long had I had no one to whom I could speak! Words had spilled from me, like tears. And tears spilled from me as well.

I had not been close to any boy, still less any man, in my life. The only man who'd ever hugged me was my dad.

There were many girls at Pennsboro High who were like me, in our discomfort with boys. My mother had said that this hadn't always been the case, when she'd been in high school she'd had friends who were boys, and boy-friends whom she'd "dated"; but times had been different then, teenagers hadn't been encouraged to spy upon each other, and inform upon each other, quite so much as they had been in the past twenty years.

And the boys at our school hadn't been sympathetic, mostly— like my brother Roderick they'd been opportunistic, untrustworthy, mean-spirited, mocking. Across a narrow abyss we'd seemed to be regarding each other—female, male. There were no "friendships" but rather "sex-contacts" that were crude and curt and likely to be ridiculed online in ugly words or photos posted by boys.

Ira Wolfman was the first man I'd *loved*. The first man with whom I was *in love*. It would not discourage me that Wolfman didn't reciprocate my feeling for him—so grateful was I that Wolfman simply *existed*.

That night he hadn't kissed me except on the forehead, and lightly on the cheek, as you might kiss a fretting child. He'd laughed saying he was *way too old* for me. He wasn't *the kind of guy who takes advantage*.

I'd wanted to beg Wolfman—*Please! Take advantage.*

In The Instructions it was clearly stated: an EI is forbidden to *procreate.* But I had no thought of such a development—*pregnancy.* No more than any other Wainscotia "coed" did I imagine that such a quandary could happen to me; and I could take solace in the fact that Wolfman was indeed older than I was, a responsible adult.

I did believe that Wolfman was my friend. I did believe— (maybe)—that Wolfman might love me, in time.

I wasn't so desperate now. I wasn't so lonely now. If Wolfman was in my life, I would not ever be lonely again. I thought.

YET STILL I COULDN'T keep from seeking out Wolfman, in public places. No one could ever suspect us, in public places!

In the week following the museum, I attended a lecture given by a visiting professor of psychology from Purdue, which Ira Wolfman also attended, and at which he'd asked questions. Not a very interesting lecture—(on a behavioral topic involving a complicated schedule of "reinforcement" in primates)—but Wolfman's questions were lively and provocative; and I thought *But you must not call attention to yourself!*—for older professors were present, one of them white-haired A. J. Axel, and they might not approve of the young psychologist's manner.

In the lecture hall I'd watched Wolfman covertly. And Wolfman was aware of my presence, I thought. But at the end of the lecture Wolfman remained at the front of the room, talking with colleagues, and I left without speaking to him.

Just being in his presence suffused me with a sense of wellbeing, happiness. I thought—*That you exist in this world, with me. That is enough for me, for now.*

In psychology it is known: a mentally ill person can have "insight" into her illness, yet the illness remains. As a physically ill person can understand the circumstances of her illness, yet the illness remains.

In love with Wolfman, how pathetic is that? When Wolfman does not love you.

THE SEARCHERS

It was a mistake. Would be a mistake. Maybe.

Attending a Friday evening Film Society screening of a "classic" western starring John Wayne called *Red River*. In the bomb shelter Wolfman had happened to mention that while he disliked the TV of Zone 9 he generally liked the movies; and so I'd gone to the Film Society showing with the hope of seeing him.

I'd arrived late. Not knowing where the Film Society was, going to another (darkened, locked) classroom building. Steep steps in a hill. And now making my slow way into the darkened room in — at last! — the right building, in a first-floor alcove. Where chairs were set up. At first, I didn't see Wolfman in the audience and hesitated, thinking I wouldn't remain — then, I saw him sitting alone, in a side-aisle seat near the front of the room.

(Had he seen me? I wasn't sure.)

(I did not sit near Wolfman. I felt that this restraint would commend me to him.)

Since I'd begun studying what was called "twentieth-century psychology" — since I'd researched the history of behaviorism, predating B. F. Skinner — it had begun to seem to me that most human situations were analogous to psychological experiments. The usual experimental subject was a pigeon or a rat but sometimes human subjects were used. You saw, or in some way experienced, a "stimulus"—the way you reacted was the "response." It was the case that, the more detailed and "objective" the description of the subject's behavior, the less the experimenter was likely to know what was happening; for one could not infer an inner life, a subjective mode of being, from mere observation. Inevitably, living things were perceived (from the outside) as resembling clockwork mechanism. You wanted to protest—*But I am me! I am unique and ungraspable.*

But now I was here at the Film Society, where my (unreciprocated, futile, sublime) love for Ira Wolfman had drawn me. Was that not predictable? Had not (probably) Ira Wolfman predicted it? In a new and unexpected variant of a "Skinner box" everywhere I went, in Zone 9, I brought this (invisible) box with me, for I was at its epicenter.

Before Skinner, but not unlike Skinner, there were leading scientists who claimed that animals were essentially machines whose behavior could be explained in simple terms, and manipulated by conditioning; yet, there were scientists, if fewer in number, who argued for a kind of *vitalism*—a "non-material" essence that suffused living things. (These were likely to be scientists discredited by their colleagues, as in the case of a German named Hans Driesch.) In my life, in the obsessive nature of my thoughts, and in the circumstances of my Exile state, I saw myself as an experimental subject of some kind, for (of course) I was being observed, and "recorded"; but at the same time, in the emotion that Ira Wolfman aroused in me, and in my yearning for him, I saw myself as unique, secret, unpredictable.

My yearning for Wolfman was leading me to places I would not otherwise have gone. As if, slowly, another being was evolving

who was both *Mary Ellen Enright* and *Adriane Strohl*. For the behaviorists also believed that the self is created out of the environment, and out of accidents in the environment, rather than out of
the rigidity of genetic determinism. *We are what we are made to
be—we must only not resist.*

Often I caught sight of myself in reflective surfaces, and was
struck—stricken—by the person I'd become, in Zone 9. For an
eighteen-year-old girl—(my eighteenth birthday had recently
come and gone unremarked: I could not bring myself to tell Wolfman), I was *not young*. My skin was ashy, my eyes were stark and
staring, my manner was vigilant, hyper-alert. I'd become one of
those lab rats that has been frustrated or frightened or shocked (by
electricity) so many times, it has lost its essential, original *ratness*
and is something else now, almost a new species: a creature waiting to be defined by the next, possibly lethal stimulus.

Yet, my love for Wolfman couldn't have been inferred or deduced from my appearance. (I was sure!) This was my secret happiness.

How unexpected, the movie!—the "western."

At first I couldn't determine if *The Searchers* was intended as
a sort of comedy, in its exaggerations, or meant to be serious—
"heroic." It was riveting and even enthralling—as a cartoon would
be to a credulous child. The Technicolor was luridly bright, the
actors awkward in their dialogue, the music accompanying every scene, from the credits onward, distracting as a clattering
of drums. John Wayne was not an actor I'd ever seen before—
conspicuously he seemed to be playing "John Wayne." The camera was fixed upon him, often in close-up, at the center of every
scene, and scenes were both melodramatic and slow-moving; you
knew that something was about to happen by the "suspenseful"
music, but it did not happen quickly. No one on-screen looked

anything like an actual person in an actual situation—it was obvious that these were professional actors in costumes reciting lines of dialogue they'd more or less memorized.

Indians were the enemy here. Menacing near-naked savages who behaved cruelly in scenes of deafening violence followed by scenes in which Indians were shot off galloping horses, to fall heavily into sagebrush. Scenes in which Indians were treated "comically" were almost worse. And—so much killing of buffalo, by the John Wayne "hero"!

Yet, at the end of the film, many in the theater applauded. Even Wolfman!

After the lights came up viewers stood about discussing the film, which they seemed to take seriously. All, except me, were adults. Most appeared to be faculty members. There was much praise for John Wayne's "performance"; there was praise for the "direction" and the "western landscape." There was a pretentious sort of talk of the "myth of the American frontier." Almost, I'd have thought that these people were joking, as Wolfman often joked, but evidently not.

Myth. American frontier. Not a frontier to those who'd been living there.

Patiently I waited for Wolfman to turn to me, or at least to glance in my direction. Patiently I waited for the people with whom he was speaking to drift away and go home. I was feeling a kind of righteous indignation for I had not liked the bombastic film and I didn't think that anyone else should like it, either.

Wolfman was friendly with these people, but they did not seem to be psychology colleagues. There was a couple, who appeared to be married; and there were two women who'd come alone to the film, but had sat together, near Wolfman. I felt dismay, that one of the women, whom Wolfman seemed to know well, was lingering in Wolfman's presence, clearly waiting for him to leave, so that she could walk out of the student union with him. There was some fuss about putting on jackets, and fur-lined hats. The woman's hair was dark, and parted in the center of her head, where her hair had begun to turn silver. Her eyes were large, heavy-lidded, staring—like my own. She was not a beautiful woman and she may have been older than Ira Wolfman but her face was sharp-boned and striking and you could see that she was edgy, and intelligent. I could taste jealousy like hot acid in my mouth. For a dizzying moment I thought—*She is myself only older. They are Exiles together.*

But this was not likely. This could not be.

Their laughter was grating, like knives and forks clattering together. I did not realize that what I felt was sexual jealousy, which strikes like a virulent illness those to whom it is previously unknown.

They were even smoking cigarettes together! I hated them.

Still I waited unobtrusively, I thought, near the back of the hall, studying posters advertising upcoming Film Society movies including several "classics of the silent era."

Nothing is so depressing as films-to-come that, if you see them, you are destined to see alone.

Seeing me alone, a lone undergraduate at the Film Society, several people spoke to me. Except that I was very quiet, and avoided looking any stranger in the eye, for fear of seeing a glit-ter of recognition, I might have been "befriended" at the Film Society—this would have attracted Wolfman's attention!

Trying to overhear what Wolfman was saying. I did not like the high-pitched laughter of the woman with the center-parted hair. (I'd seen her glance in my direction. But I had not glanced at her.) At last, the little group broke up. The woman had no choice but to depart with her friends.

It would have seemed to the casual eye that Wolfman had

been ignoring me until now, not rudely, but incidentally, as if it had to be chance merely that a faculty member and one of his undergraduate students were in the same room together; now, he acknowledged me with a frowning sort of smile.

"Hello, Miss Enright!"

My eyes were heavy-lidded, as if I'd been crying. But I had not been crying. Until seeing Wolfman with these strangers, I'd been feeling alert, excited, hopeful.

Wolfman understood, seeing my stricken expression. But Wolfman would not indulge me in any public place.

He meant to be nice. He meant to be kind. His questions to me were the questions a university instructor might make to one of his students, whose name he probably didn't recall. Politely he asked me how I'd liked the movie?—and was surprised when I told him what I thought, for I hadn't spoken at all sharply in the shelter, beneath the museum. I'd been meek and melting then but now there was a sort of adolescent vehemence in my critique of the film, and an impatience that others should pretend to admire the ridiculous "western."

Wolfman laughed—"Whoa! I guess you aren't a fan of John Wayne."

His manner, both startled and amused, reminded me of my father.

I said, I thought the movie was insulting. It was so *simple*, and so *crude*. If I'd been a Native American, I would be furious. If I'd been a woman—

But of course, I was a woman.

Wolfman regarded me with a quizzical sort of admiration. You could see he was a man who liked being surprised. In a lowered voice, though no one was likely to overhear us, he said, "You'll get used to intellectual 'insult,' my dear. If you see enough TV and movies in the Happy Place."

Not wanting to leave my ranting subject I was saying how the women's faces were so *artificial*. And the way the Indians were shot off their horses, screaming as they fell. And the awful "musical score" that ruined every scene . . .

"No one behaved in the slightest way convincing. No one even *looked* convincing."

Wolfman said, "Movies aren't about the way people look, 'Mary Ellen.' Movies are about our perceptions of them."

What this meant, I didn't understand. But I thought that I would ponder it, instead of asking Wolfman to explain.

"You could do one of your behavioral experiments, following a western. Ask people questions with multiple answers, and you'd see that their bias against Indians has increased, and their general antagonism. Seeing so many people shot off horses, the viewers would want to shoot, too."

Like Wolfman's brightest student I spoke.

Wolfman laughed as if my idea was too fanciful: How could you establish that "bias" had increased? For all you knew, bias might have decreased. And how could you establish what had caused an effect, the movie alone, or other factors? Yet, Wolfman conceded that such an experiment could be revealing—except attitudes are not "behavior" and can't be measured.

Excitedly I said, "You could administer two experiments, one a few weeks before the movie, and one immediately after the movie. And why isn't a thought 'behavior'? It takes place in the brain— you could probably see it. Some sort of X-ray."

My head was feeling blurry. I was blundering into the "censored" area, I think—remembering something that, in January 1960, did not yet exist in experimental psychological research, and could not be articulated.

"Thoughts can't be X-rayed, Mary Ellen. Not yet."

"Well—*not yet*."

We were speaking quickly, quietly. I would almost have thought that I'd clutched at Wolfman's hand, to keep him from running after his friends, and away from me.

For I was remembering—my brain had been scanned. A sequence of images had been made of my brain, to determine if I was lying or not.

(Hadn't this happened? In the Youth Disciplinary Division, Homeland Security?)

Wolfman was frowning, working his mouth in a way to signal me—*No more! Stop.*

An alarmed look had come into his face. Wolfman's brashness was calculated, and mine was—only just brash.

Quickly Wolfman walked away, without waiting for me to come with him.

I was left alone in the little movie theater, deserted except for the Film Society officers who were shutting things down. It seemed to me that quizzically—pointedly—they were observing me, but I did not meet their gaze, and quickly left.

A few minutes later I saw, outside, on the snowy walk in front of the building, the woman with the stark staring eyes, still with her friends, and now Ira Wolfman had caught up with them. In a loose group they walked off toward Moore Street for a drink in one of the pubs.

I did not follow. I wouldn't have given Wolfman the satisfaction.

THE TEST

Hurriedly dressing in the dark. I did not want to disturb my room-mates. My thoughts were preoccupied with Wolfman, which was not a good thing—my thoughts should have been preoccupied with the upcoming psychology exam.

It was difficult to eat, so early. It was difficult to force down food. But if I did not eat breakfast, by 9:30 A.M. I would be famished. My eyesight would grow blurred. My mental capacity would be affected.

I brought my notebook with me. I would study in the dining hall, as I ate breakfast.

In the brightly lit dining hall there were very few students at 7:05 A.M. These were dark-skinned "foreign" students who sat to-gether at a table near one of the windows, usually amid a sea of Caucasians except, at this early hour, the dining hall was near-deserted.

ST5 or ST6. A rare sight at Wainscotia.

The foreign students were not undergraduates but graduate stu-dents in such subjects as physics, chemistry, engineering. They were exclusively male. Their eyes, lighting onto me, registered something tentative and marginal in me, that was akin to their own sense of estrangement. Once or twice, the young men had beckoned to me to come sit with them; but I'd pretended not to have seen.

In their own countries, it was said, male and female did not so easily intermingle. Not before marriage. In beckoning to me the young "foreign" men excited themselves, testing the limits of their taboo. I felt dismay and dread, to be so considered, in their eyes, as an *object* of their (male) perception.

Especially this morning as I passed along the cafeteria line, I felt their eyes upon me. They did not seem so friendly. I fumbled placing on my tray a small glass of orange juice, a ten-cent waxed container of milk, a five-ounce box of cereal, two pieces of white-bread toast . . .

The smell of breakfast-food made my mouth water. I never realized how hungry I was until I smelled food.

Adriane! Is it—Adriane!

Adriane come sit with us.

(I didn't hear this.)

(Did I hear this?)

(No. I did not hear this.)

Pushing the tray along, and a roaring in my ears. The cafeteria worker who punched my meal ticket, a heavyset black woman with kindly creases in her face, asked if I was all right?—or maybe did I want to sit down, for a minute?

I must have seemed confused. I didn't seem to know how to answer her.

The woman took the tray from me, so that I wouldn't drop it. She set it on the nearest table, which was empty.

This all right, honey? You be fine here.

The roaring in my ears was near-deafening. I could not bear to look at the table of "foreign" students who were looking at me, and who seemed to recognize me.

They, too, were Exiles in Wainscotia. That had to be the explanation.

I drank a small mouthful of the orange juice, which tasted as if laced with turpentine. I could not manage to open the waxed milk-container. I ate some of the cereal, dry out of the box. The pieces of toast I cleverly wrapped in paper napkins to take with me, for the morning's ordeal.

I'd been in the dining hall less than ten minutes. When I left a half-dozen students were entering, stamping snow off their boots. These, too, were dark-skinned—"foreign." They were exclusively male. I saw their malicious eyes, savage-eyes, like the eyes of the doomed Indians in the western, whose fate it was to be shot off their horses. They smiled at me, and said something to me, words I couldn't quite hear, which I chose to believe were friendly.

In their wake I heard *Adriane? Adriane?*

Slightly mocking, ululating—*Ad-riane?*

So desperately I fled, I slipped on the icy pavement, and fell hard, the breath knocked out of me; but in the same instant I managed to scramble to my feet, and ran back to Acrady Cottage.

IF I'D HAD A TELEPHONE NUMBER for Ira Wolfman I would have called him. I'd have cried into the receiver *There are other Exiles here! They know me.*

But then, a few minutes later, in a less hysterical state I thought that perhaps this wasn't the case: the dark-skinned young men were not actual students but virtual images, operated by an agent in NAS-23 as military and domestic drones were operated, at great distance, to confuse and terrify me; but primarily to tempt me into acknowledging them, and identifying myself to them, in violation of The Instructions.

That is, it had been a test. The first such test (so far as I knew) since I'd arrived in Zone 9.

If I'd spoken with them, if I'd acknowledged myself as *Adriane Strohl*, this would have constituted grounds for "vaporization"—on the spot.

THE EXAM

Discuss, in 750–1,000 words, the principles, techniques, and significance of Behaviorism in 20th-century psychology, and its possible and probable social applications.

The exam was several pages long, stapled together.

More than two hundred somber-faced Psych 101 students, in rows of desks in the university gym.

Proctors prowled the large echoing space, to discourage cheating.

(Yet, cheating in such large lecture courses was known to be "rampant" at Wainscotia. Almost, there was a tradition of such cheating at Wainscotia, the more brazen the more celebrated, especially by fraternity men, alongside a tradition of good citizenship, honor, integrity, and school pride.)

(At Pennsboro High no one cheated. Surveillance cameras and monitors were everywhere. And no one wanted to get a really high grade, in any case.)

Quickly I skimmed the exam. My heartbeat was pleasantly quickened.

I thought—*Around me, the Skinner box materializes.*

My initial impression of the exam was that there was nothing here that I didn't know intimately. For I had studied, studied, and studied like a rat in a maze who keeps running in hope of a reward even after the reward has been removed from the maze.

I considered the essay question which was both disappointing and provocative. I thought—*I can write something original, and "analytical." Wolfman would wish this.*

If I answered the questions with the obvious answers, like a creature in a Skinnerian experiment, I would receive a grade of perhaps 100 percent—no one taking the exam on this day was likely to receive a higher grade. But if I experimented with the essay question, for instance, a tracing of Darwinian evolutionary thought with regard to B. F. Skinner, and something of the historical background of behaviorism, before Skinner, I could say something unexpected, and interesting.

Wolfman will recognize my writing. Wolfman will be impressed.

He'd cautioned me not to write more than I needed to write. Not to range beyond the narrow scope of the course. But I was sure he'd only meant to protect me.

I took approximately twenty minutes to answer the fifty multiple-choice questions. Following A. J. Axel's eccentric manner of testing, the multiple-choice questions were not conventional. In each, you were asked to choose the single "correct" statement—or the single "incorrect" statement—of four statements. In this way, the tester was testing not only the student's knowledge of the subject but also the student's mental agility. You could know the "correct" answer but not feel confident about choosing it because of the other, competing statements. Perhaps this was unfair?—it was distinctly Skinnerian.

Around me, my classmates were shifting in their desks, sighing, running their fingers through their hair. These were laboratory rats being run through a maze!

Yet, there was nothing in the exam that had not been set forth clearly, and repeatedly, in Professor Axel's lectures. If the students had been receptive, passive vessels for information, they'd been programmed to give the correct answers now. We were the pigeons and rats of Skinner's experiments, behaving in ways to assure rewards and not punishments.

Something in me rebelled against this limitation—*the tyranny of merely correct answers!*

In my logic course I had wanted to ask our professor if it was possible that *x* could be both *x* and non-*x* simultaneously—but I'd known that my question would have been greeted with a look of startled concern, as one might greet the ravings of a lunatic.

In life, it is (invariably) the case that *x* is both *x* and non-*x* simultaneously. But in formal logic, no.

Your sentence is to surrender to Zone 9. You must not resist.

It was Wolfman's voice, in warning.

There was Wolfman in another part of the gym. Restless on his feet. Fortunately he wasn't assigned to my quadrant of the gym.

Wolfman knew, I could write a perfect essay. But Wolfman deserved better, if he was going to read/grade my exam. He would want to think highly of me. If only in secret, he would want to feel *pride*.

Since the incident in the dining hall I'd been feeling excitement, apprehension. Like one who has narrowly escaped with her life, I felt both relief and restlessness.

I would write an "original" essay of 1,001 words! I smiled to think this.

I would write an essay that would begin with several questions: *What is "free will"? What is "The Law of Effect"? What is* entelechy? *What is "mind"? What is "the problem of imitation"?—Is there a "natural selection" of thought?* and then I would answer the questions, or rather, I would weave together provisional answers in a complex statement. I would refer to the history of behaviorism only in passing. I would not waste time stating obvious facts. I would be ambiguous, purposefully unclear. I would be highly critical of B. F. Skinner whom I would compare with the German researcher Driesch whose concept of animal *vitalism* allowed for something like a "soul" in both animals and human beings. I would write also of John Watson, Skinner's precursor, who'd made important discoveries in childhood conditioning, and whose boast of having the power to "create" any sort of human being out of rough, raw materials was thrilling, as Watson's behaviorism seemed to repudiate any sort of fatalistic/genetic determinism. And I would write about Darwin—of course. Charles Darwin was one of Wolfman's scientist-heroes.

Then—midway in the essay—a brilliant idea came to me: *I would write in the voice of a maze-rat!*

I would deal with the riddle of subjectivity from the point of view of the (hapless, powerless) subject. Whether "interior" states are in fact measurable as exterior states. How to measure the seemingly immeasurable. Can a shadow be dismissed as nonexistent, because it has no weight? Yet, a shadow has visual properties. A shadow can be *perceived.* Is there something inside the brain that receives, organizes, and interprets sensory perceptions? Is this something the "self"—is this "self" the "soul"?

I hadn't time to return to the beginning of my essay, to make the rat-voice consistent. I would hope that Wolfman would smile, and be impressed.

So deeply immersed in my maze-rat essay, like a rat in the innermost depths of a maze, I was shocked when proctors announced that there were just five minutes remaining before blue books had to be turned in—already, it was 10:25 A.M. But I wasn't finished!

Heedlessly, thrillingly, like one plunging down a steep hill, and not yet at the foot of the hill, I'd covered a dozen pages in the blue book, I'd written more than 1,001 words, I was sure; I'd intended to go back and edit what I'd written, and I'd meant to insert several important footnotes . . . Though the gym was not overheated, I'd begun sweating inside my clothes. I glanced around for Wolfman but didn't see him—he must have been at the rear of the gym, behind me and out of sight.

Three more minutes!

To my dismay I saw that my "original"—"brilliant"—essay hadn't turned out as I'd hoped. It was too ambitious—too diffuse. It was rather a précis for a fifty-page paper—a critique of behaviorism. In the rat-voice I'd digressed onto a secondary issue, the "social consequences" of behaviorism, and realized too late that I had lost precious time. I had neglected to set down the principles of behaviorism—these were so elementary, and so often repeated! I could not see the point of setting down what was obvious. My head was aching from tension, and from a lack of sleep. My thoughts were becoming jumbled like laundry in a dryer. I could not escape the accusing eyes of the "foreign" students who knew me as one of them, unless the "foreign" students were agents of the State. (Or were they but "virtual" images? I had not approached them to see if, horribly, I might have been able to pass my hand through one of them as the others looked on, snickering.) One lengthy paragraph, on page six of the blue book, a paraphrase of several passages from Skinner, had been x'd out by error, in my hasty editing; I saw now that the paragraph was required for my argument, and could not think how to eradicate the effect of the X. I'd memorized the passage from Skinner, or anyway most of the passage, which I'd read independently of our textbook, in a primary work, Skinner's *Science and Human Behavior* (1953):

> What is meant by the "self" in self-control or self-knowledge?
>
> The self is most commonly used as a hypothetical cause of action. So long as external variables go unnoticed or are ignored, their function is assigned to an originating agent within the organism. If we cannot show what is responsible for a man's behavior, we say that he himself is responsible for it. . . . The practice resolves our anxiety with respect to unexplained phenomena and is perpetuated because it does so.
>
> Whatever the self may be, it is apparently not identical with the physical organism. . . . A self is simply a device for representing a functionally unified system of responses.

In quoting and paraphrasing Skinner, somehow I'd come around to seeming to agree with him—I could not now remember what I'd objected to, initially. I was struck also by Skinner's remarks on the "absence" of self-knowledge—this, too, from the primary work, and not our textbook: *One of the most striking facts about self-knowledge is that it may be lacking.*

There was something terrifying here. And yet utterly clarifying. If I could not know myself, I could not know anything—if the lens is smudged, the vision will be smudged.

My pen fell to the floor. My blue book was filled with smudges, sentences crossed out, misspellings. But now, there was no more time—the last of the blue books were being gathered by proctors, most of the desks in the drafty gym had been vacated, a frowning proctor was approaching me with his hand extended—"Miss? Exam's over."

Like Wolfman the proctor was a quiz section instructor. Obviously, he didn't know "Mary Ellen Enright"—Wolfman had not mentioned to him that he had a brilliant undergraduate this semester . . .

With a polite smile I surrendered the blue book. My skin felt bathed in sweat. I was deeply ashamed. Yet, I was suffused with hope. *Wolfman will be impressed. He will see that I could have written an A+ paper, but tried for something more. He will—forgive me.*

When I stood from my desk my legs were weak. I looked around for Wolfman—he was halfway across the gym, stacking blue books into a pile. He and another proctor were laughing together. The exam-ordeal was nothing to the proctors as observing rats running a maze, or pigeons pecking desperately at a key that refuses to move, or a cat reacting in terror when it is shocked, is nothing to experimenters.

Throughout the two-hour exam Wolfman had ignored me. Or perhaps Wolfman had been oblivious of me. At about the halfway point there'd been a mild commotion in another part of the gym, where a student or students had been caught cheating, or (possibly) had become suddenly ill; while I'd been hunched over my essay I'd heard lowered, intense voices, and I'd heard the sounds of someone being led from the room, but I had not glanced around. I wondered if Wolfman had been the proctor involved.

I took some time to put on my fleece-lined jacket, which was heavy, and tugged at my arms. It was very cold outside: nine degrees Fahrenheit.

Strange how, in Zone 9, there was true winter. In our sector of NAS-23 there had not been "winter"—low temperatures, snow—in decades, though my parents spoke nostalgically of "winter" as of other phenomena now bygone, mere memories.

Then, I left the gym. I was one of the last students to leave the gym. A shining light fell upon me as I stepped outside.

Everywhere were banks of snow. Blinding-white.

Wainscotia Falls, Wisconsin, was a place of snowy hills, tall trees topped with snow. It was a place of beauty, I thought. Yet—it was not *my place.*

I did not want to die here. So far from anyone who knew me or loved me. Or felt responsible for me.

In the fresh frigid air my breath steamed. I realized that I was very tired though it was only mid-morning—the entire day stretched before me, unfathomable. I thought—*If I fail, I will lose my scholarship. I will be vaporized.*

Stumbling and short of breath I ran in the direction of the central campus. I did not glance back to see if Wolfman might be following me nor did I hear Wolfman, or anyone, call after me—

Adriane! Ad-riane!

THE FAILING GRADE

Wolfman said, "I told you."

"Told me—what?"

"Not to be 'original.' Not to be 'analytical.' Just give the answer—the minimum."

Stubbornly I protested, "But the exam questions were all so obvious. I could have written the answers in my sleep."

Wolfman laughed. I felt a hot flush rise into my face.

I said, "I thought Professor Axel might appreciate a response that wasn't like every other response. I thought, if you showed the essay to him . . ."

"Are you serious? Axel hasn't glanced at an undergraduate exam in twenty years. Even an A+, which you didn't get, Axel wouldn't trouble to read. Graduate students don't interest him, let alone undergraduates."

I was stung by Wolfman's casual remark. *Which you didn't get*.

Still stubbornly I said, blinking back tears of disappointment, "I wasn't thinking of a grade, Ira. I didn't write the exam for a grade. I didn't learn the course for a grade. I didn't do all the outside reading I did, for a grade. But I'd thought—"

Wolfman squeezed my icy fingers, which had the effect of stopping my mouth.

"It's over now, 'Mary Ellen.' You didn't fail, except by your own exalted sense of yourself. And I am not your instructor 'Dr. Wolfman' any longer."

WOLFMAN MY LOVE:
SELECTED MEMORIES

"'Adriane.'"

Only when we were alone did Wolfman call me by this name. He would be the only person in Zone 9 to call me by this name.

And when we were in a public place, or a quasi-public place, he called me "Mary Ellen" or, more often, "Miss Enright." It did not matter that no one was near us, as in the university arboretum where we walked sometimes, not holding hands, not touching, but close together as companions.

We hadn't returned to the bomb shelter. We'd begun going to Wolfman's apartment, after dark. Where we rarely stayed long.

As Wolfman said, now we weren't instructor and undergraduate, it was not *wrongful* for us to see each other, as it would have been otherwise.

"Still, you're too young. Or, I'm too old."

Slyly I thought—Yes. *But I can wait, to catch up with you.*

ONCE, AND THEN A SECOND TIME, I'd made my way through the Museum of Natural History, to look for the steep steps descending at the rear of the museum, and the entrance to the bomb shelter. I recalled the combination lock that Wolfman had turned carefully in his fingers, but of course—I could never have recalled the combination itself, I hadn't seen so closely.

It was strange, I hadn't been able to find the entrance to the bomb shelter.

Each time, I'd lost my courage and returned to the lighted area at the front of the museum. I'd heard Miss Hurly call my name, or a sound of voices in the library, visitors expecting to find someone on duty at the checkout desk, and hurried breathless back to my post.

In a vexed voice Miss Hurly said, "Mary Ellen, where were you? I called and called you—I went to look for you in the museum—but couldn't find you. And now—here you are."

"But Miss Hurly, I was here all along . . ."

The woman did not believe me, that was clear. Once, Professor Harrick was waiting for me also with a stack of documents to be (re)typed.

In a mild panic I thought—*But where have I been, if I haven't been here?*

"THAT'S RIDICULOUS. That never happened."

Wolfman laughed at me, when I'd begun to tell him about the "foreign" students in the dining hall, who'd seemed to recognize me as an Exile, on the morning of the psych exam.

"You were hallucinating out of exhaustion and worry. Just forget it."

"But, if I see them again . . ."

"You won't see them again, Adriane. I promise."

AND THEN: walking with Wolfman on Quad Street, south of the campus.

In this public place we were not—quite—a couple. Rather, we resembled individuals who'd happened to meet by chance and were walking together in the same direction, by chance.

A tall man of about thirty, bareheaded, with thick dark hair, dark eyebrows, a face that glowered with thought. A girl not yet twenty, in a heavy fleece-lined jacket that hid her hair and part of her face, and seemed a size or two too large for her. A girl intensely aware of the man beside her.

This was late Saturday morning, in February 1960. Following a night of snowfall. And the sidewalks, and the streets, hastily shoveled and plowed, glaring white.

We had not been together the previous night. (Where Wolfman might have been, I had no idea. Whom he might have brought back to his apartment, I didn't want to speculate.)

Let's meet. I want to see you. How have you been?—so Wolfman had contacted me.

Since the psych exam when I'd failed to follow Wolfman's advice and Wolfman had punished me with a bare blunt C on the exam, and a B in the course—"That should have been an A. But isn't."

I'd felt rebuffed. But I'd thought—*What does a grade matter? I am fortunate to be alive.*

And I thought—*No grade can matter in Exile.*

Since the psych exam there'd been a precarious sort of tenderness in Wolfman, toward me. Mingled with exasperation, and concern. I didn't want to think that this was the kind of feeling an individual might have for a younger sister or brother.

In this public place, with so much to say that could not be said safely in any public place. I wanted to take hold of Wolfman's arm, or grab his (gloved) hand, but didn't dare.

Wanting to ask him if, the night before, he'd been with any woman—the woman with the stark staring eyes, for instance, who so clearly liked him. I'd found out that her name was Cornelia—"Nelia"—she was an older graduate student in social psychology. *But no. You must not intrude. His private life is his own. He is not your lover.*

It was then that I saw, as we were crossing the street, several dark-skinned young men approaching us. These were graduate students, obviously "foreign"—their eyes moved onto me, and onto Ira Wolfman, and back onto me, with expressions of intense *interest*—almost, I'd have thought *recognition.*

In that instant I was stricken with fear. As we passed the young men I could hear them speaking in undertones to one another in their unintelligible speech. *They are talking of us—are they? What are they saying?*

In a daze I managed to cross the street but on the curb, I came to a dead stop. Like a laboratory creature that has been conditioned by a menacing visual stimulus to "freeze"—can't comprehend why he has been overcome by paralysis—I shut my eyes and stood very still until Wolfman prodded me.

Wolfman asked what was wrong?—and I told him, in a barely audible voice, "Those were the students—just now."

"What students?"

"The 'foreign' students. Who'd seen me—I think they'd recognized me—in the dining hall last week . . ."

Wolfman glanced behind us. "Who? Where?"

Were they gone? Had they never existed? I could not bring myself to look around.

"How would you know who they were? You didn't memorize their faces, did you? 'Foreign' students all look alike, in Wainscotia, Wisconsin."

Quickly I walked along Quad Street. Nearly colliding with pedestrians. Slipping-sliding on the slick sidewalk, so that Wolfman had to take hold of my arm, to steady me.

"I saw them, Ira. I did. They saw *us*."

Wolfman laughed in his way of dismissal, that was also a way of comfort.

"Maybe they were psych students. Graduate students who'd been auditing Axel's class. Maybe they know *me*—nothing to do with you at all."

We'd left the Moore Street commercial neighborhood, that was hardly more than two or three blocks. We were on the university campus now, making our way steadily uphill in the direction of the main library. Around us was a snowy expanse blazing with noontide light. Seeing that I was very quiet Wolfman took hold of my wrist that was bare between my jacket sleeve and my glove—a gesture of unexpected intimacy, in this public place.

"Didn't I tell you, Adriane—there is nothing to fear from them? There is no *them*."

I BELIEVED that Ira Wolfman loved me then. In some way, that was perhaps not a sexual way, or a way of possession, he loved me.

A BEAUTIFUL PLACE. I wanted to think, a safe place.

This was the university arboretum, north of the campus.

In the NAS-23 that both Ira Wolfman and I knew there were no longer "public" lands—desirable city and state parkland, and 90 percent of the old, national parks like Yellowstone, Grand Canyon, Yosemite had been sold to private interests, mining, fracking, logging, and vacation places now for the very wealthy.

Trespassing on these lands was now punishable by death. (The most egregious felonies in NAS were in the category VPR [Violation of Property Rights] second only to Treason and Questioning of Authority.) Beautiful privately owned shore lands along the Atlantic and Pacific coasts were protected by ten-foot electrified fences and gates manned by armed security guards. The only "parks" remaining for public use were vacant lots filled with rubble, wetlands and landfills, uncultivated land contaminated by chemicals and waste. (The Burnt Fly Bog Superfund in New Jersey, ten miles from Pennsboro, was claimed to be "safe" for picnicking and water sports but it was rare to see any visitors there above ST4.) And so it was a thrilling surprise to me to discover the Wainscotia arboretum.

Wolfman, too, loved the arboretum. It was the place that had "restored his sanity" when he'd first arrived in Wainscotia, a dazed and demoralized twenty-year-old who had been, at the time of his arrest and *teletransportation*, a brilliant undergraduate at NAS-Cambridge (formerly Harvard) majoring in computer science, math, and cognitive psychology.

Wolfman had been a runner, too. But in Zone 9, in the early 1950s, the only individuals who "ran" were athletes in training. And the only "running shoes" were ordinary sneakers.

The Wainscotia Arboretum adjoined the Wainscotia Agricultural College but extended for hundreds of acres into forestland abutting Wainscotia Bay, that emptied into vast Lake Michigan to the east—which I'd never seen except in photographs and on a map. (The lakeshore fell outside the ten-mile radius of my official confinement, as, to the west, the much smaller Lake Hallow lay outside my confinement.) By February 1960 I had not ventured more than two or three miles from the epicenter of my residence in Zone 9, which was Acrady Cottage.

Wolfman spoke disparagingly of his "epicenter"—which was Greene Hall on the university campus.

"If I want to, I'll leave—maybe. Just walk away. Or better yet, bicycle."

When Wolfman spoke in this way I felt both thrilled by his bravado and uneasy. He said:

"It has been observed, in laboratories, that animals in cages are sometimes fearful of leaving their cages, even when their doors are left open. Even when their doors have been removed."

"But—this isn't us, is it? I don't understand."

Was Wolfman saying such things to test *me*? Did he want to determine how reckless I was, or how diffident? In the desperation of loneliness I'd many times fantasized running away from Wainscotia, and violating The Instructions—but I'd never come close to such a dangerous act.

I didn't doubt that I was under surveillance in Zone 9. My figure was projected onto a screen, somewhere. If I left the ten-mile radius I might be "vaporized"—instantaneously.

For hadn't I witnessed the vaporization of the high school boy Z., in a DDS attack? Hadn't I seen the expression of incredulity on Z.'s young face, in the instant before his head exploded like a nova?

I'd begun to tell Wolfman about seeing the execution on a TV monitor in the Youth Disciplinary Division of Homeland Security, but Wolfman cut me off short—"If it was a TV monitor, you were probably seeing just a reenactment. You have no way of knowing if the execution was authentic."

"It was authentic! It was actual, and horrible. There were four of us being interrogated—high school valedictorians who'd been given 'Patriot Democracy Scholarships.' We were told to confess to being conspirators, and if we didn't, one of us would be 'disciplined' . . ."

"There is absolutely no way for an ordinary citizen to distinguish a 'virtual' staging from an 'actual' event. Especially via TV. Believe me!—I know."

Why was Wolfman not more sympathetic? I knew that the boy, whose name had begun with *Z*, had been executed—it was no reenactment, but actual death. I *knew*.

Domestic Drone Strike: a matter of less than sixty seconds to liquefy the subject, with a laser ray emitted by a flying object no larger than a robin, that self-implodes at the strike-target and is itself vaporized.

I was trembling with indignation, and anxiety. The boy's death was a sight I would not ever forget for, unlike other recent memories, it had been imprinted vividly into my brain.

"Maybe you've never seen a DDS, Ira. I have."

This was a sad little boast. Wolfman seemed about to reply to it, then did not, hiking briskly ahead of me.

On this bright sunny-snowy day we were hiking in the arboretum. Our breaths steamed for it was well below freezing. We took care not to walk too closely together and we did not ever grip hands—of course. Often, Wolfman walked ahead of me, like an expert hiker showing the way to a less expert hiker. We did not usually talk much for Wolfman preferred silence in the arboretum. He'd told me, in an undertone, hearing other hikers chattering to one another, how he'd like to *vaporize* these rude people who ruined the silence and beauty of the arboretum for others.

This was shocking to me. Just the term—*vaporize*.

How could Ira Wolfman of all people say such a thing, even in jest?

When he was alone Wolfman hiked for miles. His legs were ropey with muscle. When we were together usually we hiked for no more than two miles, along a looping trail through snowy hills. Many of the trees in the arboretum were identified with little plaques so a hike in the arboretum was educational, like prowling through the Museum of Natural History; though after a snowfall the plaques identifying the trees were likely to be obscured. Often after a snowfall the arboretum wasn't immediately plowed and so, hiking there, as Wolfman insisted, we had to make our way in knee-high boots along the trails that were difficult even to discern when covered with snow. It was my fear— (Wolfman laughed at this fear)—that we might go off-course, and unintentionally violate The Instructions, that forbade EIs to go more than ten miles from their epicenters; Wolfman thought it was ridiculous to imagine we could hike so far unintentionally, and even so, if we did, Wolfman doubted that there would be consequences.

It was Wolfman's belief that outdoor settings were "safer" than indoors, generally; the vast reaches of the arboretum were safer than the university campus; though no place was as safe from surveillance as the nuclear bomb shelter beneath Van Buren Museum, where it was certain NAS surveillance couldn't reach.

Initially I'd had no reason to doubt Wolfman. When he'd led me down into the bomb shelter I'd been so astonished, so intimidated, I had not been capable of thinking clearly. But in subsequent weeks I'd come to wonder how Wolfman knew what he claimed to know about surveillance systems in faraway NAS-23, beamed at Wainscotia, Wisconsin.

In a bemused undertone Wolfman told me: "Their security isn't perfect—NAS. We're not continuously on their radar. They want EIs to think that they know everything about us, but they can't possibly. For one thing, and this is basic, Wainscotia isn't *wired*. There's no cyberspace here. There's no grid. It's like 'atoms and the void'—preceding creation. They may have agents here, but they can't have many. Back home, in our time, it's taken for granted that everything is 'monitored'—every cell phone or computer transaction, anything electronic. We take for granted that we're being recorded—like lab animals in cages, who've been born into captivity. But Zone Nine is very different. That's why they call it the 'Happy Place.'"

But—who called Zone 9 the Happy Place? I didn't understand.

"And the connection between 'future' and 'past' is tenuous. It isn't 'Big Brother is watching you'—not at all. It's my theory that the aperture could shut down—the connection could snap like a rubber band—they could lose us, and never see us again."

This was not altogether comforting to me. If I understood Wolfman correctly.

"You mean—we would never return home? We would be permanently exiled here?"

Wolfman laughed. "Yes, it's pretty suffocating here—our 'hotbed of mediocrity.' But the alternative is not so obviously superior, is it?"

I wanted to protest—I missed my parents. I loved my parents and wanted desperately to see them again—I hadn't been allowed even to say good-bye . . .

Wolfman hadn't left behind anyone he'd loved much, evidently. Or, if he had, his Exile had been so long, his feelings had atrophied.

"Something could very easily happen in NAS, to render the Government powerless. Ordinary citizens are unaware that there are contentions within NAS—the president has his faction within the Patriot Party but there are other factions, too. There are secret dissenters, rivals. There are military uprisings that are quelled—or maybe not. In this case, it would be a 'cyberspace' uprising—whoever has control of the computers has control of NAS. The so-called leaders of NAS are invisible to us but hardly to one another. Their 'power' depends upon electric power. It's the most elemental source of power, generating a vast computer-system. It's all generated by wind now but it isn't foolproof and one day, the entire structure could just go *down*."

Such rapture in Wolfman's voice, I knew that he had fantasies of returning to NAS in triumph. The bright brash rebel-Exile who returns, overcomes his enemies, takes away their power and takes their place.

I wondered what Wolfman's role had been in NAS-23. Had he been more involved in the Government than he'd told me? Clearly he was of a caste higher than my parents'—my father was an intelligent man but lacked the confidence of Ira Wolfman, even in Exile.

Half-pleading I said, "I wish you wouldn't say those things, Ira. I—I want to go back . . ."

"Yes, they control you in exactly that way, Adriane. All 'Exiles' think they yearn for home—until they return to it."

"That's a—a terrible thing to say, Ira . . ."

"Why? Isn't it true?"

"I love my parents, I m-miss them . . ."

On the verge of tears. Throat shut up tight.

Didn't want to think that this might be true, to a degree. Painful, how much of what Wolfman said was true. To a degree.

OFTEN AS A HIGH SCHOOL STUDENT I'd been moody, sad, angry, even despondent—so *trapped* by the perimeters of my life, and my parents' lives. I'd felt sorry for my parents and (maybe) impatient with them, as a child might be impatient with her parents, unable to comprehend the complexities of their lives.

To return to a variant of that life, knowing what I knew now of the power of the State, would be difficult.

Yet, I was feeling elated suddenly. Wolfman had called me "Adriane."

I heard love in his voice, when he called me this forbidden name. I heard tenderness, respect, regard, concern—friendship and protectiveness. Of course I also heard amusement, condescension.

I heard *intimacy*. This was the most precious gift to one in Exile.

Now that I was no longer his undergraduate student Wolfman did treat me slightly differently, as if I were more adult. In the spring of 1960 I was enrolled in five new courses but I was not continuing with Psychology 102, not just yet.

Wolfman, too, thought it was a good idea that I wasn't continuing the course this semester. If I'd had a different quiz instructor Wolfman would have been overly curious about my work for this person, and how he was grading me; and if I'd been assigned to Wolfman a second time, our relationship would have been a strain to both of us.

(I did admire Wolfman for having read my final exam "blindly" and for having graded me as honestly as he had. I'd been hurt—of course—for we always want to be assured that we are special; but Wolfman had certainly done the right thing. He had not compromised his academic standards for even a fellow EI. He had not made me feel beholden to him.)

I said: "You call me 'Adriane'—you know my true name. But you have never told me your true name."

"That's right."

"But—why not?"

"My name is what you call me, Adriane. Anything you call me—that's my name. My birth name is of no significance."

"But—why can't I know it? You know mine."

"I've grown to fit 'Ira Wolfman.' I think I might prefer it, as a name to attach to my publications. And in any case it's close enough to my original name."

"Which is closer—'Ira'? Or 'Wolfman'?"

"Both."

"Is 'Wolfman' a—Jewish name?"

Wolfman laughed at this. Telling me that yes it was, but no, not, rather an "Anglo approximation" of a Russian-Jewish name back in the early twentieth century.

Despite the deep snow (which no other hikers had yet penetrated) Wolfman had been increasing his pace by degrees. It was a habit of his—I didn't want to think that it was a stratagem—walking in front of me on a trail so that my breath was taken up with the effort of following him and not with the effort of speaking with him.

Several times I staggered in the snow. My heartbeat began to hurt. I was beginning to be overwarm inside the shabby fleece-lined jacket.

Wolfman don't leave me! Wolfman please protect me.

High overhead the sky was a bright blue like china and when we passed beneath trees, dark-winged birds fluttered in the highest branches, calling to us in hoarse cries, like crows, or starlings, with indignant yellow eyes.

AND SOMETIMES, we met at the Rampike Street Laundromat.

Warm-yeasty smells of the Laundromat. Soft collapsing sounds of laundry tossing in the dryers. In this era before cell phones the place was often quiet even when relatively crowded for the clientele was mainly graduate students, who brought their work with them. (Most undergraduates had access to washers and dryers in their residences. Acrady Cottage had these, in the dank basement, but I preferred the more impersonal Laundromat where I might meet Ira Wolfman.)

This was a place of refuge, it seemed. There was a soothing and dreamlike air here. You would not ever be "vaporized" in the Rampike Street Laundromat, such a thought was preposterous.

Initially we'd met by accident at the Laundromat. But ever after that I'd contrived for us to meet.

I'd offered to iron Wolfman's cotton shirts damp from the Laundromat. This was not an era before wash-and-wear fabrics but such fabrics were considered cheap, and were in fact cheap. Cotton and linen were more formal, and more desirable. Wolfman had a half-dozen long-sleeved cotton dress shirts which he wore when he lectured. When he wore a tie he loosened it soon after lecturing and, as soon as he was off campus, pulled it off saying he felt choked—"Literally."

It was a novelty—*ironing!* I'd seen advertisements on the Acrady TV. "Housewives" happily ironing their husbands' shirts.

In Wolfman's sparely furnished three-room apartment on Myrtle Street there was, kept folded upright in a closet, an ironing board with a badly scorched covering, and a shiny iron so heavy it nearly fell from my hand the first time I lifted it. (The ironing board came with the furnished apartment.) These were artifacts of an old, vanished America of which I'd had but a glimmering awareness for I'd never seen my mother "ironing"— we'd lived in a post-cotton era, of quick-drying and wrinkle-less fabrics.

I wondered what my mother would have thought of her daughter wielding a heavy iron like a pioneer woman! Yet—I would have liked to assure her—there was some small comfort in the task.

If you love the task. If the shirts belong to someone you love. If you are not forced to "iron," but have chosen to "iron."

In NAS-23 there were households in higher-caste neighborhoods in which servants were employed. In some households, a team of servants. These were usually of the IS (Indentured Servants) caste which meant desperate individuals who, having no money, and being very likely in debt, entered into contractual relationships with employers for a prescribed number of years; many, indeed most, but not all of the IS were ST5 or lower: dark-skinned. You would not call these *slaves,* which was an offensive term, not even *indentured servants,* but simply "servants."

To this Wolfman observed *Hey! I didn't make you my friend to make you my servant. I'm not colonizing you, Adriane.*

Still Wolfman appreciated the ironed shirts. Otherwise he'd have had to take his shirts to a laundry, which was expensive and a nuisance. (It seemed that Wolfman, for all his skill as an experimental psychologist, wasn't capable of ironing his own shirts. Why was this? A masculine deficit?) And yes, Wolfman appreciated my cleaning dishes he'd left to soak for days in the sink of his "kitchenette." And to scour the sink itself, and put away things in cupboards and drawers.

Colonize. This was an ambiguous term.

For the *colonized* might be complicit in such an arrangement. In Skinner's rat mazes the rats soon learned that treats awaited them when they ran the gauntlet correctly. And so, why not run? Why prefer being locked in a cage?

And we prepared meals together, Wolfman and me. And ate meals together.

I waited for Wolfman to kiss me more forcibly, to suck at my lips so that my breath was drawn from me. I waited for him to run his hands over my body which was eel-like and yearning, that would have wrapped itself around him if he had. I waited.

Yet we lay on Wolfman's bed, in the semi-dark. Quietly talking, or not talking.

So that I thought—*This is the happiest I can be. I want this never to end.*

Listening to Wolfman's favored music, on what he called *long-playing records*: symphonies of Mozart, Beethoven, Brahms, Mahler. In his own time, growing up in New York City of the Reconstituted North American States, he'd had virtually no contact with classical music but only the electronically driven post-rap "heavy metal" music of his generation which was generic and anonymous.

It was a surprise to him, Wolfman said, to discover music that was labyrinthine like thinking and feeling commingled. Music that didn't need to be deafening to penetrate the soul.

(Did Wolfman, a scientist, really use the term "soul"? I think he did—unconsciously. For often Wolfman spoke in extravagant ways when feeling, and not just thinking, was involved.)

Wolfman had been kindly, asking me to speak about my life. Less easily, he spoke of his own.

There were secrets in Wolfman's life, I had to suppose. He was older than I was by a decade, at least—of course, more secrets would have accrued.

One evening he told me about his parents. His voice trembled with childlike wonder and another, less definable emotion—a kind of elation verging upon terror.

He had not seen much of his "famous" parents while growing up, he said. But he'd admired them—very much—for they'd been distinguished research scientists at the (former) Columbia University Medical Institute, epidemiologists whose specialties were bacterial diseases of the tropics. Unfortunately, at about the time Wolfman had begun middle school, his parents' highly publicized findings brought them to the attention of the Defense Strategies Department; soon then they were drafted into the MRP (Military Research Program) where their (secret, classified) assignment was the cultivating of strands of virulent bacteria to be "weaponized." Their work of several years was with a particular species of bacteria that could exist only within a narrow environment, to be targeted by the military in case of a declaration of war with one of NAS's many enemies. ("As you know, 'Declaration of War' is *ex post facto* since the War in the Middle East," Wolfman said dryly. "It's so much wiser to declare the war after it has begun, not before.")

At first Wolfman's parents, particularly his mother, had been upset by the nature of their new research; then by degrees they were drawn into the competition and excitement of the work-team at MRP, which employed the most prestigious research scientists in the country. "My parents received all sorts of special perks including a rent-free condominium overlooking the Hudson River, a car and a driver, joint election to the National Academy of Sciences, and virtually unlimited funding for their research; also, for me, their only child, admission to the most elite prep schools though I'd never been a 'good citizen' as a student, and had actual demerit marks on my record. As it was, I went to three private schools before I finally managed to graduate. I'd told you, Adriane, that I'd been a computer hacker in middle school. I also created video games, several of which were bought by Nightmare Works, Inc., when I was twelve, and became modest best sellers, and several of which were considered so politically dangerous, and so obscene, they were banned. I was the Mystery Teen who'd penetrated security at Congress in December, NAS-11, and sent a flotilla of toy 'drones' into the legislators' midst as they were debating one or another lobbyists' bill—you probably wouldn't remember, you were too young at the time. But it was everywhere in the media. It was a real scandal—if a teenaged kid could breach congressional security with toy domestic drones what about our enemies with their 'sophisticated weaponry'?—everywhere online and on TV. "

Wolfman laughed. He was drinking beer from a can, set atop a copy of a journal called *Brain* on his bedside table. I was feeling dismay at what he'd told me about his parents but said nothing to Wolfman.

"The legislators panicked of course—stampeded and trampled one another in their desperation to get out. The few females in the room, legislators and aides, were stomped and injured. Fortunately, no one was killed. I'd planned to perfection the onslaught of the toy drones—but hadn't given much thought to what would follow. That's a kid's brain, typically—no matter how bright, the kid is immature. Immediately, I stopped hacking for months, not wanting to get caught. My parents knew nothing about my 'experiments'—they were wholly innocent. I wouldn't have been tracked by FCS—Forensic Cyberspace Security—except a 'friend' informed on me. By this time I was a freshman at NAS-Cambridge—formerly Harvard. My parents tried to intervene for me but failed. Unlike you, I was incarcerated in a Youth Facility for a while—the Government was hoping to co-opt me in one of their 'defense security' programs but I refused." Wolfman swallowed hard, and fell silent.

Lying beside Wolfman, just lightly touching him, the side of my upper body, the side of my legs, I held my breath wondering what he was not saying.

I asked Wolfman about his parents: Had he been close to them? Had they confided in him? What had he said to them, the last time he'd seen them? (For in my case the last time I'd seen my parents I had had no idea that it would be the last time. I was sick with regret that I'd probably been my usual moody self-absorbed self, and could not even clearly remember what we'd said to each other. Oh, terrible!) Did he know if they'd been arrested too, in connection with his arrest? Did he know what—if anything—had happened to them after he'd been arrested?

Wolfman didn't answer immediately. Wolfman gave the impression of one who is searching his mind, trying to recall, like a man with a wide rake dragging the rake along the ground.

"I—I think—they are all right. I think so."

Then, "Well—my mother suffered some sort of breakdown, and had to be hospitalized in a NAS facility in Bethesda, Maryland. This was after my arrest. While I was incarcerated."

I asked Wolfman if he'd seen his mother, since his arrest? If she was still hospitalized?

"I—I'm sure my mother isn't hospitalized after so long. I've never heard—never been informed—exactly what happened to her, or even if—if she's still alive."

I asked Wolfman if the interrogators had tried to make him confess to having been "influenced" by his parents?—"That's what they tried with me."

"No."

Wolfman's answer came too quickly. I could feel heat lifting from his skin as he lay beside me, only part-consciously aware of me, staring at the ceiling.

"I—I do miss my parents. I guess. Just now, I could 'see' them— almost—as I hadn't seen them in years." Wolfman began trembling. I was shocked, but I held him tightly. It had not been the case in my life, that I'd had the opportunity to comfort others.

I thought—*I can love him more than he knows. I am strong enough to love for him, too.*

Wolfman said, "I was 'drafted' too—but managed to be phased out. Not at DSD but CSD—Computational Strategies Division— a younger generation, with some extreme personalities. Of course, they'd all been co-opted. No one so vindictive as those who've been co-opted by the enemy. For much of their energy is fueled by *shame*."

I wondered what Wolfman meant by this? But I didn't feel comfortable about asking him.

"What happened was—it was a stupid mistake on my part. Inadvertently I'd texted a friend, using a code we'd devised, which aroused suspicion . . ."

Wolfman was saddened. Wolfman couldn't continue.

Wolfman pressed his face against my neck and we lay together in silence comforting each other in our separateness, our ignorance.

"I love you."

I spoke softly, tentatively. So quietly, Wolfman could pretend not to hear.

SANE

The voices were pleading yet bold: "Come join us! March with us."

And: "Save your life! The lives of your loved ones! Save the life of the Earth! March with us."

A chorus: "Come join us! March with us. Come join us! March with us. *S-A-N-E*—now!"

But I held back. I was very frightened.

I remembered—my father had been arrested at a public demonstration when he'd been in his mid-twenties. He'd paid for the remainder of his life for that single impulsive act of curiosity and sympathy.

And my uncle Toby—he'd paid with his life.

A large rowdy crowd had gathered in front of the university chapel, spilling onto the snowy campus green. Was this a spontaneous sports rally? A parade? The mood of the crowd didn't appear to be cheerful or festive but rather angry, jeering. Were they heckling marchers? But why were they heckling marchers? In Wainscotia everyone was so *friendly*.

Dirt-tinged snow lay in heaps beside the walks and drives. It was an overcast afternoon in March 1960. On my way to class I'd been hearing voices—shouts—a sound of agitation and derision that was upsetting in this place, for it was unusual. Only on Saturday afternoons in the football stadium, in football season, did such excited calls and cries waft across the campus, or on weekend nights on the Hill, where fraternity and sorority houses glared with festive lights.

My instinct was to turn the other way. I could take a backway to Masson Hall, to my literature class, and avoid the commotion.

Yet somehow I found myself pressing forward. Curiosity drew me, and a sense that, if something was happening on campus, I would want to know what it was, to discuss it with Wolfman. Through each day I accumulated incidents, episodes, paradoxes and riddles to present to Wolfman for his reaction. For I wanted to entertain and intrigue Wolfman, badly I wanted to be essential to the man's life.

Go home! God-damn Commies go back home to Russia!

I saw that the loose, rowdy crowd had surrounded a small group of marchers who'd been making their way from the chapel to the administration building across the quad. Just as I'd arrived the marchers had had to halt midway. There were about thirty of them, ringed in by as many as two hundred individuals of whom most appeared to be vociferous male undergraduates. The marchers were carrying picket signs, swords in vivid red letters—

SANE

NATIONAL COMMITTEE FOR A

SANE NUCLEAR POLICY

SANE

NUCLEAR TESTING—

STOP STOCKPILING WAR

Demonstrators for peace! They were protesting U.S. nuclear experimentation in Nevada and in the South Pacific—protesting war. I had heard of this newly formed national organization, in fact Wolfman had spoken of it with reserved admiration, but I'd never met anyone who belonged to it. I had never seen a "peace march" before.

The hostile jeers of the crowd, raised fists and angry faces— these were shocking to me. *Commies! Traitors! Go back to Russia, you don't like it here.*

How contorted with hatred, the usual bland-white midwestern faces, and how shiny their eyes with rage!—you could not comprehend the power of so small a gathering of protesters, the effect of their handmade picket signs, to rouse such emotion.

Campus security guards were holding back the hecklers and simultaneously urging the SANE marchers to leave the area. Security vans had arrived to carry the marchers away in safety. Yet, the SANE protesters resisted. Bravely and stubbornly, they resisted. They were strangers to me though it would be revealed, in the student newspaper, that several protesters were on the Wainscotia faculty, and several were graduate students; most of the protesters were from Milwaukee and Chicago. They ranged in age from mid-twenties to mid-sixties or older. Many white-haired, and a scattering of bearded men. The elders appeared to be the leaders. A revelation to me—at least one-third of the protesters were female.

Eagerly I looked but saw no one I knew—at first, I'd thought that Ira Wolfman might be among them. (He wasn't. I felt relief yet also disappointment.)

The crowd wasn't dispersing, and the marchers remained where they were, prevented from marching across the quad but brandishing their handmade signs aloft. Their chanting was drowned out by the shouts of the crowd. When they passed out pamphlets most of the pamphlets were taken rudely from them, torn and tossed down.

I thought—*How can they take such risks? How are they so brave?*

No one would be shot down or "vaporized" in Zone 9. Yet surely there were federal agents, F.B.I. informers, in Wainscotia. I'd learned from Wolfman of the Cold War, the anti-Communist hysteria, Senator Joseph McCarthy's campaign of slander, innuendo, and persecution. To be anti-war, anti-weapons, anti-nuclear was to risk being persecuted as a Communist in 1960 though there was, legally, *freedom of speech* and *freedom of assembly* in the United States at this time.

Subsequently, sometime in the troubled years post-9/11, these freedoms were curtailed, or banished outright. It had happened easily, Dad had said—at the time, few had seemed to notice.

Many of the hecklers were fraternity boys. A scattering of agitated girls. To my dismay I saw my roommate Hilda among them. And others from Acrady Cottage. Their faces were baffled, stern, indignant. They were "scholarship girls"—how did they perceive the SANE marchers as threats? Several young men in ROTC uniforms were shouting at the marchers, approaching them in a threatening manner.

Badly I wanted to support the SANE marchers. I felt sympathy for them, and shame that the Wainscotia students were so hostile, and so ignorant. I thought that the marchers must be nervous or anxious, in this chaotic place, yet they appeared calm, and they were certainly courageous.

A voice bellowed through a bullhorn—*This is University Security! We are asking you to evacuate the quadrangle! Evacuate the quadrangle immediately! Everybody—no exceptions!*

The noisy crowd was beginning to back away if not disperse.

The marchers were smiling in relief. Some, in giddy relief. Several young women sighted me, and must have seen the sympathy in my face for they began chanting, "Come join us! March with us! Save your life! March with us!"—appealing to those of us who hadn't shaken fists at them or turned away in disgust.

Yet I held back. I had not the courage, or the recklessness . . .

"March with *us*! Save your *life*!"

As the marchers began moving away, tramping through the dirtied snow, I followed after them. No one could hear me, I think, but I was calling to them, trying to tell them that I was sympathetic with their ideas for war was "evil"; I told them that they were very brave and that I respected them—"Where I come from, people can't protest the way you are doing."

(Oh, what was I saying? Fortunately in the din and confusion no one heard me.)

One of the male marchers approached me. Was this someone I was supposed to know, from one of my classes? A burly youngish man built like a wrestler with coarse flyaway hair, a scruffy beard, urgent eyes. Evidently he was one of the SANE leaders—I'd seen him being confronted by the head of the campus security police. Unexpectedly he smiled at me—"Hey! Hello! You and I know each other—yes?"

"We do?"

"What's your name?"

"My name is"—I faltered, for the name seemed so obviously fraudulent—"'Mary Ellen Enright.'"

"'Mary Ellen'—whatever. Come with us! Here's a sign."

"But I can't. I have a class . . ."

"Hell with your class, 'Mary Ellen'! Come with us! Save the Earth."

"I—I don't think there will be a nuclear war . . ."

Weakly I stammered this. What was I thinking of!

The burly young man stared at me bemused.

"'Don't think there will be a nuclear war'—what the hell, you *don't think* there will be a nuclear war? How in hell can you say such a thing? You're a prophet? You can see into the future? Nuclear weapons have already been used—atomic bombs dropped on Hiroshima and for no God-damned reason anyone can explain, Nagasaki. Why not again? Why not many times? The president is a U.S. war general, the U.S. Congress are rabid anti-Communists, there is plenty of profit to be made from the Cold War, so why not a nuclear war? How can you make such an irresponsible remark, 'Mary Ellen'?"

I was mortified, the burly young man stared at me with such disdain.

I stammered an apology. I said that, even if there wasn't a war, it was important to educate people against nuclear war. It was important to support SANE.

"Yes! Right! Come with us! March with us!"—the young man grabbed my hand and pulled me after him; I was too surprised to resist, and he was too strong in any case. He gave me a picket sign which I brandished as the others did. I was very excited by this time, suffused with emotion. *These are my friends. My family.*

I asked the young man what his name was and he said what sounded like *James, Jamie.*

Now came a resurging of hostility. A swarm of yelling frat boys storming the quad from a new direction. *Commie traitors! Bastards! Get the hell out of Wainscotia!* The picket sign was wrenched from my fingers. Someone collided with me, hard. I found myself on the ground, in the snow. I saw only legs and feet—I heard only shouts and screams. Then, I saw the SANE regrouped marchers being escorted off the quad by security police. One of the white-haired leaders was bleeding from a cut in his forehead. The burly young man who'd given me the picket sign was running as a football player might run, aiming his right shoulder at—whoever it was, I couldn't see.

The crowd was breaking up another time. Sirens were deafening. Emergency vehicles were driven onto the quad. Hecklers were being pushed back. Marchers were being escorted into vans, unresisting now. Picket signs lay strewn in the snow, many of them broken. Pamphlets lay scattered and torn. The SANE demonstration of March 11, 1960, at Wainscotia State University, Wisconsin, the first such demonstration at the university in its history, had ended in a sort of stalemate after a tumultuous forty minutes.

Back at Acrady Cottage girls looked at me in surprise as I limped inside.

"Mary Ellen!—what on earth happened to you? Did those awful 'SANE' people knock you down?"

WISHING I COULD tell my parents about the demonstration—the "march for peace."

No one was arrested.

No one was beaten or (badly) bloodied.

Had the march been a failure, or a (qualified) success?

That there'd been a demonstration for peace at all seemed to me a wonderful thing.

My shoulder ached where someone must have run against me to knock me down, the way guys knock one another down in football. My right knee ached from the fall. And there were scratches on both my hands. But I was elated, and anxious—as if a border had been crossed, that could not be recrossed.

I'd limped away from the scene. I was thirty minutes late for my class. Yet I would go to class, I would not miss a single class for I was determined to be an A student here in the Happy Place, Zone 9.

The rest of that day my eyes repeatedly glanced up, narrowed. Steeling myself for the quick-darting appearance of *domestic drones* en route to vaporize us all—though the year was 1960, and the setting was idyllic Wainscotia, Wisconsin.

NUCLEAR PROTESTERS ROUTED FROM CAMPUS—next day's banner headline in the student newspaper. Photographs of outraged Wainscotia frat boys shoving back protesters. (The burly young man built like a wrestler might have appeared in one of the photographs, seen from behind: I looked.)

In the local newspaper, the *Wainscotia Falls Journal-American*, the banner headline was less friendly:

**COMMIE PICKETERS ROUTED
FROM WSU CAMPUS**
"Outside Agitators" Blamed

THAT EVENING, Wolfman telephoned me.

For the first time, Wolfman telephoned me at Acrady Cottage.

Did the girl who picked up the phone in the parlor understand what an extraordinary event this was, that Mary Ellen Enright had received a telephone call?

Any telephone call. From anyone.

Most remarkably, from her psychology instructor.

Benumbed I came to the phone. I had no idea who would call me and could only think that I was being informed of something unhappy.

I had long given up the foolish fantasy of my parents calling me.
Dad's voice—*Hiya, kid. How're you?*
Mom saying—*Oh, Adriane. Oh* . . . Bursting into tears.
"'Mary Ellen'? Hello."
No need for Ira Wolfman to introduce himself. Of course.
"Ira . . ."
"What did you do, this afternoon? What the hell did you do?"
"You mean?—the SANE protest . . ."
"How could you do such a reckless thing? Participate in a mob scene? Call attention to yourself in an act of 'civil disobedience'? The SANE people didn't have permission to march on campus—they were defying orders. Their request had been turned down. They're very lucky they weren't all arrested, taken away in vans. Their heads broken." Wolfman paused, breathing audibly. I knew better than to try to interrupt. Relenting he said, "It's good of them to come here. It's foolhardy but—noble. I admire them but—look, you can't possibly get involved. I can't possibly get involved. There will be no nuclear holocaust—not as they are imagining it. No Russian-provoked conflagration. So, that's that. They don't need us." Again Wolfman paused.

Then: "You know there are spies here. You know that 'Mary Ellen Enright' is under surveillance. Everyone who participated in that demonstration will be noted in FBI files. And probably in university files."

"But, Ira—it isn't against the l-law here . . ."

"It isn't against the law—legally. But since when does the FBI or the federal government give a damn about 'legally'? Seeming to side with communism is perceived as treasonous."

"These people aren't 'Communists'—they are protesting nuclear weapons . . ."

"I know what they're 'protesting'—for Christ's sake. And I know that *you are not Adriane Strohl here,* if you want to survive."

Wolfman sounded disgusted with me. I could not believe how harsh his voice was.

Stammering I told Wolfman that I hadn't thought of all those things—I hadn't had time. One of the protesters had handed me a sign. In fact I'd expected to see him among the SANE protesters . . .

Now Wolfman became vehement.

"Are you crazy? Join the marchers? Of course not! I'm up for adjudication in two years, I'd never risk resentencing."

"But it's the right thing to do, isn't it?"

"The 'right thing to do' is to survive, Mary Ellen. You know that."

"I—I'm not sorry, Ira. The SANE people are very nice. I really liked them. They were courageous, and everything they said made sense about nuclear 'stockpiling'—the 'Cold War'—the 'tragic lesson of Hiroshima' . . ."

"I told you: the *nuclear holocaust* never happened. You know that, so fuck this crap. Just—fuck this."

Now Wolfman had truly shocked me. And hung up.

THE LONELY GIRL I

Ira, forgive me! Please don't cut yourself off from me.

You know, I love you so much. I can't bear to live in this place without you . . .

These words I wrote in blue ballpoint pen on a sheet of notebook paper.

Pleading and without dignity. Words spilling like blood from a slashed wrist.

Dear Ira, I am sorry for having behaved "recklessly"—as you'd charged.

Please forgive me for behaving foolishly. You are right, I should have thought more carefully of the consequences . . . that is, the consequences that might have been.

(I'd begun to worry: What might the consequences be? *Would* there be consequences for "Mary Ellen Enright"—sometime in the future?)

I don't think it is fair to break off our friendship over this matter—I will never behave so foolishly again.

Please don't punish me! I think that I am "punished" enough . . .

Though I don't really think that I did anything other than what was the right thing to do, by showing my sympathy with the marchers—I hope you will forgive me.

These words I wrote in days subsequent to the SANE march. Trying to compose a letter for Wolfman—an actual letter to be addressed to *I. Wolfman*, 433 Myrtle Street, Wainscotia Falls, Wainscotia.

Impossible! I could not.

Could not send such childish pleas, Wolfman would be repelled. Wolfman would laugh at me.

Had I lost my only friend in Wainscotia? Like a sleepwalker I continued with my "life"—attended classes, wrote papers and took exams—worked fifteen hours a week in the Museum of Natural History. Would not have wished to confide in anyone how Wolfman's harsh words had hurt even if there'd been someone in whom I might confide.

One evening, Miss Steadman invited me to dinner in her sitting room, with another girl—another girl without friends. What did we think about the SANE protesters?—Miss Steadman asked; the other girl had much to say ("My father knows all about Communists—he's a Korean War veteran") but Mary Ellen Enright only just listened, politely.

So lonely, several times I had meals in the dining hall with my roommates. Since Betsy had (mysteriously) withdrawn from school we had a new roommate named Millicent who'd had a single room in Acrady since the start of the fall term, and had grown to hate being alone; and so Miss Steadman had allowed her to move upstairs into our room.

Yet, Millie was still lonely. Sulky, easily vexed, and lonely. Millie was a plain, somber, stern-faced girl from a Wisconsin dairy farm, a scholarship holder like the rest of us; she'd hoped to major in something called agriculture education but her fall semester grades had been mostly C's, which put her scholarship at risk. Often when I returned to our room there was Millie sitting at her desk, blankly staring at an opened textbook. Sometimes she sat slouched on the edge of her bed, eyes welling with tears.

Millie was enrolled in Psychology 102 which, she complained, was too hard, and boring. Her quiz section instructor was Dr. Wolfman—"He doesn't like me. He looks right through me. He's a hard grader—I think he's unfair. He's *sarcastic*. He's *mean*."

When I didn't respond Millie said, hissing, "People say—he's a *Jew*. From New York City."

When I still didn't respond with anything more than a vague sympathetic *Uh-hmm* Millie said, with a harsh little laugh, "—What Daddy calls Jew York City. *That's* where."

Never would I ask Millie about "Dr. Wolfman" for I would not betray my interest in Wolfman, and I did not want to encourage Millie to talk to me.

Yes, it was cruel of me. Yes, I am sorry.

For I might have been the one to tell the new, so very unhappy roommate the somber wisdom I'd acquired at Wainscotia/ Zone 9: *You can live a life even if it is not the life you would have chosen. You can live breath by breath. You can live.*

THE LONELY GIRL II

Returned to the Film Society. Not that I expected to see Ira Wolfman there but that in my loneliness I had to be *somewhere*.

This time, a "classic" Alfred Hitchcock film, *Rear Window*.

And this time too, though it was billed as a suspense film, the
film moved with excruciating slowness. The actors were so obviously *acting*. The film was so obviously a *film*. Serenely blond
Grace Kelly was very beautiful and not so made-up as actresses in
the John Wayne western, and James Stewart was sympathetic and
winning, but it was impossible to take them seriously as anything
other than glamorous movie stars going through the paces of an
improbable story, again to heavy-handed background music that
made me so restless, I had to press my fingers against my ears.

Midway in *Rear Window*, I had to slip away. My head had begun to ache, I felt so—*exiled*.

How was it possible, the other viewers were totally absorbed in
this movie! I could concede that it was superior to the western,
and not so garishly colored, but each scene ran for a predictable
length, always too long—my mind skittered ahead of the dialogue,
and it made me restless and resentful to wait for the dialogue to
catch up. Bewildering to me, how in the flickering light of the

screen the viewers' faces were rapt as children's faces . . . I felt trapped, as in a child's cartoon-world. I felt as I did those several times when I'd tried gamely to watch television at Acrady Cottage, not sitting on the sofa with the other girls but standing close by, tentatively; wanting to be drawn into the stilted comic sketches, or the melodramatic scenes.

Wolfman had had a name for it—*intellectual insult*. Yet he'd been entertained by the John Wayne western and would have liked the Hitchcock suspense film even more, as it was more cleverly plotted.

There is terror in such revelations. *You can't be deceived. You can't "suspend disbelief." You are trapped inside your own head.*

Later that week I went to a poetry reading in the university chapel given by the distinguished Wainscotia poet Hiram Brody on the occasion of his seventieth birthday and the publication of *The Collected Poems of H. R. Brody.* At first, because poetry is a more subtle art than Hollywood movies, I was intrigued by the poet's words, and by the musical cadences of his poems, which rhymed in the way of Robert Frost's, whom we'd been studying in my American literature class; there were echoes too of Archibald MacLeish, another revered American poet whom we'd been studying. At the front of the large audience H. R. Brody performed brilliantly, you would have thought for the first time though in fact (it seemed) he had committed much of his poetry to memory and recited it, rarely glancing down at the page. For his age he was youthful, silky-white-haired, with a white beard that looked as if it had been lovingly brushed and combed; his manner was sly, self-effacing and elfin. The audience loved his poems and applauded after each as I did as well, though I began to realize, after about fifteen minutes, that the poems were essentially the same poem, of familiar "rural" subjects, presented with a sort of self-righteous coercion, and invariably rhyming—like a music box that strikes only the same notes, repeatedly. You could tell when the poet was concluding a poem by the expression in his face, and by the cast of his voice.

There were echoes of Robert Frost, I had to think others heard—though possibly not—for the applause continued with uncritical ardor, like a narcotic that must dull the poet's senses.

Overhead the snow-boughs sweep the sky.
Our promises to keep are bye-and-bye.

And:

Nature never will betray
A heart that cherishes each living day.

At the end of the hour-long reading the poet's eyes shone with tears. Several times H. R. Brody bowed low, his silky-white hair falling into his face so that it had to be brushed away. The crowded university chapel was deafening in applause. The audience sprang to its feet with uplifted hands, excitedly applauding.

I felt so sad! I'd tried to be swept away with these others, but had failed.

Clapping until the palms of my hands smarted. And thinking—*If I am being watched they will see how I belong here.*

SOON THEN I ATTENDED a lecture given by a historian of science from Michigan State University, introduced by Professor Myron Coughland; the thesis of the lecture seemed to be that *Homo sapiens* is comprised of distinct races and these races are genetically determined in all things, including I.Q. According to "historical data," certain races (Caucasian, Asian) are superior to other races (African, Australian aborigines).

I tried to listen but for the distracting words of Ira Wolfman muttering in my ear—*Nothing they do will have the slightest value to anyone. No one is "original" here—no one is "significant."*

Wolfman wasn't at the lecture. Wolfman would have held the lecturer in contempt, perhaps.

Gamely I tried to take notes. All serious students take notes. For it seemed to me an important gesture, *to try to learn something— anything.*

SO LONELY FOR WOLFMAN!

Even for Wolfman's disapproval, contempt.

In my restlessness drifting through the Fine Arts Building, peering into the high-ceilinged studios whose doors to the hall were all flung open—studios of painting, figure drawing, sculpting. There was a sharp smell of paint here, and turpentine. There was an air in this building of risk-taking, and bravado.

I was envious of the art students, who didn't have to concern themselves with "ideas" or "data"—who weren't doing work that could be shown to be "mistaken." I resolved that, the following year, if I was still in Exile, and still at Wainscotia, I would take a course in art.

Yet, the paintings which the students were working on, and the canvases displayed on the walls, were not very promising. Most of the paintings were meant to be "realistic"—awkward attempts at landscapes, sunsets, old barns, covered bridges; awkward attempts at sketching the human figure, or the human face. It did not seem to occur to the artists that these subjects had been done before—many times before. A few of the more daring paintings were "abstract"—bold swirls and streaks of color in the style of European expressionists of the early twentieth century, or maybe it was the large paint-splattered canvases of the late Jackson Pollock whose radically experimental work had been featured in *Life* magazine.

It seemed to be true, as Wolfman had said: there were no new ideas here in Wainscotia. No originality, no—surprise. Everything they undertook had been done already, and done more beautifully by others. Or it was wrongheaded.

Yet everyone worked energetically, enthusiastically. Everyone was convinced that whatever work they were doing was important; if you'd inquired they'd have surely said with modest bravado—*My work will outlive me . . .*

The last studio in the Fine Arts Building into which I peered was a sculpting studio. Here some ten or twelve students were working earnestly in modeling clay, trying to capture the likeness of a model that was in fact a life-sized female mannequin in emulation of a Greek statue, draped in a white sheet. The mannequin had a perfect, symmetrical face with sightless eyes, its arms were broken off at the elbows, and its legs were missing. A quick glance into the sculpting studio told me that the students, mostly male, were not very gifted as sculptors; their dwarf-sized "Greek statues" were subtly misshapen, each with something flawed though it would have been difficult to say what it was. Yet, their instructor appeared to be intensely involved with their work, and spoke passionately to them—not uncritically, but not unkindly—for you could see that he *cared*.

"Good work, Mark! Good work, Jonny! Very impressive."

In the doorway I stood hesitantly, hoping not to be seen. But no one was interested in me, for the mood of the studio was elec-

tric; the instructor spoke rapidly and knowledgeably, making occasional corrections in his students' clay figures. I wondered what
the man had found to say over the years about these diligent amateur efforts—his voice was both critical and comforting. For he
did care. You need not be exceedingly talented to feel welcome
here.

Again, Wolfman's pronouncement murmured cruelly in my
ear—*Hotbed of mediocrity!*

And then I was astonished to see that the instructor was the
broad-shouldered youngish man who'd been one of the SANE
leaders, who'd pressed the picket sign into my hand. In a sweeping attack of fraternity brothers I'd lost sight of him.

So the SANE marcher was a member of the art department
at Wainscotia! Evidently he hadn't been fired by the administration. (Maybe arts faculty were regarded differently from other
faculty?) It was touching to see what respect the students felt for
their instructor.

Indeed, James, or Jamie, was an appealing figure, if somewhat
ridiculous (as Wolfman might have thought) in bib-overalls soiled
with clay, and a tattered T-shirt, and crude sandals worn with
gray wool socks; his hair was coarse and gnarly, straggling onto
his shoulders, and his beard was scruffy, not at all "sculpted" like
H. R. Brody's beard.

Quickly I looked away, before he could turn to see me.

APRIL

"Hello, Adriane."

Adriane! In this place where, not so many feet away, in an adjoining office, Miss Hurly was speaking on the telephone.

Wolfman had returned, unpredictably. After three weeks of estrangement he turned up in the Museum of Natural History at closing time.

So often I'd imagined glancing up to see Wolfman, when he did appear it was with the unsettling authority of a vision, or a hallucination.

He was smiling at me. He extended a hand, to take my hand. "Ira! Hello."

Outside, it was April: rain-lashed, chilly as winter.

NEVER WOULD WOLFMAN SPEAK of SANE, nor would I speak of SANE to him. I came to think that he'd been embarrassed not to have joined the marchers, and this had made him unreasonably angry with me.

I thought—*Maybe someday. We will both belong, and we will both march for peace.*

"TERMINATED"

Now, I was not sure that I could trust Wolfman.

I loved Wolfman more than ever.

SOMETHING HAD GONE WRONG in Wolfman's life. Or perhaps in Wolf-man's career, for Ira Wolfman scarcely seemed to have a life.

He was preoccupied, moody. He frowned a good deal, until the furrows in his forehead didn't fade. Then, his laughter was quick, loud, and inappropriate. As Millie had said, he was *sarcastic*.

Compulsively he lit a cigarette and took several quick puffs then left the cigarette burning in one of the ashtrays scattered through his apartment.

I waved away smoke. I tried not to cough. I thought—*Has he forgotten all that we know about smoking, and cancer? Has he forgotten who he is?*

When I was in the kitchen preparing a meal or cleaning after a meal I was likely to hear Wolfman, in an undertone, speaking on the phone in his bedroom. And if the phone rang, he asked me please to wait in another room while he answered it.

That woman, Cornelia. "Nelia."

I had to think that she was as desperate for Wolfman as I was. And that she knew much less about Wolfman than I knew.

More frequently now, I was the one to hold Wolfman in my arms. As he lay on his back, on his bed, smoking, staring at the ceiling, wracked with thoughts like sharp-serrated waves that worked their way through his flesh.

Given to saying recklessly, he might just *walk away*.

(From Wainscotia? But where would Wolfman walk *to*?)

Given to saying he had new, radical ideas for experiments, of which A. J. Axel would not approve.

For much of the time, A. J. Axel was an (invisible) presence in Wolfman's life. In the sparely furnished apartment, at Wolfman's table, there were three chairs of which one was (invisibly) occupied by the white-haired professor.

As if pleading his case Wolfman said, to me: "I'm not making a break with behaviorism. I want just to stretch the concept. I want to postulate a 'mind.' I want to manipulate and expose a 'mind.' This is a post-Holocaust era, you know, Adriane, though no one at Wainscotia is likely to talk about it—the Nazi death-camps, the 'Final Solution' of the Jews, experimentation with human beings. Eichmann—the seemingly ordinary man who was 'only following orders.' None of this is taught in the history department there, still less is it spoken of. America is founded upon amnesia—denial. Conscience cannot keep up with acts. I want to investigate the mind of the 'ordinary' man—and woman. I will devise a series of experiments that replicate, in miniature, the Nazi experiment— for Nazism was a social experiment of a kind unique in history. Skinner has led the way—now, others will continue. There is a 'cognitive'—and 'moral'—revolution soon to erupt. I envision experiments that are, to the uninitiated, theatrical events. Experimenters would wear uniforms of authority, like Nazis. But more clever than Nazis, we would wear white uniforms that suggest medical or clinical authority—professorial authority. We would involve our experimental subjects in decisions of a moral nature. Not the equivalent of rats running mazes, or pigeons pecking for seeds, but the equivalent of a 'good German' complying with the Nazi agenda. In the experiment there will be a 'captain,' and there will be a 'private.' The figure of authority, which will be the psychology professor, will instruct the 'captain' in the necessity of punishing the 'private' who is slow to learn, or doesn't obey orders quickly enough. The 'captain' will administer to the 'private' a sequence of escalating punishments, as the figure of authority bids him. The figure of authority says he will accept 'all responsibility.' The intention of the experiments will be to investigate to what length a seemingly ordinary, 'moral' human being will go, if he's told what to do by a figure of authority. I envision punishments that escalate, ending in 'death' . . . of course, it's just an experiment: no one is punished, no one dies. But the subject will be revealed to himself, irrevocably. When I publish these results, American psychology will be changed forever."

Wolfman spoke excitedly, as I had rarely heard him.

How complicated, Wolfman's proposed experiment! I had to wonder uneasily if in fact it was an experiment that Wolfman was remembering, out of his past; an experiment, or a series of experiments, devised by another psychologist; if the filtering device in Wolfman's brain had failed to block this forbidden knowledge . . . The Instructions forbade all EIs to "remember" anything that had not (yet) occurred in their places of exile. Wolfman was putting himself at risk with such ideas.

Yet, his experiment was not so very different, in essence, from the experiment I'd suggested to him after the Film Society evening. For it was the pursuit of the elusive "consciousness" that underlay behavior, whose shadow behaviorists shunned.

I did not know what to tell Wolfman. Since the SANE episode he'd become edgy and impatient; feeling the need, it seemed, to explain himself, defend himself in a way he'd never done before. Almost, another Wolfman was emerging, insecure, argumentative.

In a faltering voice I told Wolfman that the experiments sounded "very exciting"—"important"—but would be controversial. There was the "moral question"—for instance.

"What 'moral question'? Jesus!" Wolfman laughed derisively.

Diligently I continued: Was it fair, was it ethical, to so deceive the subject—the "captain"? After the experiment was over, this person would feel very bad about himself, and bitter at being used . . .

"Also, from what I know of behavioral psychology, which admittedly is a small fraction of what you know, Ira—Professor Axel will certainly not approve."

There. I had said it. The obvious: Wolfman's mentor would not approve.

"But I will be measuring behavior, Adriane," Wolfman said irritably. "That is—exactly—precisely—what I will be doing. Do you not comprehend?"

"But—"

"*Do* you not comprehend? How important this work will be? In the history of psychology? 'Obedience to authority'—a uniformed patriarch. Please help me, Adriane. Don't discourage me. You could be my lab assistant, as well as my friend and companion. My *kinsman*. I could teach you all that I know, we could work together, a series of brilliant psychological experiments . . . Like John Watson and his graduate student, whom he eventually married."

For a moment I could not speak. The heat of elation lifted from Wolfman's skin. His face was slick with perspiration. I could hear only the words—*eventually married.*

"But, Ira—is this experimental work that has already been done, that you are remembering? The Instructions forbid us to—"

"No. It is not work that has 'already been done.' It is, it will be, *original work*."

Wolfman was vehement. I could not challenge him.

"Where would we do this work? Here at Wainscotia?"

"No. Not here. Don't be obtuse. We would need to be independent researchers. We can make our way to California, or Oregon. I have contacts there. I mean, as 'Ira Wolfman.' Of *that other*, no one but you knows. We would have to leave here. My time here is finished, in any case. In the department."

"But we can't leave. We're Exiles. We each have sentences, we can't travel more than ten miles from . . ."

Wolfman shrugged impatiently. Wolfman threw off my hand, where I'd been stroking his twitchy arm.

"Fuck 'Exile'! Fuck 'NAS'! It's here and now I'm concerned with. All that—phantasmagoria"—Wolfman waved his arm dismissively—"the hell with it. I think it has all evaporated—it's *gone down*. The fact is, Axel isn't happy with me any longer. He knows beforehand that I'm the son who will betray him. He's shrewd—you can't run rats through mazes for most of your life and not know how they will run in the future. And pigeons, exactly how they will peck. Accordingly, Axel has 'axed' me—as he has axed other protégés—I've been denied tenure. The departmental meeting was last week. They won't even be sending my name to the president's committee. Axel invited me to dinner at his home, to inform me. And for the last time to dinner, I'm sure—'It was a tragic mistake on their part, Ira. I tried to intervene on your behalf, but I was outnumbered.'"

I tried to comprehend this. Wolfman, the most promising of assistant professors in the department, the young man who'd been selected by A. J. Axel to substitute for him in his lecture course, and to work closely with him in his laboratory, had been fired from the Wainscotia psychology department?

"Not *fired, terminated.* Supposedly there's a distinction. I'm not being promoted, and I'm not getting tenure, because they perceive me as rebellious, which happens to be true. Axel has arranged for me to have two more years, however. (Just coincidentally, the terms of my so-called sentence.) But we can leave earlier, Adriane. At the end of this term."

All this was stunning news to me. Though I could understand Wolfman's harsh-bitten words, I could not comprehend their meaning.

Tried to explain to Wolfman, I didn't think it was possible for us to leave Wainscotia. He had to know that it was forbidden to leave the "epicenter"—"It must be the same sentence, for you as for me."

Wolfman flung off my arm, and sat up. He didn't want to be held now. He didn't want female comfort, or solicitude.

He swung his legs off the bed. We had not—yet—become lovers, and now I understood that we never would be. At least, not in this place. Not in Wolfman's apartment on Myrtle Street.

"If we violate The Instructions . . ."

Wolfman laughed harshly. He was lighting another cigarette, his fourth or fifth of the evening. He flung the wooden match in the direction of an ashtray, and it fell to the floor.

"You believed that crap? 'Exile'? 'Teletransportation'? 'Zone Nine'? None of this is real, Adriane. It's a construct."

"A 'construct'? What do you mean?"

Weakly I sat on the edge of Wolfman's bed as Wolfman paced about the room, sucking in smoke, coughing. Rain continued to lash at the windows of the room, so thickly that a streetlight just across the street was virtually invisible except for a nova-like halo of light at its top.

"As I've told you, I don't think that NAS is monitoring EIs as they'd once done. It might be that the Government is under attack—there could be a new president, in fact—a new administration. Or a new war. You've been out of contact for at least eight months." Wolfman laughed. "Maybe NAS has lost a war? Maybe one of the 'terrorist' enemies has conquered North America? Maybe there has been a 'reform' overthrow, and the old regime is gone. We wouldn't know—maybe, we will never know. And what has it to do with *us*?"

Seeing the look of confusion in my face Wolfman said, extravagantly, "Look: it's possible to see that this is all 'virtual'—Zone Nine. The 'Happy Place.' I'm your friend Ira but I'm also a researcher at CSD—Computational Strategies."

Now Wolfman spoke more calmly, matter-of-factly.

He might have been explaining something to his students that was both an astonishing revelation (to the naïve) and the most obvious common sense (to one like himself).

"You believed it! All Exiles do! Such credulity is based upon your—our—sense of guilt and inadequacy. Zone Nine is 'virtual'—it isn't real. As a CSD researcher I did VVS work—'Visceral Virtual Settings'—my most ingenious project has been a replica of a state university in a small town in Wisconsin, 1959 to 1960—Wainscotia U. It doesn't exist on any map, except the CSD map. I think it's pretty damned impressive—built absolutely to scale, in space and in time. Did you discover any glitch? Any anachronisms? You did not—because the person who created Zone Nine is a genius in his highly competitive field—(you should see the skills of the kids coming up, as young as thirteen—killer sharks!)—and because you didn't expect to see them. In fact, Adriane, you're still in custody in Youth Disciplinary. You've never left the state of New Jersey. You've been there—I mean, here—for approximately eight months, in a comatose/hypnotized state. They're feeding you through a tube, and emptying you out through a catheter and no, your parents never did get notified. *They* feel guilty, thinking it's something they did to have a child taken from them."

I was staring at Wolfman. Speechless.

Not just struck dumb but struck blank. Brain-blank.

Wolfman, smiling at me slyly. Defiantly. Thinking maybe he'd gone too far?—but no, Wolfman never went too far.

"Hey. Listen. I'm a staffer at YD, part-time. They lend me out from CSD. The challenge was to create a virtual reality appropriate for EIs in which to incarcerate you, via the microchip. The microchip *is real*. But nothing else is. You've been dreaming that you were 'teletransported' to a past time, absurdly—how could a past time *exist*? Time travel is a preposterous notion, my dear girl. How could you, so skeptical about B. F. Skinner, fail to understand that simple concept? There is no *past*. There is no *there*. For instance—if we were to 'travel back' in time thirty seconds, how could this be? Would we 'travel back through' our own bodies? And what would we encounter? Our own bodies, at an earlier age? If we traveled back an hour, eleven hours, a month, a year— our own bodies? Would we see these bodies, or would we inhabit them? Which would be 'us'? You've fallen for all of this phantasmagoria, because our virtual creations are so exquisite. I've been particularly praised—my code name is 'the Wolf.' I'm especially pleased with the Van Buren Museum of Natural History and the 'nuclear bomb shelter' below it. Quite a joint achievement! And the pompous professors of Wainscotia—Axel, Morris, Coughland, Harrick, Stein. I know them intimately, since they are my creations. You didn't get to meet them all yet. It did take me a while to create Zone Nine. But I'm patient, as I am skilled at my craft. *That* was why I was drafted into 'special forces' within CSD." Wolfman smiled at me, much pleased with himself. "And, absurd as it sounds, I truly feel protective of you, 'Mary Ellen'—almost, at times, I've come to feel that I've created *you*."

I was stunned by Wolfman's remarks. I could not think how to respond.

I felt like a soft, winged thing, a moth that has been batted out of the air. Not hard enough to break its wings, but hard enough to knock it stunned to earth, and the wings slow-moving, wounded and mute in wonderment.

In a teasing voice Wolfman said "You will never know, 'Mary Ellen,' will you?"

"I—I don't understand . . ."

"But I've told you: Zone Nine is a virtual construct. Not one thing here except you, and me, is *real*."

"Ira, are you serious?"

Wolfman spread his fingers like a magician intent upon showing that he has no hidden tricks.

"Do you think that I'm 'serious'? Or—what is the opposite of 'serious'?"

I stared at Wolfman who smiled at me with maddening insouciance. As if to prompt me, he said: "As a dreamer doesn't know she is dreaming, so you can't know your circumstances."

"But—you know, don't you? Please . . ."

Wolfman lay his hand on my shoulder to comfort me. In his face was an expression of both kindliness and exasperation.

"Yes, I'm joking, Adriane. Of course. I was just running a maze—that is, exploring a possibility. If I were a cybertech staffer at CSD, instead of a teletransported EI, what I've told you might be plausible. Or plausible in some way. But the painful fact is,

I'm at a loss to control my life in Zone Nine, just like you." Wolf-man laughed, baring his teeth in a grimace. "Maybe I did cre-ate Zone Nine but my supervisor appropriated the program and trapped me inside it! Like my parents who'd invented a biological weapon who might then have been infected by it as a part of the research . . . possibly by their own supervisor, to get rid of them once they were no longer useful. We're trapped in our own exper-iments. Please don't look so dismayed, Adriane—I'm telling the truth now. The bare, plain, irremediable truth. If Zone Nine were my invention I'd have arranged for my own freedom—and now, for yours. But I'm trapped here, in the epicenter—just like you."

Wolfman held out his arms, to comfort me. Sobbing I pressed myself against him.

I could not trust him. His words had been devastating to me, even those I could not truly comprehend.

Yet, I had no one else.

ELOPEMENT

Wolfman insisted that we leave Wainscotia at the end of the term.

"No one can stop us. No one will even notice. We'll 'elope'—and begin a new life."

Elope. So long I'd waited for Wolfman to seriously love me. Or, at least to say that he loved me.

He had not uttered the word, yet. But if he wanted me to leave Wainscotia with him, and to begin a "new life" with him, it must be understood between us that he did love me, so far as Wolfman could love another person.

Wolfman's plans were vague but thrilling. We would live together under assumed names, somewhere in the West. In April 1960 the United States was not yet the (Reconstituted) North American States in which all citizens were under continuous surveillance. Much of it, in the western states, was unpopulated, virtually uncharted. It had not yet been *wired*. There was no *grid*.

"Our enemies thought to exile us in this 'prehistoric' place," Wolfman said, "but in fact, the absence of technology will be our salvation. There are vast empty zones where human consciousness doesn't intrude. We can get work of some kind when we need money. I can use my hands as well as my wits. I have friends in

Berkeley who will take us in. We can prepare new documents—
birth certificates, ID's. No one will know who we are, or once
were."

Wolfman spoke with such conviction, I could not doubt him.

It was the end of the spring term. "Mary Ellen Enright" had
completed her first year at Wainscotia, with a near-perfect grade-
point average.

Impulsively Miss Steadman hugged me when she'd received
the news from the dean's office. "All of us in Acrady Cottage are
so proud of you, Mary Ellen!"

I was embarrassed by Miss Steadman's words. And I felt a
vague shame, for my high grades had been a consequence of desperation and loneliness, and few other students could have been
so motivated. My very misery had given me an unfair advantage.

Overnight, Acrady Cottage had emptied. My roommates had
departed with hugs, kisses, promises to see me in the fall—though
there were no plans for us to room together again.

In the summer term I would continue to live on campus, but
in another residence. I would work at the university library five
days a week. During the university's summer session, which was
an accelerated six-week period that began in late June, I would
take a course in mathematics, which was a weak subject for me.
(So Wolfman had advised me, months ago.) A life of routine lay
before me like an unmarked calendar. I felt the comfort of such
emptiness—no emotion, no anxiety. If Wolfman departed Wainscotia as he'd been threatening to do.

Without Wolfman, I would live my life from day to day, until
my sentence was completed. Could I endure this life?

In despair I said to Wolfman, "How can I come with you, when
it's forbidden?"

And when you've never said you love me.

"I've told you, Adriane. No one is watching us."

"But—how can you be sure?"

"Trust me."

Badly I wanted to trust Wolfman. His teasing words about Zone 9 as a virtual construct in which I'd been placed unknowingly had had a profound effect upon me—I woke in the night, frightened that Wolfman had been telling the truth, and that he was an agent of NAS; at other times, I was certain that he'd been joking.

I said to Wolfman, "What is the meaning of our lives? Does psychology tell us?" and Wolfman said, "Psychology is a mirror into which we look, and see our own faces. No science can tell us *why*."

Though Wolfman had spoken disparagingly of the psychology department at Wainscotia, I understood that he'd been deeply wounded by his failure to be granted tenure, and promoted. And I understood that the reason was probably the reason he'd told me—his ideas were unacceptable to the older faculty members, particularly A. J. Axel.

Wolfman had the idea of packing a selection of our belongings and mailing them c/o General Delivery, Berkeley, California. A half-dozen cardboard boxes filled with clothes, books, personal items, neatly taped and stamped for third-class mail.

"But—how will we leave?"

"We'll leave unobtrusively. We'll hike in the arboretum—and we'll never return."

So Wolfman convinced me. Ever more, Wolfman was affectionate with me, and appealing to me as he'd rarely been in the past. His disgust with me over the matter of SANE seemed to have been totally forgotten.

So it happened, I brought my things—a small selection of my

mostly unwanted things—to Wolfman's apartment, where we packed them into boxes as he'd planned. Wolfman was a perfectionist in taping boxes shut, making sure they were secure and snug. Then, we took them to the post office to mail to Berkeley, California.

Did I ever think, as we hauled the boxes to the post office, that I would open them? Remove my folded clothes, my books?

Did I ever think that Ira Wolfman and I would journey to California together, to begin a *new life*?

In the days before our planned departure, I was very excited—frightened, elated. I felt both reckless and resigned.

And Wolfman, too, seemed both reckless and resigned.

For every time I expressed doubt of what we planned to do, Wolfman cut me off: "Adriane! Have faith in me."

I was very frightened at the prospect of violating The Instructions. But I was more frightened of losing Wolfman and finding myself again alone in Exile.

I thought—*I have lost my parents, but I have Wolfman.*

Wolfman's plans were becoming more specific. The latest news was that we would definitely have jobs in a research institute at UC Berkeley, where Wolfman had friends. These friends would provide legal documents for us, as well.

"They think it's romantic, that we're 'eloping,'" Wolfman said happily. "And they have no idea who we are—of course."

On May 19, at 8:00 A.M., Wolfman waited for me just inside the arboretum. We wore hiking clothes and boots. On Wolfman's head, a baseball cap. "It's important that we look like anyone else," Wolfman said. "The more ordinary, the better."

We were both wearing backpacks. Wolfman stipulated that we were to carry several bottles of water apiece as well as food for several meals, a change of clothing, warm socks. Wolfman would supply a Swiss army knife.

Wolfman had charted our route. We started out on our usual trail but, instead of taking a loop that would return us to campus after two miles, we took another fork, that led us farther from campus, in a region of pinewoods that stretched for miles to its northern boundary at a state highway in the adjacent township of St. Cloud.

Wolfman had calculated: we would take a Greyhound bus from St. Cloud to Minneapolis; from Minneapolis, we would take another Greyhound bus to Denver, Colorado; and so to San Francisco, California, and Berkeley. Wolfman knew the departure times of these buses, and Wolfman had money for our tickets.

Briskly Wolfman was hiking into the arboretum. I had to half-run to keep pace with him.

It was early enough in the morning that no one else was around.

It was a partially cloudy day in spring, not warm, but clear, still. My heart was beating in that quick erratic way it had learned to beat, when the arresting officers had come for *Strohl, Adriane* and Mr. Mackay had so eagerly handed me over to them.

This was not a good memory. Blindly I clutched at Wolfman's sleeve. "Will I ever see my parents again, Ira? How will I return to see my parents?"

"Please don't cry."

I hadn't known that I was crying.

Why, when I was so happy?

THE BAT

We set our faces forward. We did not look back.

Soon we were out of sight of the university campus and ever more faintly came the chapel bell like something heard undersea.

The trail ascended. The trail was soft with pine needles yet there were outcroppings of rock that mimicked steps, you had to be careful not to turn your ankle as you climbed.

The trail was steep enough yet would have been steeper, near-vertical, were it not for numerous switchbacks.

These switchbacks cast us back upon ourselves, it almost seemed. Our progress was slow for the trail was shaped like a great snake with a rippling body. If Wolfman plunged ahead, the switchbacked trail cast him not so very distant from me, approaching and then passing me, as I followed trying not to stagger beneath the weight of my backpack.

Wolfman had said—*A backpack can mean the difference between death and survival. A backpack can save your life.*

High overhead in the trees, quarrelsome birds.

The trail became fainter. The trail was lost in the pine forest.

Yet, Wolfman pressed forward. And I followed.

Miles of forest. Climbing what must have been the side of a

mountain. And then, beginning a sharp descent that was as difficult as the ascent.

Small avalanches were loosed by our feet.

We were panting, from the ascent. We were wary of the descent for descents are always dangerous.

The sun had shifted in the sky. The sun had been blinding, our eyes filled with tears.

We were very happy! Soon, we would be free of Wainscotia.

Yet, after six hours of hiking Wolfman cried out in surprise and disappointment—"My God!"

At first, I didn't understand.

Then, I saw.

Though we'd been walking for six hours, following the trail that Wolfman had carefully chosen for us, to lead us to St. Cloud, it seemed that the trail had looped back upon itself, and had brought us again to the entrance of the arboretum, close by the university campus. The tolling sound we'd been hearing was the chapel bell.

Wolfman said, "This isn't possible . . ."

I could not believe that we'd returned to the campus, yet it appeared to be so—the trail had betrayed us, leading us in a large looping circle. We had climbed up the side of a mountain and we had descended the side of a mountain. We had believed that we were miles from Wainscotia and yet—we had covered less than a quarter mile.

Somehow, the trail with its many tortuous switchbacks had turned us around, though on the maps it led out of the arboretum, unmistakably.

Wolfman was staring at the map, trying to determine where he'd made a mistake. But I could see that Wolfman hadn't made a mistake, except in his conviction that we could leave Wainscotia.

"We're trapped, Ira. We can't leave. I knew this."

"Go back, then! Go back alone. I'm not coming back with you."

Wolfman's voice shook with fury.

Wolfman pushed away my hand. He'd turned and was half-running, panting, back along the trail.

A roaring rose in my ears. I wanted to follow him, to stop him. But I could not move any farther, I was so very tired.

Six hours and more I had been hiking in this accursed place. Six hours and more, and the trail had brought us back to our beginning.

I called after Wolfman, who ignored me. Wolfman, my love!

Overhead, birds were fluttering in the trees. Was it dusk, so soon? A nervous chattering of birds, a flurry of wings.

In pine branches overhead a small bird, possibly a bat, was circling strangely, as if rabid. Then, as I stared, the thing—black, swift, unerring—swooped down to rush at Wolfman, struck him on the side of the head and entered his head, suddenly aflame, engulfing him in flames, within seconds turning the man to vapor only a few feet from me.

There had been time for only a brief high-pitched cry of horror and loss—I think it must have been mine.

For in that instant Wolfman fell, Wolfman died, Wolfman vanished from Zone 9.

III

WAINSCOTIA FALLS

HIKER STRUCK BY LIGHTNING, WAINSCOTIA ARBORETUM

18-YEAR-OLD WAINSCOTIA STUDENT HOSPITALIZED

Found Unconscious on Trail by Hiker, Dog

Wainscotia Falls Journal-American, MAY 20, 1960

SAVED

A warm soft muzzle against my face that had gone stiff with cold.

For in shock the blood had drained from my brain.

My breathing had ceased. Like a living creature that has been pinched cruelly in two, and its life snuffed out.

Where I'd fallen heavily on the trail. Outcropping of shale sharp against my stunned face which bled from abrasions I could not see and could scarcely feel.

Could not move not even to cast my eyes into the place where Wolfman had died.

What had been Wolfman, destroyed.

Shock like icy water into which I'd been flung. The shock of an arrested heartbeat. Eyes blinded by light.

Then came a sudden warmth against my face. The panting-hot breath, a dog's quick damp astonishingly soft tongue.

There came a cry—Rufus! Where are you?

A hiker on the trail. Running up behind us where we'd fallen.

Then—Hello? Are you hurt? What has happened . . .

Through the haze riddled with pain came a dog to lick my face to revive me. Eagerly, anxiously. Emitting sharp-distressed cries so close to human, I wept to hear and to know that I had been saved.

THE MIRACLE

For a long time then I was sick.

It might have been weeks. Months.

For there were intervals of "wakefulness"—followed by periods of "forgetfulness."

For there was not what could be called a *steady recovery*. There was not a *steady progress*.

More like switchbacks on a trail. With painful slowness you make your way forward—only to find yourself switched-back, facing the direction you'd come from.

In such a way time turns back upon itself. You believe that you are making progress, but it is an illusion.

Yet, this is *progress* of a kind.

At first I wasn't sure where I was. These (austere, white-walled) rooms to which I was taken (gurney, wheelchair) were different places serving different functions, purposes. Different medical workers, in different uniforms. Often I would open my eyes—(my aching eyes with their blurred vision)—to find myself in a place that was new to me.

Possibly, this was a hospital. Or a rehabilitation clinic.

Some place in which there are degrees of "hospitalization" relating to degrees of physical injury, psychological deficit.

Do you know your name?

Do you know where you are?

Do you know the date?

Do you know who is president of the United States?

Like a small child who is eager to talk but doesn't yet have the power of speech I was eager to give the correct answers to these questions but I did not feel confident that I knew the correct answers. I understood that it was wiser to give no answer than to give an incorrect answer, that might be held against me.

For I was remembering an elaborate examination I'd once taken—when I'd been a student—(I think this was recent, since I've been told that I am a student at the present time, at Wainscotia State University in Wainscotia Falls, Wisconsin)—in which it hadn't been enough to know the correct answers to questions, you had to know which "correct" answer was the "most correct" choice of multiple choices.

Your name is "Mary Ellen Enright"—do you remember your name?

And do you know where you are, Mary Ellen?

Do you know why?

SOMETIMES I WOKE, and the voice(s) continued.

As if there'd been no interruption, the voice(s) continued.

Do you know what happened to you, Mary Ellen?

The doctors believe you were STRUCK BY LIGHTNING.

And, you survived!

It's very amazing, Mary Ellen! It's considered a "miracle."

Your picture has been in the newspapers, Mary Ellen! At least, in Wisconsin papers and on local, Wisconsin TV.

You will be a sophomore at Wainscotia State in the fall. Do you remember your freshman year?

Do you remember—you are an honor student? You are the recipient of a University Scholarship?

You are recovering, Mary Ellen. Dr. Fenner—your neurologist—says the prognosis is GOOD.

All of your doctors say the prognosis is GOOD.

You were found on a trail in the arboretum—a hiker found you—in fact his dog found you.

You were unconscious. You did not seem to have a pulse nor did you appear to be breathing.

The hiker's dog led him to you. By his quick actions he saved your life.

Mary Ellen, you are a very lucky young woman.

You were in an acute state of shock but—you are recovering.

Your blood pressure had plummeted. Your left eardrum was perforated.

You are recovering steadily—you are ALIVE!

A MIRACLE—*people have said.*

STRUCK BY LIGHTNING *in the arboretum where you might not have been found for hours and already it was believed to be approximately ninety minutes you'd been unconscious, before you were found.*

Hiking alone in the arboretum. In the more distant parts of the arboretum, it can be dangerous.

It is always recommended that no one hike alone. Even an experienced hiker should not hike alone.

Fortunately, you were discovered. In time.

Your lips were blue. You appeared to have ceased breathing. You appeared to have no pulse.

Fortunately, the hiker knew CPR.

Mouth-to-mouth resuscitation. He'd been trained in first aid as a Boy Scout.

STRUCK BY LIGHTNING *and survived—that's very rare.*

That's a "first" for Wainscotia Falls!—we are proud of you.

If you had family, Mary Ellen—they would be so very grateful that you are ALIVE.

MARY ELLEN—why are you crying?

Are you in pain, Mary Ellen? Where is the pain?

You can point, Mary Ellen, if that's easier.

Your heart? Your heart hurts?

Or do you mean—your heart is broken?

GRIEF

Can you explain but I could not.

Why you are crying as if your heart has been broken.

SO MUCH TO BE grateful for! I knew.

For slowly I was regaining the use of my legs. And the coordination of muscles that in normal people is taken for granted.

If for instance you decide to stand, your legs will not meekly buckle and cast you to the ground.

If you decide to lift a glass in your hand, your fingers will not release the glass to fall clattering onto the floor.

If you open your mouth to speak, you will not begin shivering, shuddering, weeping inconsolably instead.

Weeping as if I'd lost something—or someone—and could not remember what, or who.

Nor could I remember who the person had been—*Mary Ellen Enright*—(the name on the hospital bracelet on my left wrist)—though I stared into a mirror at the unfamiliar face, shaping the unfamiliar name with numbed lips.

Mary Ellen Enright. A riddle that, no matter how long I puzzled over it, I could not solve.

For I knew—(but how could I *know?*)—that it was a false name.

As I knew—(but how could I *know?*)—that I had not been alone on the trail when the lightning struck.

As if I'd been flung down from a great height onto the needle-strewn trail. And the breath had been knocked from me. And when I shut my eyes there came rushing at me a mysterious object the size of a bat—in panic I ducked, cried out in terror, covered my head with my arms, tried to scream but could not.

No no no no no.

THEY BELIEVED: the patient was reliving the lightning-strike. The powerful charge of electricity that threw her to the ground, and the deafening thunder that followed, that perforated her eardrum.

They believed: the patient was an orphan, had been an orphan since early childhood, and had lost her adoptive parents as well, in an automobile accident. These traumatic losses the patient was mourning in her weakened and semi-delirious state.

OFTEN I WAS very happy! With each improvement in my "condition"—I tried to rejoice that I was *very lucky.*

Yet in the midst of, for instance, physical therapy—(which included swimming)—the terrible tears came, and would not soon cease.

Like a convulsion of the body, such grief.

Like a great snake rippling inside me, that cannot be contained or controlled and whose outlet is tears.

Why did I cry so helplessly, and bitterly—I did not know.

(It was not pain. Or it was not solely pain. The many pains in my legs, my backbone and my neck, my head, my eyes—these I could bear without crying, for they were only *physical symptoms.*)

Slowly my ability to *think* was returning. My ability to *concentrate*, and to *remember*.

The neurologist explained to me that my "short-term" memory had been affected by the trauma of the lightning strike. There had been, he believed, some (temporary) brain damage, judging by my symptoms.

Memories stored (temporarily) in the hippocampus before being stored (permanently) in the cerebral cortex had been lost, and could not be retrieved.

It was a normal reaction to a traumatic brain injury, Dr. Fenner said. I had not lost my grasp of language—how to speak, read and write—and certain skills—like walking, climbing stairs, swimming—but it was evident that I'd forgotten much in my recent life.

As if someone with a giant sponge had vigorously washed, scrubbed, wiped a part of my brain clean.

I asked Dr. Fenner if there had been a CAT scan of my brain? Or—what was the term—"MRI"?

Dr. Fenner smiled quizzically, cupping his hand to his ear.

"What was that, dear? Something about a—cat?"

"A CAT scan. Or an MRI."

"But—what are you saying, dear? 'MIA'?"

Dr. Fenner was an older gentleman, kind and solicitous to his patients but with a very limited patience for foolish questions; I had noticed that the nurses were careful in his presence, behaving more like servants than trained staff workers. He wore a stiff-starched white shirt and a necktie beneath his physician's white coat, that bulged over his small high belly. Involuntarily I'd shrunk from his touch, as from the nurses' touches, which were with bare hands, where I would have expected thin rubber gloves, when I'd first become conscious of my surroundings; but by this time, after several weeks in the hospital, and now in a rehabilitation wing adjacent to the hospital, I'd become accustomed to the hospital staff working with bare hands. I told myself—*But they are probably washing their hands all the time. When they leave a patient's room.*

Dr. Fenner's necktie fascinated me also. For it was just perceptibly mottled with stains—(food stains?)—and when Dr. Fenner bent over me, it swung at my face.

"'MIA'—'Missing in Action,' I think?—why are you asking about that, Mary Ellen?"—Dr. Fenner was utterly bewildered.

Rapidly I tried to think. But my thoughts came slow and halting, like my walking—sometimes with a spurt of strength, sometimes a flurry of weakness.

"I—I don't know. I guess I don't know what I mean, Dr. Fenner."

It was so: I had no idea what I'd meant. The terms *CAT scan, MRI* were bewildering to me too, like the arcane Latin terms I'd typed in the dim-lit university building whose name I couldn't quite recall—a museum of natural history with which I had an association both vague and exhilarating, like a tumultuous dream that has faded except for its emotional residue.

"Well! You've had a brain trauma, my dear. But you'll recover—I've seen miraculous recoveries in the past."

So often the word *miracle* was voiced in my presence, I was beginning to believe—*miracle!*

(Yet I wondered what percent of the neurologist's patients recovered. And what percent of the neurologist's patients were untreatable.)

Again, when Dr. Fenner left, I burst into tears.

Out of nowhere a terrible sadness swept over me, and I began to cry helplessly, like an abandoned infant.

One of the nurses asked what on earth was wrong?—Dr. Fenner had been *so positive.*

VISITORS

Few visitors came to see me. And each was a surprise to me, as if stepping out of a pitch-black part of my brain.

The first identified herself as *Ardis Steadman,* who'd been the resident adviser of Acrady Cottage, my freshman residence.

Did I remember Miss Steadman? Did I remember my roommates? Did I remember Acrady Cottage—"Such a spirited group of freshman girls this year! And you, Mary Ellen, pulled our grade-point average 'way up.' We were all very grateful for *you.*"

I told Miss Steadman that yes, I did remember her—of course.

We'd gone to a music recital together, I thought. Or—we'd watched TV together in the sitting room of the residence.

As Miss Steadman spoke, reminiscing of events I could not recall, and had no interest in recalling, I began to feel the sensation of terrible loss wash over me, and began to cry.

"Oh, Mary Ellen! What's wrong? Have I said something—?"

I tried to think *why.*

"It's like—I've lost something. But I don't know what it is."

"But when did you lose it, Mary Ellen?"

"I—I don't know . . . Maybe in the arboretum."

"From what I've read and heard, there was nothing on the trail that you'd left behind, evidently. You were hiking alone, and you were wearing a backpack—that was all."

"I remember the arboretum—from before. I'd gone there before. But the last time—" Quick sharp pains struck my eyes, from inside. My vision, that had been improving, became misty now, so that I could barely see Miss Steadman's concerned face. "—the last time, I don't remember."

Miss Steadman clasped my hand. "That may be a blessing, dear. No one would want to remember being struck by lightning!"

And there came, with some embarrassment, Miss Hurly—she'd been astonished to read about me in the papers, and to hear about me on the radio, and to realize that the girl-struck-by-lightning was *Mary Ellen Enright*.

"Such a shock! Dr. Harrick sends his regards and very best wishes for your recovery. He said—when I'd told him about you, and showed him the news stories—that he had once come close to being struck by lightning himself, boating on Lake Michigan as a teenager—when a sudden storm came up. It was quite a suspenseful tale Dr. Harrick told, really quite emotional—how terribly close Dr. Harrick had come to being killed, at such a young age; and only think, all the great work he has done in science would have been lost . . . He did remember you, Mary Ellen, though at first he'd mixed you up with our other girl Lorraine who'd been working on Mondays and Wednesdays."

And there came a woman whom I didn't know, whom I had never seen before, who introduced herself as *Cornelia Graeber*— "Please excuse me, Miss Enright, but I saw your picture in the newspaper, and I—I felt that I knew you, somehow—and I wonder if you know me? In some way? I'm sometimes called 'Nelia.'" This individual was edgy, nervous; as she spoke to me, she picked at her fingernails, and tugged at her hair; she seemed quite anxious in my presence, as well as bewildered—she had no idea why she'd been drawn to see me, but the idea had obsessed her, and at last she'd come. She was about thirty years old. She reminded me of myself—somehow. Her eyes were disconcertingly intense, fixed upon me in a way that suggested that yes, she knew me; but she wasn't sure why. She explained that she was a Ph.D. candidate in psychology, working with A. J. Axel. She'd looked up my course work, out of curiosity, and discovered that I'd taken Axel's popular lecture course Psychology 101 the previous fall, and that I'd had an instructor named *Ira Wolfman* as my quiz section instructor— "Do you remember him? 'Ira Wolfman'? He left Wainscotia abruptly, at the end of the term. He didn't say good-bye to any of us. Not even A. J. Axel. It was very—upsetting. Of course, Ira may have had reasons—professional reasons. But to leave without saying good-bye to any of his friends and colleagues—that wasn't like him . . ."

The sharp shooting pains behind my eyes were making me blink rapidly. Tears flooded my eyes. I could not hear the woman's words clearly for a roaring in my ears like a freight train. Badly I wanted to tell the woman to go away—I didn't know her, I didn't know what she was talking about, the names of most of my professors had vanished from my memory, and could be retrieved only with great difficulty. And it wasn't clear why this stranger was addressing *me*.

Drawn by the sound of my sobbing, one of the nurses intervened. Quickly the woman apologized, and departed.

The fit of convulsive sobbing was so extreme, I could not catch my breath and began to hyperventilate. I had to be rushed to the ER and given oxygen and an IV medication to bring down my suddenly accelerated heartbeat, that was racing at 260 beats a minute.

THEN, THERE CAME Jamie Stiles to see me.

"Hello? Mary Ellen—"

At first, I didn't remember him. In ungainly stained bib-overalls and no shirt beneath, with bristling dark whiskers, and sandals on his large gnarly feet, he was a startling and intimidating sight.

Telling me, with awkward humor, that "Rufus" wanted to see me also but hadn't been allowed inside the hospital—his leash was tied to a bicycle rack outside.

For it had been Jamie Stiles's dog Rufus that had discovered me fallen on the trail, and rushed barking to me.

It had been Jamie Stiles who'd been hiking on the trail—(not the trail that I'd been on but one close by)—and who'd applied mouth-to-mouth resuscitation to make me breathe again.

Jamie Stiles, who'd been one of the SANE protesters. Who taught sculpting part-time in the fine arts department at Wainscotia State.

Jamie had remembered me, he said. Seeing my picture in the newspaper he'd remembered me from the SANE march.

By this time he'd stepped inside the room. For a big man he was shy. Nurses were glancing at him curiously as they passed by in the hall. A flush rose in his face. I saw now that in his right hand he clutched a bouquet of flowers which looked as if they'd been picked in haste in a field.

There was some fuss about getting a vase for the flowers. When I lifted the flowers to smell them, the scent that lifted from their petals was very faint, yet sweet. Confusedly I thought—*Is this a long time ago, or now? Or has it not happened yet?*

In fear of shutting my eyes. For there was the possibility that, if I did, something small and dark would rush at me and cause me to scream and when I dared to open my eyes again, I would be alone; or worse yet, one of the nurses would be bending over my body.

As Jamie Stiles spoke haltingly to me I thought that yes, I would have remembered this man—I wanted to remember him. For his face was familiar, in a way that a face is familiar which you have been seeing all your life.

Jamie Stiles was deep-chested, with a thick neck, muscular arms and shoulders, jaws covered in bristling dark whiskers that looked as if they would be wiry to the touch; yet his eyes were kindly and concerned, and puzzled. For he'd seemed to know me, he said. When he'd seen me on campus at the time of the protest he'd felt this strongly—"No reason, I guess. The phenomenon is called *déjà vu*."

(*Déjà vu*. The term was familiar to me. Though in the psychology textbook in which I'd encountered it the phenomenon was discounted as usually mistaken, delusional.)

Jamie Stiles was soft-spoken at my hospital bedside but I recalled from his behavior at the protest march that he was a man of strong feelings, impulsive, courageous.

He'd spoken scornfully to me, I recalled. Oh but why?

But then, he'd spoken tenderly to me. He'd forgiven me for my ignorance.

Despite the size of his hands, and his stubby clay-stained fingers, there was gentleness in Jamie's handclasp. In his voice, sympathy and hope.

I felt tears threaten me. Though I tried very hard to keep them back tears spilled onto my face like acid and I began to cry as Jamie looked on in distress.

Unlike the hospital staff Jamie Stiles did not ask why I was crying. He did not remark that I was *very lucky* to be alive.

He said very little on this or subsequent visits. For Jamie Stiles is not a verbal person.

He did not say *What have you lost, why are you grieving. Why, when you should be grateful to be alive!*

He did not say *In another story, you were electrified to a crisp. A smear and a smudge and a smelly hole in the snow, that's all of you.*

So that I knew, and there was great comfort in the knowing—*He will be the one.*

"UNCLE"

His name was Cosgrove, David R. He introduced himself to me as *an old-fashioned family doctor.*

He didn't know Dr. Fenner, or any of the physicians at the hospital. His practice wasn't in Wainscotia Falls but twenty miles away in St. Cloud. He'd been "intrigued" by my picture in the newspaper. He'd come to visit me just to say hello and to ask a question or two.

"Like Benjamin Franklin, I have an abiding interest in electricity. It's a hobby. The history of 'reanimation'—for instance."

Reanimation? I tried to think what this might mean.

Dr. Cosgrove was a lean wiry man of some mysterious age— fifties?—sixties? Slightly ill at ease, though boyish, and very friendly; thistle-colored hair had nearly vanished from his head, and his eyes were couched in shadowy puckered skin. He had a long thin nose with a tiny bump in the bone and he was smiling hard at me as if to impart some secondary meaning to his words, totally lost on me. In his left cheek was a tic or a twitch, which was distracting. He was carrying a brown leather bag much creased from years of wear.

"Unless you'd rather not talk about it? The 'lightning strike'?"

Vaguely I shook my head, no. Meaning—yes.

Meaning—yes, I could talk about it. Though I didn't remember anything helpful at all.

Dr. Cosgrove continued to smile at me in that curious way. A strange sensation came over me—*I know this man! I have seen this man before.*

Dr. Cosgrove chattered about lightning, electric current, those instances of individuals who'd been "struck" and yet lived, until one of the nurses who'd been in the room left; then, he ceased speaking abruptly.

He went to the door, and shut it carefully.

Doors to hospital rooms are rarely shut, still more rarely are they shut so carefully, so that the lock clicks.

"Very good to see you, 'Mary Ellen,' dear! I've been told that you will be discharged in a few days, you've made a miraculous convalescence."

Dr. Cosgrove continued to smile, even as he removed an object from the brown bag. It looked like a wand, or a small phone—a small, flat phone?—a shape that was familiar to me, that fit in the palm of the man's hand, but which I could not have identified.

Ah, a cell phone! I had not seen a cell phone in—how many months? Years?

But no, the object was not a cell phone after all.

Dr. Cosgrove was frowning, turning a dial between his thumb and forefinger, that caused the small flat object to stir, to emit a low buzzing sound like a hive of wasps.

Seeing the perplexity in my face he said quickly, "Just a little 'baffle'—of my invention."

"'Baffle'?"

"A kind of 'white noise,' in case anyone is listening to us. Advertently, or in—"

Dr. Cosgrove adjusted the buzzing-level of the object in his hand. Now it sounded like honeybees at a little distance, a friendly sound. But all this was utterly mysterious to me.

I wondered—why would anyone be interested in our conversation? I could not imagine who these people could possibly be—no one on the hospital staff, certainly.

Satisfied that the small flat object was operating as he wished Dr. Cosgrove pulled a chair close beside my chair—(for I'd been sitting up and reading beside my bed; I'd become very tired of lying flat and helpless in bed)—and smiled at me with an air of complicity.

An eccentric individual, I thought—but a gentleman.

Dr. Cosgrove asked if indeed I did remember anything about being struck by lightning; and when I shook my head no he said, frowning, "But think again, 'Mary Ellen.' Try to remember."

I had tried, many times. This request had been put to me frequently. Most recently by a *Wainscotia Falls Journal-American* reporter who'd been assigned a feature article with the sensational title THE GIRL WHO CAME BACK FROM THE BEYOND. The disappointed reporter had had to interview several doctors, nurses, and a professor of physics from the university, since I hadn't been able to supply him with much information.

"Did it feel as if you'd been 'transported' from—somewhere? I mean, when you woke up."

"I—I don't know. What do you mean 'transported'?"

"Or, a more accurate term might be 'teletransported.'"

Teletransported. What did this mean!

(I wished that my new friend Jamie Stiles were here. It had become a habit to tell Jamie most of what happened during my days in rehab, as Jamie told me what happened during his days, when he came to visit me in the evening; but this exchange with Dr. Cosgrove was so peculiar, like something in a dream, I despaired that I would be able to convey it successfully.)

In a yet more cautious voice Dr. Cosgrove asked if the name "Eric Strohl" meant anything to me

Eric Strohl. I wasn't sure.

"'Eric Strohl.' 'Madeleine Strohl.'" The doctor spoke slowly, and quietly, just loud enough for me to hear him over the buzzing sound of the flat little object in his hand, even as he held his hand over his mouth as if to shield his lips.

Eric Strohl. Madeleine Strohl. I began to tremble. I had no idea why.

"Does the name 'Adriane Strohl' mean anything to you?"

My heart was beating rapidly. I was frightened that I would begin to hyperventilate again. The pain in my eyes throbbed.

Dr. Cosgrove reached out to take hold of my wrist. Gently, he pressed his forefinger against my pulse.

"Calm, now! Remain calm."

"I—I don't—"

"Calm, my dear! Just breathe normally. You may want to count your breaths."

I counted my breaths, to ten. By ten, I was not so agitated. Dr. Cosgrove released my wrist.

"'Adriane Strohl.' I'm just curious—if you have ever heard this name before."

"I—I think—" I was straining, trying to remember. I felt as if I were about to stumble over a curb, as in a dream. "I don't *know.* What is it again—'Adrian'—"

"'Adriane.'"

Very deliberately Dr. Cosgrove pronounced this name. It did not mean anything to me—did it?

Dr. Cosgrove and I stared at each other as if across an abyss. Not a wide abyss, but very deep. I felt that familiar weakness in me, as if my bones were turning to water.

From some long-ago time I remembered being told, unless I'd read these words—*The striking thing about self-knowledge is that it may be lacking.*

When I'd awakened from my comatose state in the Wainscotia hospital I'd felt this weakness through my entire body—no way to describe it except as terrifying, appalling.

That sense that the body is a precarious entity comprised of numberless atoms, that might disintegrate at any moment.

And beyond the body, the world itself—the very Universe—poised at the brink of detonation.

Leaning close to murmur in my ear Dr. Cosgrove said, "It may be, dear, that I used to know—your parents . . . In fact, I have reason to believe that I am related to your father. And so, to you."

This was astonishing! For a long moment I stared at the bald-headed doctor with the earnest eyes, and had no idea how to reply. "I don't mean to surprise or shock you, dear. I realize—you've been told that you were 'adopted.'"

"I—I don't really remember my parents, I'm afraid. The parents who adopted me, and the parents who . . ."

Dr. Cosgrove regarded me thoughtfully. He reached out to take my hand in his, my hand in both his hands. I thought—*He knows me.* The absurd thought came to me—*Maybe he was the doctor who delivered me.*

"Your father Eric was—is—my older brother. I have reason to believe. For you look very like him—unmistakably. I saw the resemblance in the newspaper photograph—I was sure." He paused, wiping at his eyes. "But we've been separated, your father and me. We have not seen each other in nearly twenty years."

Have not? This man was speaking as if my father were alive at this time.

I was very confused. Pulses pounded in my head, exacerbated

by the drugs that coursed through my veins, seeping into my blood.

"You're my uncle? But—where did you—where did my parents—live? How is this possible?"

It did seem to me now that Dr. Cosgrove looked familiar. The uncanny dark glisten of his eyes, the tiny bump on his nose . . .

"I think that it is possible, 'Adriane.' You won't remember—probably—because you were very young when I'd last seen you. About two years old, I believe."

"Oh—where was this?"

"In another part of the country."

"But—which state?"

"In New Jersey, I believe."

But I have never lived in New Jersey—have I?

"Have you ever heard of Pennsboro? New Jersey?"

"N-No . . . I don't think so." I could not think at all.

"Maybe—*y-yes* . . ."

I was shivering badly. Wiping tears from my eyes. Dr. Cosgrove apologized for upsetting me.

He held my chilly hand in his and stroked it, for some time.

"Adults have not treated you well, my dear girl. I would not exacerbate your confusion and grief. But let me ask—do you know the name 'Tobias'?—'Toby'?—'Uncle Toby'?"

I did not know how to reply. If this man was my uncle, I wanted to say *yes*.

"I—I'm not sure. 'Uncle Toby.'"

"That was my name once—'Tobias.' Before I was sent to Wisconsin to complete my medical degree at U Wisconsin–Madison; then I settled north of Wainscotia Falls, and haven't left St. Cloud since. I'm married—have been married for a long time. I married a dear, kind, beautiful girl from here—Zone Nine. I have children—that is, you have cousins . . . But we may not see each other again, I think. It's too risky for us both." Dr. Cosgrove blinked tears from his eyes, though he was smiling. "I've been concerned about coming to see you—I wasn't sure if it was a good idea. But of course, I wanted to see *you*—my dear niece Adriane. I wanted to ask about your father—your family. How they are, how things are in that other—dimension. But you can't answer me, I think. You can't 'remember.'"

"I—I can almost—remember something . . ."

When Dr. Cosgrove gazed at me with his dark-brown glistening eyes, and as he smiled at me in so kindly a way, almost I did remember—something. But at once it evaporated like a dream in bright daylight.

"You seem to be doing well, 'Adriane'—that is, 'Mary Ellen.' At least physically, now that you're recovering. Young people are so much more resilient than older people—you're still young, though you've traversed decades."

"But—how old are *you*? When were you my uncle?"

"Ah, dear—it hasn't happened yet! And when it happens, when I am a young man, in my twenties, and you are newly born, some years from now—we won't remember a thing. Amnesia is all that saves us from the abyss."

Seeing that I was looking confused Dr. Cosgrove said quickly, "All we can do is persevere in our own time. No one has to deal with more than one day at a time. That's the blessing of our temporal universe—time is spread out horizontally, you might say; it doesn't all happen at once, as at the instant of the 'big bang.' Some of us are political beings, and some of us are not. But it can't be the case that any of us—in 1960, in the United States—or elsewhere, at another time—can be wholly *well*."

I didn't understand this. I knew nothing about politics—

(though I knew, now, who was president of the United States: Dwight D. Eisenhower, an actual person, a former general, and not a genial smiling emoji). I knew that the United States had recently been "at war" with a faraway foreign country called Korea and I knew that World War II was a "fresh memory" to many.

"The new decade will bring us true 'newness'—even revolution. But it will be revolution mixed with tragedy, and not a little farce. And beyond that—!" Dr. Cosgrove laughed, with a shudder. "But we must not look ahead, you know. That is the essential lesson we Exiles learn. We must look to save ourselves."

I told Dr. Cosgrove that I wanted to know more about my parents—please! And how it was that—somehow—Dr. Cosgrove had known me, as a small child; and he'd had another name at the time: Uncle Toby . . .

A sensation almost of delirium overcame me. What would it mean to me, what happiness, to have an *Uncle Toby* in my life!

"No. We must live *here*, Adriane. I've learned that, after having made painful blunders. The 'future' exists only in the way that the other side of the earth exists, whether we can see it or not. Or," Dr. Cosgrove said, indicating the sky outside the window of my third-floor room, "those clouds in the eastern sky that look so sculpted—they exist in our immediate future because they are being blown in this direction, from Lake Michigan. In some time—it might be an hour, or several hours—the clouds will be overhead, the sunshine will be obscured. But now, we can see the clouds approaching, we can see into the distance and so, in a sense, we can see into the future. Most things are too distant for us to see, however, and we have to surrender the effort. 'One breath at a time'—your father used to say, wisely; though I was too young to listen to him at the time."

With a snap Dr. Cosgrove turned off the small flat object that had been buzzing and slipped it into his breast pocket.

"Dear Adriane—good-bye! Maybe we will meet again, if you wish—you know my name, and you know that I live in St. Cloud, just over the mountain."

Smiling Dr. Cosgrove squeezed my hand tenderly in parting. Badly I wanted to say *Good-bye, Uncle Toby!*—but the words could not come.

HERON CREEK FARM

"Rufus! Time for our walk."

Rufus is a six-year-old mixed-breed dog, part boxer, part retriever, part border collie. His fur is sand-colored and coarse. He weighs about thirty-five pounds—eager, clumsy, hot-panting. His ears are alert and his eyes are alive with happiness when his name is called. If he rushes at you, tail thumping, he's heavy enough to cause you to stagger.

He swathes your face with his damp soft eager tongue.

You do not want to think what that tongue has touched. You do not want to think of a dog's myriad loves of which you, as yearning as any dog, but determined to disguise that fact, are but one.

Rufus and I walk together in the fields behind the farmhouse. There is a bond between us, I want to think: this is the creature who saved my life.

An animal is a kind of machine, the behaviorists teach. Yet Rufus is hardly a machine. Rufus's soul shines in his eyes, when he is in the presence of someone he loves.

Rufus's great joy is fetching a stick. Especially from out of the pond in which Jamie stocked trout a few years ago, and where a colony of noisy frogs live.

When I hug Rufus, and when Rufus licks my face, I feel tears stinging my eyes for this reason—*I am so happy.*

SOMEDAY, I MAY HIKE over-the-mountain to seek out Dr. Cosgrove in the place with the beautiful name—St. Cloud. But I have not, yet—my new life is too demanding.

THERE WAS NO NEED for Jamie Stiles to ask me. Somehow it seemed to have been decided between us that when I was discharged from the rehabilitation clinic I would come to live with him in the farmhouse on Heron Creek Road.

You know—you have a home with me.

You know I love you.

The first sight I had of Jamie's house I fell in love with it—a sprawling clapboard farmhouse someone had painted the startling hue of goldenrod, with dark blue shutters. Other farmhouses scattered along the rural Heron Creek Road were a uniform weatherworn white.

The farmhouse is set back from the road. The (dirt, rutted) lane must be a quarter of a mile long. In the front yard are a 1949 Ford pickup truck with a smashed windshield and no tires, the remains of an ancient International Harvester tractor, an eviscerated 1947 Buick convertible, a child's sled with badly rusted runners: these are not remnants of discarded, formerly functional objects but an elaborately constructed scrap-metal sculpture titled, by Jamie Stiles, *Hazards of Time Travel.*

On the front porch, in addition to rattan sofa, chairs, and a Schwinn bicycle leaning against the wall, a display of—quilts? But these are not ordinary quilts for they have been fashioned of

Plexiglas, heavy brocade, and aluminum, in stark metallic hues—
Quilt Legend Diorama 1958.

Behind the house, a large hay barn that has been painted
brick-red. On the peak of the roof is a weather vane bearing the
heraldic figure of a stag. On the front of the barn is a large copper
sunburst-face, like the face of a (benign) ancient deity. Beside the
barn is a crumbling stone silo, and close by are several smaller
outbuildings.

And behind these, the pond Jamie has stocked with trout.

In the red barn is Jamie Stiles's sculpture studio. Several feature articles for Wisconsin newspapers have been published
about the sculpture of Jamie Stiles, as well as interviews on TV
arts programs, and these always include coverage of the studio
which resembles, at first glance, something between a mortuary
and a junkyard—the myriad, very *physical* materials of a sculptor's trade. In the smaller outbuildings are studios used by artist-
friends of Jamie's, some of whom have had the rent-free studios
for years while others are "just visiting."

As he has inherited and accumulated people in his relatively
young life—(Jamie Stiles is thirty-one)—so Jamie has inherited
and accumulated farm animals. These include a chestnut-red
mare named Hedy, retired from harness racing at the punishing track at Traverse City, Michigan; Leila and Lee, two
goat-companions who graze with Hedy; a dozen dingy-coated
sheep—(for sheep are nothing like storybook pictures of them,
I've discovered: they are not naturally *white*); a half-dozen cats
of various sizes, ages, and colors of whom some are affectionate
house-cats and others, more feral, live in the barns; and Rufus
our ever-vigilant guard dog. Also, a flock of chickens—white-
feathered, Rhode Island reds, Plymouth Rock barred—to provide
eggs for the large household. (Chicken-tending, egg-gathering,

are tasks that quickly fell to me, with help from my young step-niece Chloe and step-nephew Tyler.) On the pond behind the barn, often squawking and contentious, are Canadian geese, domestic (white) geese, and a shifting company of wild ducks.

Sometimes, a pair of swans arrives at the pond from Heron Creek a short distance away. Vivid-white and usually silent amid the smaller waterfowl the swans are beautiful like figures in a dream, that seem to represent something for which there are no adequate words.

The oldest part of the farmhouse was built in 1881, by a long-ago relative of Jamie Stiles's; it had come into Jamie's grandparents' possession, and they'd left it to him a dozen years ago, along with forty acres of mostly uncultivated land. Along a circuitous rural route the house is approximately five miles from the vast Wainscotia campus where Jamie Stiles teaches, or has taught, sculpture in the fine arts department for the past nine years, and I am a student enrolled in the liberal arts college—my twin majors are biology and art.

I am still a University Scholar. I don't any longer work part-time on campus, since I am no longer living in a university residence.

A census taker would be frustrated seeking to determine exactly who lives in the goldenrod-colored farmhouse on Heron Creek Road, Wainscotia Township. For the place is always open to Jamie's friends, acquaintances, fellow-artists and fellow pacifists/anti-nuclear protesters; frequently they are strangers passing through on their way elsewhere, who are known to Jamie, or recommended to him by mutual like-minded acquaintances. Also there are likely to be Stiles relatives on the premises, temporarily or permanently: a male cousin of Jamie's who'd dropped out of the agriculture school at WSU years before; a morose older half-brother who works in a local stone quarry, and who weighs three hundred pounds; a ravaged hard-drinking stoic uncle who'd had a twenty-year career in the U.S. Marines and who'd been severely wounded in the closing week of World War II, in Germany.

And, to my surprise, two children aged five and eight, not Jamie's children but the abandoned children of an older sister of Jamie's who'd departed several years before, leaving nothing in her wake but the children and their clothes and toys—"They're a little sad-hearted but very sweet kids, Mary Ellen. You'll love them."

It is true, I will love Chloe and Tyler—in time. I am sure of this.

The significant fact about the children, I guess you could say, is the color of their skin: a rich mocha-brown.

In all of (rural) Wainscotia Township there is probably no one else with mocha-brown skin like my little step-niece Chloe and my step-nephew Tyler.

Jamie and I will have our own children sometime soon. This is our hope.

Much of the farm property isn't tillable soil but there are a number of acres that are fertile enough to be leased by neighboring farmers, which provides a much-needed source of income for the household, and there is a half-acre behind the house where we can plant a few things—tomatoes, green beans, sweet corn, carrots, cucumbers, melons . . .

In time the care of this ambitious vegetable garden will fall to me, I think. (As the care of Chloe and Tyler will fall to me.) By the time Jamie brought me to the farmhouse in late summer everything had been planted but there were weeds choking the smaller plants, and the corn and melons had been ravaged by deer and raccoons; what flourished most was a bed of basil, catnip, and mint, which grow like weeds, and a small jungle of hollyhocks and wild rose.

I was eager to clear away weeds and thistles. But Jamie laughed at me.

"Sometimes you come too late to a garden, the way you might come too late in another's life. Best just to take things as they are. 'One breath at a time.'"

Walking through the overgrown garden, with Rufus at my heels, sniffing and bounding into the dry-rustling corn, I am suffused with happiness. I think—*All this was waiting for me. I didn't know.*

In my former life—(which I can remember only vaguely, like something glimpsed through frosted glass)—I don't believe that I lived on a farm, worked with the soil, grew things. Yet, I have confidence that I can learn.

The smells of a garden, in late-afternoon sunshine, or after a quick flashing rain—so beautiful, they leave me feeling faint.

Jamie and his friends are continually repairing the house— shingled roof, shutters, rotting porch, steps. Jamie has a plumber friend, a friend with a backhoe, a friend who digs cisterns and wells. He has (commercial) painter friends as well as (artist) painter friends. His closest sculptor-friend is a welder. Jamie himself is very handy, and strong; he has to be prevented from lifting things that might throw out his back, or break it altogether. For a reticent person he seems to thrive in a whirl of communal activities. Meetings of the local branch of SANE are usually held at Heron Creek Farm. Though sometimes Jamie drives to Madison, or farther still to Chicago. (Chicago! How very far away that seems to me. Since being struck by lightning, I tell myself that my traveling days are over.) Jamie's artist-colleagues work in their studios and eat with us, most nights. And their wives, girlfriends, children eat with us. Parents arrive, and stay for the night, or several nights. Elderly grandparents too are welcome. (But please don't die on the premises! Jamie jokes about such things, which don't seem funny to the rest of us.) Of course, there are poetry evenings—many (of us) are poets. (H. R. Brody is a friend of Jamie's.) There are musical evenings—(Jamie plays drums). Even the young niece and nephew have school friends, who are invited to the house often, brought by their parents who stay for dinner as well. Because WSU has seemed to be exploiting some instructors and staff workers Jamie has been trying to organize a union without having realized how much time this would consume, and how quarrelsome people can be when you are trying to help them. One night at dinner, which had sprawled out of the house, onto the porch and onto the grass, I counted twenty-six adults and children before I gave up.

Thinking—*There is no time here for sorrow. I am too busy!*

Jamie identifies himself as a "multi-media" sculptor. His great model is Rodin. He has fashioned strange and striking works of art out of scrap metal, including wrecked cars and tractors; he utilizes iron (which he allows to rust, in a "natural progression"), stainless steel, aluminum, brass, wood, clay, fiberglass and other materials, even papier-mâché; yet he's a traditional sculptor too, whose most successful commission has been a Korean War memorial for Wainscotia Township.

At the university Jamie Stiles is highly regarded as an original and important artist; at the same time, his political involvement with SANE, and his general pacifist/anti-war statements, have made it difficult for his departmental chairman to promote him, still less to give him a permanent appointment with tenure at WSU.

Jamie is kept on, however—"kept on." Semester after semester, year after year.

"The department likes me well enough—I mean, most of the guys on the faculty are my friends. I've known them for years.

I've helped them with their work. But the dean of the college, and the president—they're afraid of my 'notoriety.' They've been hearing that there's a 'rabid Commie' in the art department. One of the trustees thought that I'd been arrested in 'some sort of protest riot.' So part-time employment is the best I can hope for."

Wistfully Jamie spoke, and yet boastfully. I had to rush to him then, to kiss him.

Soon after I was discharged from the rehabilitation clinic and had come to live at Heron Creek Farm, Jamie took me to see the Korean War memorial in front of the county courthouse in Wainscotia Falls. Though made of stainless steel the figures of the eleven soldiers, ground troops on parole, were unnervingly lifelike; you could almost imagine that they were breathing, and that their stony skin was actual skin. They loomed above the viewer slightly larger than life-size, at about seven feet in height. Their faces were both youthful and ageless. Their hands—their fingers—were particularly lifelike. Names of the Wainscotia Township dead had been carved in stone on a parapet surrounding the sculpture, and with this too Jamie had taken a good deal of care, rejecting several fonts before he chose the one he finally used.

There was a good deal of positive publicity in local newspapers regarding *Ground Patrol: Korea 1950–1955*. Relatives of soldiers who'd died in the war were deeply moved by Jamie Stiles's memorial and wrote letters to him, each of which he answered. (Jamie had not yet become involved in anti-war protests, fortunately!) His model for the memorial hadn't been Rodin but a revered midwestern sculptor of the early twentieth century named Harry Hansen, at one time promoted as the midwestern Rodin, who'd executed more than two hundred memorial sculptures during his fifty-year career. Jamie seemed embarrassed by the sculpture though I praised it as moving, tragic, beautiful; he said, humbly, "It wasn't my idea to do a 'realistic' memorial, but that's what they wanted. I tried to explain to the township board that what they wanted was an old-fashioned kind of sculpture that had been valued in the past, before photography; the rendering of human faces and figures so close to human that they appear human. Sculpture today is more likely to be abstract than realistic. I tried to explain, but . . ." In the end if he'd wanted to accept the commission he'd had to create the kind of memorial the citizens of Wainscotia Township wanted, which he did.

"But it's very powerful, Jamie. It is."

Seeing the figures of the eleven soldiers transformed by Death was deeply moving. I could not speak for some time as Jamie circled the memorial, staring at it. An artist is one who never believes what others say about his work—perhaps that was it. Jamie could not see what I saw and could not believe what I believed about his sculpture.

On the soldiers' heads and shoulders were chalky-white bird droppings. This, we tried not to notice.

Though finally I wetted tissues in a nearby puddle and tried to wipe the droppings away, with limited success.

When we returned home Jamie showed me his framed award from the Wisconsin Council of the Arts for Sculpture, 1957. He seemed both embarrassed by the reward, and proud of it.

"Jamie! Congratulations."

Only three years before but Jamie had looked younger then. He'd been thinner, his face leaner, and his jaws clean-shaven.

Jamie Stiles without a beard! The young man in this picture wouldn't have approached me on the university campus, possibly.

Seeing his young face I felt tears gather in my eyes. For when this photograph had been taken I hadn't known Jamie Stiles. He hadn't known me. And it was quite likely that we would never meet.

Indeed, how was it possible that we'd ever met? Ever glimpsed each other? It was not possible. Yet it had happened.

Tears ran down my cheeks. I was overcome with joy indistinguishable from grief. At such times Jamie comes to me, and silently gathers me in his arms.

Jamie's strong arms. Jamie's heavy body like a fortress.

His comfort is immediate and unquestioning.

I have you, I will protect you. I love you.

IN JAMIE'S STUDIO, in the old red barn. We've fashioned a small studio for me in the hayloft, to which I climb by a ladder. It's a private space that looks out over pastureland and down into Jamie Stiles's workspace. I can watch him below but he isn't likely to watch me.

My artwork, so-called, is of a much smaller scale than Jamie's. I have no interest in heroic or monumental works. I'm content to wander outside sketching for hours with pencil, charcoal, pastel chalks, then return here and work on my sketches. I've experimented with portraits—my little step-niece and my little step-nephew, residents and visitors, Jamie's ex-Marine uncle who calls himself, with some measure of irony I am not able to gauge, "Captain Shalom."

In the loft I have a workbench of about six feet in length, that Jamie made for me. Jamie has stretched canvases for me and encouraged me to experiment with paints.

Often, I gaze over the edge of the loft floor, at Jamie Stiles below. He's a restless figure, burly, muscular, yet like an athlete he is agile on his feet, alert and curious. The sliding barn door to his studio is usually kept open except on cold days. He works with fire, sometimes. He works with spray paint. He has fashioned sculptures out of old broken floor lamps, discarded baby carriages, bullet-ridden STOP signs; scrap metal, fiberglass, window glass, aluminum and brass rods. I think that his work is not only strangely beautiful in its own way but "profound"—"important." I don't think that it's naïve of Jamie Stiles to be thinking in terms of Rodin, not Harry Hansen of Whitefish Bay.

Jamie has the ability to totally lose himself in his work. No matter how much he's worried about, for instance, the financial upkeep of Heron Creek Farm, or the insanity of U.S. nuclear testing in the Southwest, or whether he will have an instructorship at WSU in the fall, his concentration on his work is complete as that of a lone child who has lost himself in play. He's pitiless on himself when his work isn't going well, which is fairly often; he's stubborn, easily discouraged; it makes me sad that he sighs often, runs his fingers through his hair in a gesture of exasperation, or despondency; that he pulls at his beard, that seems to me a beautiful bristling beard, of the hue of mahogany, curlier than the hair on his head.

Yet, so much of James Stiles seems beautiful to me! I could stare and stare at the man, even in his soiled bib-overalls worn without a shirt beneath, and his weatherworn sandals.

When Jamie makes love to me he is awkward, tender, hesitant—he fears hurting me, or crushing me. And it is true that Jamie's considerable weight takes the breath from me, and makes me worry that my ribs might crack. The fierce thrusts of his body cause me to shudder in pain, which Jamie interprets in another way. I never betray the slightest discomfort for I think only of Jamie. I think of my need to love, and to be loved.

Never in my life before this have I loved any man—(I am

sure)—yet by instinct I've known that I must not hurt Jamie's feelings. The slightest tone of reproach or criticism of his work—never. I would never undermine Jamie Stiles's sense of himself as a man, an artist or a sexual being.

The truths I reveal to Jamie Stiles are those "truths" that will nurture Jamie's love for me. For only Jamie Stiles's love for me can validate the love I feel for him, so powerful it leaves me faint and breathless.

It must be—*I came so close to dying. Now, nothing matters except this life.*

SOME EVENINGS, we watch TV.

On a sofa we sit holding hands. We are open about our affection for each other—(this is Jamie's way, in the expression of affection generally)—and not at all embarrassed about seeming sentimental, even when "Captain Shalom" grunts wryly in our direction, passing with heavy steps through the living room on the way to his quarters at the rear of the house.

We're rarely alone watching TV. The prime hours are 8:00 P.M. and 9:00 P.M. Jamie laughs as loudly as the children laugh at the foolish pranks of Milton Berle, Lucille Ball and Desi Arnaz; Ozzie and Harriet are household favorites, as well as Arthur Godfrey, Lawrence Welk, and Phil Silvers. Jack Benny, Sid Caesar and Imogene Coca require more thought, like Jack Paar, *Truth or Consequences*, and *What's My Line?* Sometimes, Jamie falls asleep while watching TV, exhausted from a full day of work, and I don't wake him, but continue to grip his hand tight. TV images wash over us. Our most torturous thoughts are obliterated in the bluish-flickering TV light.

In our lumpy brass bed in Jamie's longtime bedroom on the

second floor of the house we lie entwined in each other's arms. Here we talk, and kiss; we kiss, and make love; sometimes a sensation comes over me, that I am in the arms of someone else and not Jamie Stiles—someone whose name I have forgotten. I shudder with dread, but don't cry out, and manage not to weep.

For life is *now*. Life is not thinking, not reflective or backward-glancing; life is forward-plunging; life is the present moment as, on TV, it is always *now*.

And I think—*I am in the right place, at the right time.*

TO CELEBRATE OUR WEDDING in late October, Jamie's poet-friend Hiram Brody gave a large, festive party for us in his Victorian house in the Faculty Hills neighborhood of Wainscotia Falls, and wrote a "sprung love sonnet" for the occasion. At the party, in addition to Jamie's many friends, was a dazzling mix of writers, artists, sculptors, musicians, university professors, and their wives; here were such Wainscotia luminaries as Amos Stein, Myron Coughland, my former employer Morris Harrick, Carson Lockett III, and A. J. Axel—all of whom were friends of H. R. Brody's and, I wanted to think, admirers of Jamie Stiles's sculpture. (The most distinguished of these professors, judging by the deference paid to him, was Professor Axel of psychology, who made only a brief visit at the gathering, and left after the first champagne toast. Mr. Brody said proudly of his friend that A. J. Axel had just received the "largest federal government grant ever given to any research scientist in the State of Wisconsin" to help establish the Wainscotia Center for Social Engineering, of which Professor Axel is the founding director; the Center is to pioneer in the behavioral conditioning of anti-social, psychopathic, and subversive personalities.) To this gathering, white-haired Mr. Brody read his "sprung sonnet" in a richly dramatic voice; it was a poem containing an echo, he said, of a sonnet of Shakespeare's—*Let us not bar the twining of true love but celebrate Love's fixed star* . . . Everyone applauded loudly; Jamie wiped tears from his eyes. He didn't understand most poetry, Jamie said, but he often cried just the same when he heard it.

I thought Mr. Brody's poem was strange, and beautiful. I did not quite understand it, either, but it brought tears to my eyes.

Jamie and I had been married that morning by a justice of the peace at the Wainscotia Falls courthouse, with just a few witnesses from Heron Creek Farm. The fatherly justice had expressed surprise and a little concern that the bride seemed to have no family, or at any rate no family members who'd come to the wedding, but I smiled and assured him that Jamie Stiles was all the family I needed.

(One of many kindnesses Ardis Steadman had done for me, in addition to boxing those possessions of mine which I'd left behind in Acrady Cottage, was to provide me with a birth certificate out of my university file, which I needed in order to be married. This birth certificate, with an ornate gilt seal from the State of New Jersey, was not a document I could remember having seen before; it stated that *Mary Ellen Enright* had been born in *Pennsboro General Hospital, Pennsboro, New Jersey,* on *September 11, 1942,* and that her parents were *Constance Ann Enright* and *Harvey Sterns Enright.* Were these my birth-parents? Or were these simply fictitious names someone had provided, for a birth certificate? The names meant nothing to me—not a stir of emotion. But I remembered Dr. Cosgrove speaking of *New Jersey.*)

In black ink, with a fountain pen, H. R. Brody copied "Wisconsin Epithalamium" inscribed to *Mary Ellen* and *Jamie,* signed and dated in the poet's flowing signature, on a stiff sheet of parchment paper, which Jamie has framed to hang in our bedroom. "It's like having a handwritten poem by Robert Frost, or T. S. Eliot," Jamie has said. He is deeply moved by the poem, as I am. Sometimes one of us reads it aloud to the other, as we prepare for bed.

I thought—*I have always loved this person. I have always known this person. Before I was born, I loved him.*

SHORTLY AFTER THIS, a strange and disturbing thing happens.

I am not sure how to speak of it. So much in my life has floated beyond language, like a high scudding cloud so distant it can't be identified, I have lost confidence in my ability to comprehend many things, let alone explain them.

I have not tried to avoid "Captain Shalom" for I have not wanted to hurt the man's feelings, nor do I want to hurt Jamie's feelings; but there is something unnerving about Jamie's ex-Marine uncle, that hasn't only to do with the poor man's ravaged face, his hairless battered head, and unblinking rheumy eyes; or his breath that smells of something metallic, like coins held hotly in the palm of a sweaty hand. In our household, in which there are so many individuals coming and going, in the kitchen, on the stairs, in the living room and hallways, not to mention in and out of bathrooms—(there are just two bathrooms for all of us, one on each floor; but there is an outhouse between the house and the hay barn, that Jamie says had been in use when he'd been a boy, not so long ago)—you are constantly encountering the same people yet you're usually on your way elsewhere, hurrying past them with a murmured *Excuse me!*—or no words at all. The intimacy born of sheer proximity is a curious phenomenon—there is something mocking in it.

Jamie's middle-aged ex-Marine uncle uses crutches some-times, though not always; often he can make his way upstairs by hauling himself hand over hand along the banister, and his method of descending stairs is a kind of free-falling plunge. The one thing you must not ever do is offer to assist him: this error, I made shortly after coming to live at Heron Creek Farm.

The man had stared coldly at me. His eyes were fierce and glaring. Along the left side of his face was a zipper-like scar, and a portion of his upper lip was missing. His teeth were grayish, like the teeth of a malnourished child. His breath was coppery-hot. The ex-Marine who called himself Captain Shalom knew how to render me helpless, by not speaking as I stammered an apology.

Finally he said in an ironic, gravelly voice, "When I need your help, 'Mary Ellen,' I will request it. Thanking you beforehand."

The way in which Captain Shalom pronounced *Mary Ellen* allowed me to know that he did not think much of the name, or the subterfuge which such a name can validate.

Jamie has been concerned about his uncle's "mental health" but—what can he do? Captain Shalom refuses to see any doc-tor, even local doctors; he flies into a fury at the suggestion that someone might drive him to Milwaukee, to see a VA doctor (that is, a psychiatrist). Jamie says that unless he overpowers his uncle and ties him up, and bundles him out to the pickup, it isn't likely that anything can be done for the veteran.

"Does he have a gun? Guns?"—this was my innocent query.

"No firearms are allowed on this property. That is understood."

How was this a satisfactory answer? Jamie was incensed, that I should even suggest this.

I think it's likely that Captain Shalom keeps a gun, or guns, in his room. (This room, awkward for the disabled man to access, was his choice when he'd moved in, Jamie has explained.) But I think it's likely that, if Captain Shalom feels the impulse to fire one of his guns, he would not kill any of us—(out of contempt, or indifference)—but only himself.

For Captain Shalom is heroic, in his tortured way.

I have tried to "sketch" him—though only by memory. I would like to take photographs of his face when he isn't aware of me—but that isn't likely.

Captain Shalom is both jocular and despondent, by turns; his moods are not so very different from Jamie's but more frequent, and unpredictable. By mid-afternoon he's likely to be moderately drunk, which gives to his manner a playful, bitter-ironic air; as he is a wreck of a man, an object of pathos, he isn't given to hypocrisy, if for instance someone tells him, as visitors to our house invariably do, meaning to be kind, that he is *looking good* he will say dryly, "Really? In whose eyes?—yours, or mine?"

Or he will say nothing but grunt in a way to convey disgust, amusement, contempt; and lurch his way past, with the gleeful rudeness of the disabled for whom condescending kindnesses from the *abled* are particularly insulting.

Between "Captain Shalom" and "Mary Ellen Enright" there is an uneasy sort of truce, I think. As I am Jamie Stiles's wife, Jamie's uncle believes that he should respect me; he's dependent upon Jamie for a place to live and a household, for his own marriage ended shortly after he'd been discharged from the VA hospital in Milwaukee, and had returned to his wife and children in Racine with both physical and psychiatric problems. Yet, as I'm a young woman of only nineteen, an undergraduate at WSU, and, since coming to live at Heron Creek Farm, as Jamie Stiles's dear companion, a fairly attractive young woman with a quick bright friendly smile, it is quite possible that Captain Shalom resents me, as men often resent women who are unattainable to them, as women. It is the case that when Captain Shalom and I are alone together in a room or in a hallway, we move past each other with averted eyes, and indrawn breaths. Captain Shalom is exceedingly polite with me at mealtimes, and often volunteers to help with kitchen cleanup, a chore Jamie mostly avoids with the excuse that he must return to his sculpting studio for an hour's work or so before quitting for the day; at such times, I'm grateful if others in the household help out in the kitchen, for being alone with the ravaged and embittered ex-Marine is painful, and makes me very self-conscious.

Here is a man who sees through you. Your happiness, your relentless smile, even your "love."

Yet, Captain Shalom is an obsessive reader, and has accumulated a library of secondhand books in his room; unlike the rest of us, he rarely watches TV, and never without snorting in derision and distaste. (Anything on TV that has to do with soldiers, armed forces, veterans, "war" he particularly scorns; but also pacifists, anti-war protesters, and SANE—to Jamie's disappointment.) Often I see Captain Shalom limping outside with a book from his library, in good weather; he has found a place to read overlooking the pond, and has strung up a hammock there, for his own, private use. (Yet Captain Shalom has invited me to lie in the hammock any time I wish, an offer I would never take up for it seems to hold a veiled threat.) One warm afternoon when Captain Shalom went out to read in his hammock by the pond, I hurried upstairs to his room, to examine his books; though I knew that they were mostly historical books about war, as Jamie had told me. (I thought of searching for his guns, but could not so violate Jamie's uncle's privacy. I could not bring myself to look, for instance, through the man's bureau drawers, or between the mattress and box springs of his bed.)

Captain Shalom's room was sparely furnished, with no carpet on the floorboards; there was a table, a single chair, a floor lamp that looked as if it had been salvaged from a junkyard. Surprisingly, the room was relatively neat, for Captain Shalom had made his bed as if he were in a barracks—tight-drawn, corners tucked in, the single pillow perfectly positioned. (I smiled to think how Jamie would never think of making any bed in which he'd slept. He kicks his covers off, leaves the bedclothes rumpled and churned-looking.) There was only one actual bookcase in the room, that stood about five feet high; but everywhere were books, some of them oversized picture or photography books, on the floor and table, on a windowsill. Hesitantly I pulled out a book from the bookcase, noting its place so that I could return it, and Captain Shalom would never know, but when I opened the heavy hardcover book I saw to my surprise that there were *no words on the page*—no printed words.

I turned pages, and all were the same: blank.

On the book's spine there was nothing as well. On the book's cover.

Shaken, I replaced this book and opened another at random. And this book did have printed pages, but the print was blurred and incomprehensible as if it had melted; and a third book I opened, now quite frightened, and this too had pages that were incomprehensible to me, not like words in a foreign language but hieroglyphic-like words that did not use familiar letters. And the thought came to me, chilling yet somehow calm—*That's because you are dreaming. In a dream, you can never read print.*

Quickly I replaced the books, and quickly retreated downstairs to the first floor of the farmhouse. I have never returned to Captain Shalom's room since.

MY INJURIES WILL NEVER disappear entirely, I've been told—they are "neurological deficits." Always I will be susceptible to migraine headaches. I can't "play catch" as most others can in the household—playing with my little step-niece and my step-nephew, for instance, is embarrassing, since I so often fumble the ball. In chill, damp weather, both my knees ache. My eyesight begins to fade at sundown. My eyes are weak, and water easily. A mild agitation will precipitate heart palpitations even when I am otherwise serene.

And I still cry, for no evident reason.

And Jamie comforts me at such times, without asking me what is wrong. And Rufus too, if he hears me.

This afternoon, visitors are arriving. I think they are friends of Jamie's from Madison—anti-nuclear activists, who are also artists. I have no idea how many people will be eating meals here for the next several days but I will have help in preparing these meals, and cleaning up in the kitchen afterward. Strangely, there is a kind of calm amid so many people, and commotion—and then, I can retreat to my loft in the hay barn anytime I wish, or nearly.

Always room for one more at Heron Creek Farm—this is Jamie Stiles's dictum.

Anytime you are in the neighborhood of Heron Creek Farm, or anywhere in the vicinity of Wainscotia Falls, Wisconsin, you too are welcome to drop by here with all the others—of course.

Please come! I would so like to meet you. Stay with us as long as you like.

ACKNOWLEDGMENTS

Special thanks to Greg Johnson for reading this manuscript with his usual care, thoughtfulness, and sympathy; and for my husband Charlie Gross for his continued support.